D0077540

STUDIES IN GEOPHYSICS

Climate in Earth History

Geophysics Study Committee
Geophysics Research Board
Commission on Physical Sciences, Mathematics, and Resources
National Research Council

NATIONAL ACADEMY PRESS
Washington, D.C. 1982

NOTICE: The project that is the subject of this report was approved by the Governing Board of the National Research Council, whose members are drawn from the Councils of the National Academy of Sciences, the National Academy of Engineering, and the Institute of Medicine. The members of the committee responsible for this report were chosen for their special competences and with regard for appropriate balance.

This report has been reviewed by a group other than the authors according to procedures approved by a Report Review Committee consisting of members of the National Academy of Sciences, the National Academy of Engineering, and the Institute of Medicine.

The National Research Council was established by the National Academy of Sciences in 1916 to associate the broad community of science and technology with the Academy's purposes of furthering knowledge and of advising the federal government. The Council operates in accordance with general policies determined by the Academy under the authority of its congressional charter of 1863, which establishes the Academy as a private, nonprofit, self-governing membership corporation. The Council has become the principal operating agency of both the National Academy of Sciences and the National Academy of Engineering in the conduct of their services to the government, the public, and the scientific and engineering communities. It is administered jointly by both Academies and the Institute of Medicine. The National Academy of Engineering and the Institute of Medicine were established in 1964 and 1970, respectively, under the charter of the National Academy of Sciences.

The Geophysics Study Committee is pleased to acknowledge the support of the National Science Foundation, the Defense Advanced Research Projects Agency, the National Aeronautics and Space Administration, the National Oceanic and Atmospheric Administration, the U.S. Geological Survey, and the Department of Energy (Grant # DE-FG02-80ER10757) for the conduct of this study.

Library of Congress Cataloging in Publication Data

Main entry under title:

Climate in earth history.

(Studies in geophysics)
Papers presented at the American Geophysical Union meeting, held in Toronto, May 1980.
Includes bibliographies.
1. Paleoclimatology—Addresses, essays, lectures.
2. Historical geology—Addresses, essays, lectures.
I. National Research Council (U.S.). Geophysical Study Committee. II. American Geophysical Union. Meeting (1980: Toronto, Ont.) III. Series.
QC884.C574 1982 551.69 82-18857
ISBN 0-309-03329-2

Available from

NATIONAL ACADEMY PRESS
2101 Constitution Avenue, N.W.
Washington, D.C. 20418

Printed in the United States of America

Panel on Pre-Pleistocene Climates

WOLFGANG H. BERGER, Scripps Institution of Oceanography, *Cochairman*
JOHN C. CROWELL, University of California, Santa Barbara, *Cochairman*
MICHAEL A. ARTHUR, University of South Carolina
WILLIAM A. BERGGREN, Woods Hole Oceanographic Institution
ARTHUR J. BOUCOT, Oregon State University
GARRETT W. BRASS, University of Miami
DAVID L. CLARK, University of Wisconsin
HANS P. EUGSTER, The Johns Hopkins University
J. FERRER, Exxon Production Research-European
ALFRED G. FISCHER, Princeton University
W. LAWRENCE GATES, Oregon State University
JANE GRAY, University of Oregon
ANTHONY HALLAM, The University of Birmingham, England
BILAL U. HAQ, Woods Hole Oceanographic Institution
JON HARDENBOL, Exxon Production Research Company
WILLIAM W. HAY, Joint Oceanographic Institution
W. T. HOLSER, University of Oregon
JOHN IMBRIE, Brown University
L. D. KEIGWIN, JR., Woods Hole Oceanographic Institution
THEODORE C. MOORE, JR., Exxon Production Research Company

W. H. PETERSON, University of Miami
NICKLAS G. PISIAS, Oregon State University
JAMES B. POLLACK, National Aeronautics and Space Administration
RICHARD Z. POORE, U.S. Geological Survey, Reston
E. SALTZMAN, University of Miami
SAMUEL M. SAVIN, Case Western Reserve University
J. L. SLOAN II, University of Miami
J. R. SOUTHAM, University of Miami
HANS R. THIERSTEIN, Scripps Institution of Oceanography
PETER R. VAIL, Exxon Production Research Company
JAMES W. VALENTINE, University of California, Santa Barbara
FRANKLYN B. VAN HOUTEN, Princeton University
JACK A. WOLFE, U.S. Geological Survey, Menlo Park

Geophysics Study Committee

CHARLES L. DRAKE, Dartmouth College, *Chairman*
LOUIS J. BATTAN, University of Arizona, *Vice Chairman*
JOHN D. BREDEHOEFT, U.S. Geological Survey
ALLAN V. COX, Stanford University
JOHN C. CROWELL, University of California, Santa Barbara
HUGH ODISHAW, University of Arizona
CHARLES B. OFFICER, Dartmouth College
RAYMOND G. ROBLE, National Center for Atmospheric Research

Liaison Representatives

BRUCE B. HANSHAW, U.S. Geological Survey
GEORGE A. KOLSTAD, Department of Energy
MURLI MANGHNANI, National Science Foundation
NED OSTENSO, National Oceanic and Atmospheric Administration
WILLIAM RANEY, National Aeronautics and Space Administration
CARL F. ROMNEY, Defense Advanced Research Projects Agency

Staff

THOMAS M. USSELMAN

Geophysics Research Board

Commission on Physical Sciences, Mathematics, and Resources

Studies in Geophysics*

ENERGY AND CLIMATE
Roger R. Revelle, *panel chairman*, 1977, 158 pp.
CLIMATE, CLIMATIC CHANGE, AND WATER SUPPLY
James R. Wallis, *panel chairman*, 1977, 132 pp.
ESTUARIES, GEOPHYSICS, AND THE ENVIRONMENT
Charles B. Officer, *panel chairman*, 1977, 127 pp.
THE UPPER ATMOSPHERE AND MAGNETOSPHERE
Francis S. Johnson, *panel chairman*, 1977, 169 pp.
GEOPHYSICAL PREDICTIONS
Helmut E. Landsberg, *panel chairman*, 1978, 215 pp.
IMPACT OF TECHNOLOGY ON GEOPHYSICS
Homer E. Newell, *panel chairman*, 1979, 121 pp.
CONTINENTAL TECTONICS
B. Clark Burchfiel, Jack E. Oliver, and Leon T. Silver, *panel cochairmen*, 1980, 197 pp.
MINERAL RESOURCES: GENETIC UNDERSTANDING FOR PRACTICAL APPLICATIONS
Paul B. Barton, Jr., *panel chairman*, 1981, 118 pp.
SCIENTIFIC BASIS OF WATER-RESOURCE MANAGEMENT
Myron B. Fiering, *panel chairman*, 1982, 127 pp.
SOLAR VARIABILITY, WEATHER, AND CLIMATE
John A. Eddy, *panel chairman*, 1982, 106 pp.
CLIMATE IN EARTH HISTORY
Wolfgang H. Berger and John C. Crowell, *panel cochairmen*, 1982, 198 pp.

*Published to date.

Preface

In 1974 the Geophysics Research Board completed a plan, subsequently approved by the Committee on Science and Public Policy of the National Academy of Sciences, for a series of studies to be carried out on various subjects related to geophysics. The Geophysics Study Committee was established to provide guidance in the conduct of the studies.

One purpose of the studies is to provide assessments from the scientific community to aid policymakers in decisions on societal problems that involve geophysics. An important part of such an assessment is an evaluation of the adequacy of present geophysical knowledge and the appropriateness of present research programs to provide information required for those decisions. Some of the studies place more emphasis on assessing the present status of a field of geophysics and identifying the most promising directions for future research.

This study was initiated in response to the recommendation in the report *Geological Perspectives on Climatic Change* (National Academy of Sciences, Washington, D.C., 1978). The recommendation called for the following: "(1) to assess the state of the art, . . . and to stimulate progress in geological aspects of climate research; (2) to evaluate comprehensively and in detail the research opportunities in, the operational needs of, and the scientific and societal relevance of geological and geophysical processes affecting the understanding of climate and climatic forecasting; and (3) to recommend the appropriate geological content of a national and global climate program." In considering this recommendation, the Geophysics Study Committee felt that the major geologic and geophysical questions concerning climate were those found in the pre-Pleistocene (older than 2 million years) record.

The study was developed through meetings of the Geophysics Study Committee and the Panel on Pre-Pleistocene Climates. The preliminary scientific findings of the panel were presented at an American Geophysical Union meeting that took place in Toronto in May 1980. These presentations and the essays contained in this volume provide examples of current basic knowledge of the climate in the geologic past. They also pose many of the fundamental questions and uncertainties that require additional research. In completing their papers, the authors had the benefit of discussion at this symposium as well as comments of several scientific referees. Responsibility for the individual essays rests with the corresponding authors.

The Overview of the study summarizes the highlights of the essays and formulates conclusions and recommendations. In preparing it, the panel chairmen had the benefit of meetings that took place at the symposium, the comments of the panel of authors, and selected referees. Responsibility for the Overview rests with the Geophysics Study Committee and the chairmen of the panel.

Contents

Climate in Earth History

Overview and Recommendations

INTRODUCTION

The long record of climate on the Earth is valuable in understanding how the climate system works. The study of past climates can conveniently be viewed on three time scales: (1) the last 10,000 years; (2) the period back to 2 million years ago (Ma)—the Pleistocene Epoch, which witnessed the rise of man; and (3) the pre-Pleistocene period prior to 2 Ma. Much attention has already been devoted to the records of climatic change in the first two intervals; this study is devoted to the third, the vast (billions of years) pre-Pleistocene period of Earth history. The climate record is contained in strata and rocks of the continental crust and in sediments of the oceans. By examining this record we can learn much about the long-term changes in climate, what the state of climatic normalcy has been back into remote periods of geologic time, and what some of the factors are that perturb this state and by how much. We offer here a summary of achievements in understanding ancient climates and of current research problems and make recommendations concerning profitable avenues for future research.

Climate is the result of flow of the air and ocean system on the rotating Earth in accordance with the laws of physics. These fluids flow and interact within a geographic setting determined primarily by the arrangements of land and sea, the orientation of mountains and lowlands on continents, the depths of the oceans, and the location of ocean gateways. The total system is complex, and climate is the result of the interplay between the air, ocean, and land on a range of time scales. These scales range from the short period, dealt with in weather forecasts, to long-period changes resulting from changes in atmospheric composition and the drift of continents on the mobile lithosphere.

Meteorologists are currently developing computer models of the atmosphere that give increasingly accurate descriptions of the global circulation. Through the use of such models in the decades ahead, they will explain the flow of air and the movement of weather patterns. In addition, oceanographers are attempting to understand the way the ocean works. In time their models can be combined with atmospheric models leading to general-circulation models of both the air and the ocean waters. Most changes within the ocean system take place over many centuries, so the study of oceanography aimed at such time scales as well as over shorter times must be melded into the framework of the weather and climate models, which deal with changes over days, weeks, months, and years. Paleo-oceanography, concerned with describing the ocean system back farther into time, can be expected to reveal how longer-term components of the ocean play their part in the way the climate system works.

Geologists and geophysicists, employing the concepts of plate tectonics and continental drift, are now reconstructing past arrangements of lands and seas for the last several hundred million years. These reconstructions are providing boundary conditions for investigations by paleo-oceanographers and paleoclimatologists. In the near future they will lay out the framework for computer modeling of ancient climates. With enough information from sediments on the seafloor and strata within the continents, an iterative approach promises to disclose how changing land-sea patterns and topography and bathymetry influence the flow of ocean waters, which, in turn, strongly influence the circulation pattern of the atmosphere. By comparing differences in the factors controlling past climates we will learn of the checks and balances in the system and of the complex feedbacks, overshoots, and dampenings. We will learn what physical mechanisms bring the perturbed climatic system back to a near-steady state. The climatic record contained in strata will provide information on what past climates were like, on how quickly they changed, and on how large these changes were. Nature has performed a large number of experiments during the course of Earth history, and the geologic record contains the outcome of these experiments.

In this Overview, we summarize key factors for the understanding of climates in particular periods in the pre-Pleistocene and various approaches (synoptic, time-series, and event analyses) for investigating ancient climates. We also make several specific recommendations for future investigations. Detailed discussions of recent advances in understanding ancient climates appear in the chapters that follow this Overview.

Our principal goal is *to encourage specialists in all the earth sciences to find ways that their disciplines can contribute to understanding the complex atmosphere-ocean-land interactions that constitute the climate system and its variations through time.*

WHY STUDY ANCIENT CLIMATES?

Climate, its variation, and its change are closely linked to the growth of our crops, to our well-being, and to the economy. Consequently, predictions or forecasts of short-term climate changes promise substantial benefits. Recognizing this potential, Congress recently enacted the National Climate Program Act, which calls for "a well-defined and coordinated program in climate-related research, monitoring, assessment of effects, and information utilization" (U.S. Congress, Public Law

95-367, 1978). Improvement in the understanding of climate has become urgent in view of the potential for inadvertent climate modification by our industrial society. Heat and particulates are released to the atmosphere and affect climate both locally and regionally, and perhaps even globally. The greatest potential global impact is from carbon dioxide release (NRC Geophysics Study Committee, 1977; NRC Climate Research Board, 1979; NRC Climate Board, 1982). At present, the rate of carbon dioxide added to the atmosphere is about 0.7 percent per year of the atmospheric carbon dioxide content. The input of industrial carbon dioxide has been called a "geophysical experiment" of unprecedented scope—an experiment whose course and outcome is unplanned and unknown. On the whole, a global warming over the next century is indicated: a doubling of atmospheric carbon dioxide would raise the global temperature by some 2 to 3°C (Hansen *et al.*, 1981; NRC Climate Board, 1982).

To obtain the necessary perspective on present climatic conditions, we study ancient climates. How stable is the present climate, and how has it changed with time? How fast might the present climate respond to pertubation? What might be the effects of rapid change of climate on the biosphere? Any changes in the biosphere, consisting namely of changes brought about by human activities in agriculture and in the forests, will have profound economic and political consequences. Because modeling of the climate system over time spans more than a decade is still rudimentary, appraisal of rates of climatic change need to be extracted from the historical and geologic record.

Many important materials owe their origins to the interplay of climatic variables both in the oceans and on land. Petroleum and natural gas, for example, have at times and at places originated where wind-driven upwelling of deep ocean water encouraged the flourishing of microscopic algae, resulting in organic-rich muds on the seafloor. Marine phosphate deposits, important as fertilizers in crop production, were also largely associated with regional upwelling. Bauxite and laterites, prime sources of aluminum and nickel, respectively, resulted from extensive deep weathering of crystalline rocks under warm and humid conditions. Manganese deposits and coal also require specific climatic conditions for their origin. Coal deposits derive from swamps, generated in a humid climate. Much of our knowledge about past climates results from the search for such economic deposits, and our ability to locate additional resources will be improved by an increased understanding of climates throughout Earth history.

PAST CLIMATES

The geologic record shows that surface temperatures on Earth have not been too different over most of the Precambrian and Phanerozoic times from those of today (Frakes, 1979). Sedimentary rocks deposited 3500 Ma and since show that water has been predominantly present in its fluid state in the past as it is today. The air-ocean system has been driven by the Sun's energy as it is today. Life flourished and kept evolving within the seas and later upon the lands. Climate has been remarkably stable when viewed in this time perspective.

Paleoclimatic investigations can be grouped into three general time periods on the basis of the precision with which the climatic record can be read, the tools and techniques available for study, and the applicability to future climate forecasts. Studies of the most recent period make use of instrumental, historical, and high-resolution proxy records and are the most definitive. Proxy data, such as tree rings, glacial records from ice cores, and pollen distributions, have extended our detailed climate knowledge back some 10,000 years ago. The goals and challenges of these

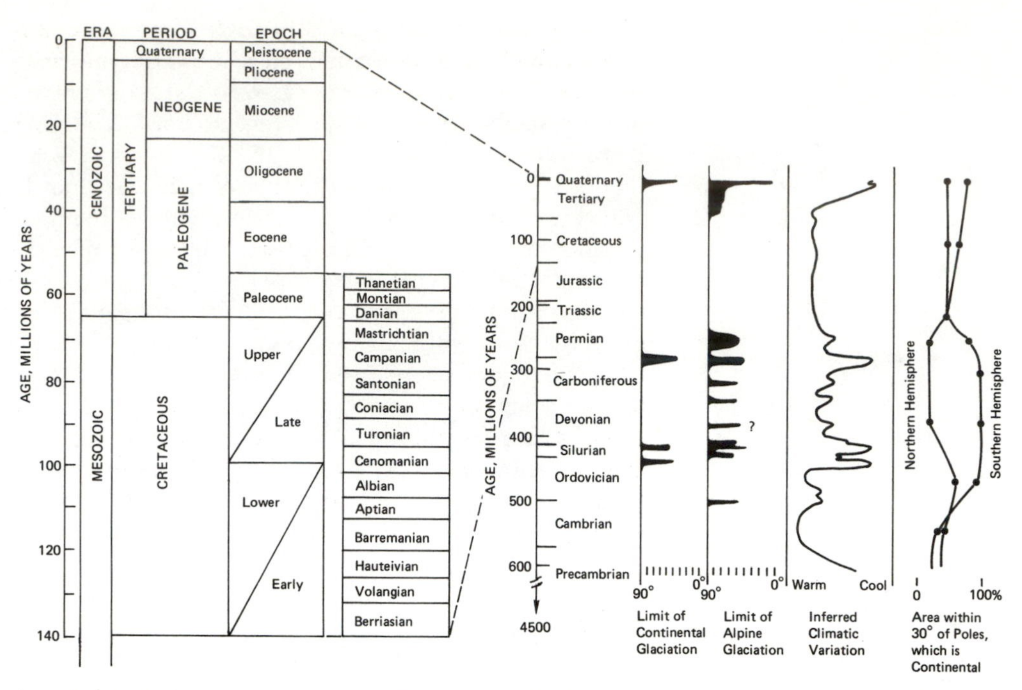

FIGURE 1 Reference geologic time scale and generalized climatic trends (Seyfert and Sirkin, 1979) for the last 600 million years.

types of investigations have been covered in previous reports, for example, *Understanding Climatic Change* (NRC U.S. Committee for the Global Atmospheric Research Program, 1975).

The second period is the Pleistocene Epoch, a time of alternating glacial and interglacial stages that extends back to about 2 Ma. Major advances in understanding the climate of this interval have resulted from quantitative approaches similar to those used in the CLIMAP (Climate, Long-Range Investigations, Mapping and Prediction) program. Recent results of such investigations are summarized by Imbrie and Imbrie (1979).

The subject of this report is the third period, the pre-Pleistocene (Figure 1), a period that encompasses the entire geologic record older than 2 million years (m.y.). The goals pursued through the study of pre-Pleistocene climate are as varied as those of historical geology itself, as indicated by the range of the specific contributions in this study.

PRE-PLEISTOCENE CLIMATOLOGY: A MATTER OF SCALE

The central task of paleoclimatology is to describe the climatic patterns throughout geologic time and to understand the trends, cycles, and discontinuities in these patterns—in short, to find out how the climate system works.

Weather elements include temperature, precipitation (as either rain or snow), wind strength and direction, cloud cover, and humidity; climate consists of weather patterns in time and space. In human affairs, climate usually refers to the sum of weather patterns over periods of months to a few years. Changes in some weather elements from one decade to the next, such as changes bearing on the occurrence of droughts, are of course significant. In reconstructing past climates from the geologic record, however, we are forced to integrate over longer and longer intervals as we go back in time. In general, the further back we go, the less sure we are of the ages of our sample and of the spans of time they represent. Because of the increasing number of unknowns in progressively older records, the type of questions we can hope to answer must change with the age of the record. We can conveniently consider the available record of pre-Pleistocene climates on five different time scales.

1. The most detailed climatic records can be constructed for the time during which the continents were essentially in their present positions and, as today, the polar areas had snow and ice. The onset of northern continental glaciations, about 3 Ma (Berggren, 1972; Shackleton and Opdyke, 1977) marks the beginning of this period. Essentially, the climatic mechanisms of the Pleistocene (1.7 Ma to the present) apply.
2. The second scale applies to the time during which we have an adequate sedimentary deep-sea record. This period spans about the last 100 million years. Paleontology and geochemistry of the deep-sea sediments provide detailed climatic signals on a global basis. In addition, the positions of continents and the morphologies of ocean basins—both of which affect the atmospheric and oceanic circulation—can be reconstructed from seafloor paleomagnetic information. Because there are more uncertainties on this scale than on that in 1, above, our understanding of climatic change within this period will necessarily be less complete. However, some of the questions that can be asked regarding changes, variations, and repetitions over relatively long periods will lead to new insights not derivable from Pleistocene studies.
3. The third scale applies to the last 200 Ma or the time since the disruption of Pangaea, the early Mesozoic supercontinent that made up of most of the Earth's land masses. Although there are several reconstructions of Pangaea, the differences refer largely to detail: the broad outlines of paleogeography are reasonably well known (McElhinny and Valencio, 1981). A deep-sea sedimentary record exists for some of this period but is much less complete than that of the 0-100 Ma record.
4. The fourth scale includes the entire Phanerozoic, nearly 600 m.y., the time for which we can read climatic zonation from biogeography, aided by geology and geophysics (positioning of the continents), geochemistry (mapping of climate-sensitive deposits), and interpretations based on data from strata incorporated in the continents. It includes all of Paleozoic time.
5. The fifth scale is applicable to Precambrian time and includes the entirety of Earth history except for the last one seventh. Climatic information exists for this scale but is scarce and commonly imprecise. On this scale, we see the chemistry of the atmosphere change in response to the evolution of organisms. On the whole, oxygen content increases and CO_2 content decreases.

On each of the time scales, we are interested in learning what were the inputs to the climate system, its boundary conditions, its inner workings, and the way it expressed itself in the geologic record. The levels of detail of description and of understanding differ considerably for the various scales considered. For the most remote intervals, such as those within the Precambrian, statements about typical temperature ranges on the surface of the Earth and about limits on atmospheric composition and on solar radiation may be all that can be hoped for. For the Paleozoic Era, fossils provide clues, and it is possible both to recognize climatic zonations and to estimate the rates of change of such zonations. Yet even within these remote times, there may be glimpses into climatic cycles, based on continuous sequences of finely layered rocks. We may be able, for example, to test the concept of the Sun being a stable star for thousands of millions of years.

Continental configurations of the Mesozoic Era are reasonably well known. However, the geologic-geochemical setting during this era is different enough from today's so that we cannot unambiguously reconstruct pole-to-equator temperature gradients or humidity patterns. From the middle Cretaceous on, we have the advantage of extensive deep-sea records of the necessary time constants that contain detailed climatic signals of global significance. The record of the last 100 Ma will therefore be most productive in our search for the sequences of climatic change and their causes.

The Tertiary Period, and especially the Neogene part, is well represented by deep-sea sediments, and it is possible to extract information on details such as temperature anomalies within latitudinal bands.

SYNOPTIC STUDIES

Paleoclimatic investigations can be conveniently characterized by three approaches: synoptic studies, time-series studies, and event and episode analysis, as summarized below.

One way to study ancient climates is to select convenient intervals of past geologic time and to assemble all pertinent information bearing on climate within that interval, that is, to study synoptic intervals or time slices. The length of the time slice selected is determined by the time control available, and it should be short enough so that climate change is minimal through its duration. Difficulties of correlation and the fact that the record is more complete for younger periods than for older ones mean that time intervals are longer and less truly synoptic as we go back in time. The crucial ingredients of synoptic analyses are reliable reconstructions of continental positions and accurate stratigraphic correlation from one continent to another and from continents to ocean basins.

The comparison of reconstructed climatic zones with present-day belts has been a traditional approach in such studies. For example, interpretation of the latitudinal distribution of coral reefs, evaporite deposits, coal deposits, and glacial moraines has relied on present-day analogies; this is also true for studies of the biogeography of plant and animal fossils. This analog approach involves the establishment of boundary conditions such as the distributions of continental landmasses, continental shelves, and ocean basins; the presence and extent of snow and ice (Figure 2); the overall pattern of surface albedo (insofar as it is dependent on snow and plant cover and the area of the seas and oceans); and the sizes and fluxes among the important carbon reservoirs influencing the carbon dioxide content of the atmosphere. The distribution of land heights and the location of mountain chains are also part of the boundary conditions. The distribution of climatic indicators and of theoretically expected belts of temperature and humidity can then be compared and the reasons for mismatches within a given synoptic interval discovered. Once such understanding is achieved, synoptic maps acquire predictive value. Not only can climatic zones be completed by interpolation, but their probable extent and associated sedimentary deposits of zones, not yet discovered, may also be predicted.

In studying the older geologic record we are commonly hampered by the lack of present-day or Pleistocene analogs. In the absence of a physicochemical understanding of ocean-atmosphere-land interaction, we need such analogies to provide guidance. Quantitative simulation of paleoclimates proceeds from using the present as an analog rather than from first principles of fluid flow. To understand the full functioning of the system, we need a greater array of climate conditions than the Pleistocene can yield. Nor does the Late Cenozoic yield enough analogs for more remote times. Although a synoptic analysis of the early Pliocene and late Miocene can be constructed from existing data, it is insufficient as an analog for earlier times. Oceanic circulation was thermally driven in Cenozoic times, but there is evidence (Chapter 7) that it was driven by salinity gradients in the Mesozoic Era.

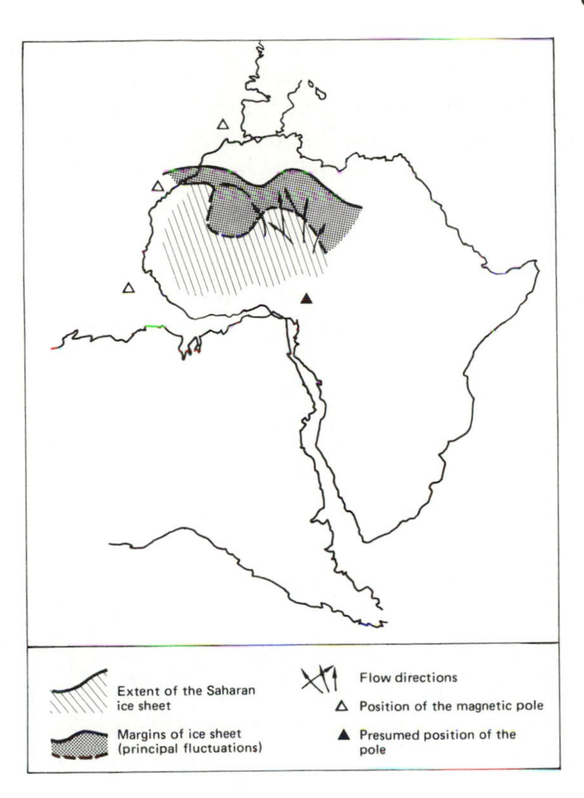

FIGURE 2 Late Ordovician ice flows in the Sahara (after Beuf *et al.*, 1971).

Precambrian Time

For late Precambrian intervals, intercontinental correlation is imprecise and continental masses are difficult to define and to place relative to one another and to the equator. For older periods within the Precambrian, the time slices become ever thicker and the identification of sites of sediment deposition become less precise. Investigations in the Precambrian must therefore focus increasingly on the statistical significance of climatic indicators. As an analogy to such research in the fragmentary Precambrian record, consider what might be learned of the Earth's climate today if it were visited by an unmanned automated spacecraft and lander. Random samples from the present Earth's surface would be revealing. We probably would be able to reconstruct (1) the relative abundance of ocean, shelf, and land environments; (2) the fact that we live in an ice age but with warm tropics; and (3) the presence of considerable diversity in soil types. We might even be able to arrange these soil types in a series from those for which mechanical processes of soil formation predominate to those for which chemical processes prevail.

In the collection of such data from the Precambrian sedimentary record, the overrepresentation or underrepresentation of certain regions or certain rock types needs evaluation. Such distortions come both from differential preservation of rock types and from accidents of exposure and location. Improvement is also needed in understanding the systematic physical and chemical changes that can modify the sedimentary record after deposition and the effect of these postdepositional changes on the distortion of the climatic information. In addition, depositional conditions may not have analogs in the present because the chemical composition of the oceans and atmosphere was different from now. We suggest that studies in organic geochemistry, with its abundance of compounds and pathways, have been underutilized in providing understanding of the postdepositional changes of climate indicators. The diagenetic processes themselves may be climate controlled.

The composition of the early Precambrian atmosphere was significantly different from the composition of the present atmosphere. From astrophysical theory and by comparison with the inferred evolution of other stars, it is assumed that the Sun was somewhat less luminous when the Earth was young (3000 or 4000 Ma). The concept that the Sun has continuously increased its heat output (by about 30 percent) has led to a search for a balancing factor to keep the Earth's temperature hospitable. If the Earth's atmosphere had its current composition throughout geologic time, calculations (Budyko, 1969; Schneider and Dickinson, 1974) predict that the planet would have been permanently glaciated since near its beginning and would not have thawed yet. If the early atmosphere were rich in carbon dioxide, an enhanced greenhouse effect would have kept the mean surface temperature of the Earth above freezing.

The composition of some Precambrian strata indicates that the early atmosphere was poor in oxygen (Cloud, 1965). This has been inferred from study of the banded-iron formations found in ancient rocks over the world (about 2200 Ma). The transport of the iron without the concurrent movement of other materials (e.g., Al_2O_3) to its depositional sites implies a carbon dioxide content of the Precambrian atmosphere approximately double that of the present atmosphere (Rubey, 1951; James, 1966). A similar conclusion for atmospheric composition has been drawn from the Precambrian gold-uranium-pyrite occurrences typified by the Witwatersrand (South Africa) and Jacobina (Brazil) deposits. These have been interpreted as placers. However, the detrital uranium-bearing mineral and the pyrite would have been oxidized and made soluble in the surface waters if the atmosphere were rich in oxygen as it is today (Hutchinson, 1981).

The Paleozoic and Mesozoic Eras

For each geologic epoch within the Paleozoic and Mesozoic Eras, several synoptic intervals hold promise for satisfactory worldwide climate correlation. Within such time slices, those data on the geochemistry of sedimentary rocks and fossils that bear on climate need to be systematically collected and synthesized. For each time slice the former position of continents needs to be reconstructed, incorporating paleomagnetic and other data. Such reconstructions should be augmented by geophysical, geologic, and pertinent compositional information from sedimentologic investigations. The sizes of the error margins also need evaluation. If this is done properly, each reconstruction can start where the previous one left off, thus adding further confidence. Estimates of the latitudinal gradients of temperature and humidity, for each time slice, will emerge from interpretation of the distribution of climate indicators with the continental reconstructions.

Such synoptic analyses are currently being attempted for several periods in the Paleozoic and Mesozoic. See, for example, time-slice analysis in Chapter 21 for Silurian-Devonian times and in Chapter 17 for the Jurassic. Similar analyses have been attempted by Gray and Boucot (1979), Habicht (1979), and Ziegler *et al.* (1979).

The Cenozoic Era

As we move toward the present, the requirements of reconstruction of place and time are met with comparative ease. The reconstruction and interpretation of climate, and especially of ocean-atmosphere interactions, can now move from the level of narrative toward one of numerical simulation. Such simulations have been attempted for the last glacial maximum in the Pleistocene with significant success (CLIMAP Project Members, 1976).

We need synoptic intervals for an Earth with no (or minor) northern ice caps. One time slice closest to the present and useful in this regard is a warm peak in the early

Pliocene. Further back in time, another instructive interval occurred within the early Miocene, when the Antarctic ice cap was small. Stratigraphic correlations are probably adequate within Miocene and Pliocene sequences, and data coverage is good because sediments of these ages are available over wide areas of the seafloor, as well as on continental shelves.

TIME-SERIES STUDIES

Thick continuous stratigraphic sections consisting of numerous thin beds may record climate variability if the bedding is rhythmic or repetitious and the thickness, or some other characteristic, is the response to a seasonal climate signal. Statistical analysis of the recurring signal may disclose variations in climate on a seasonal, annual, or even longer basis. Well-dated continuous sequences invite such analysis, and the investigation of several nearby sequences may enable the sorting out of local or regional responses from those of broader or even global significance.

Time-series analysis can also employ data from many localities if the rocks or sediments can be accurately dated and stratigraphically correlated. Ancient ocean temperatures may be deduced from oxygen isotopic analyses (Figure 3) and are the type of data appropriate to time-series study. Such analyses show the gross variations of a climatic parameter with time, and continuous stratigraphic sections show more detailed variations, which may reflect differences in climatic driving forces.

With long quasi-rhythmic sequences in hand, variability can be extracted both from complete time series and from random sampling within a given section. Where there is order along the time axis, so that a pattern or repetition is recognized, climate cycles may be present. For example, the extraction of such cycles from the Pleistocene record of deep-seas cores by frequency analysis has shown that climate has responded to orbital variations in the recent geologic past (Imbrie and Imbrie, 1980).

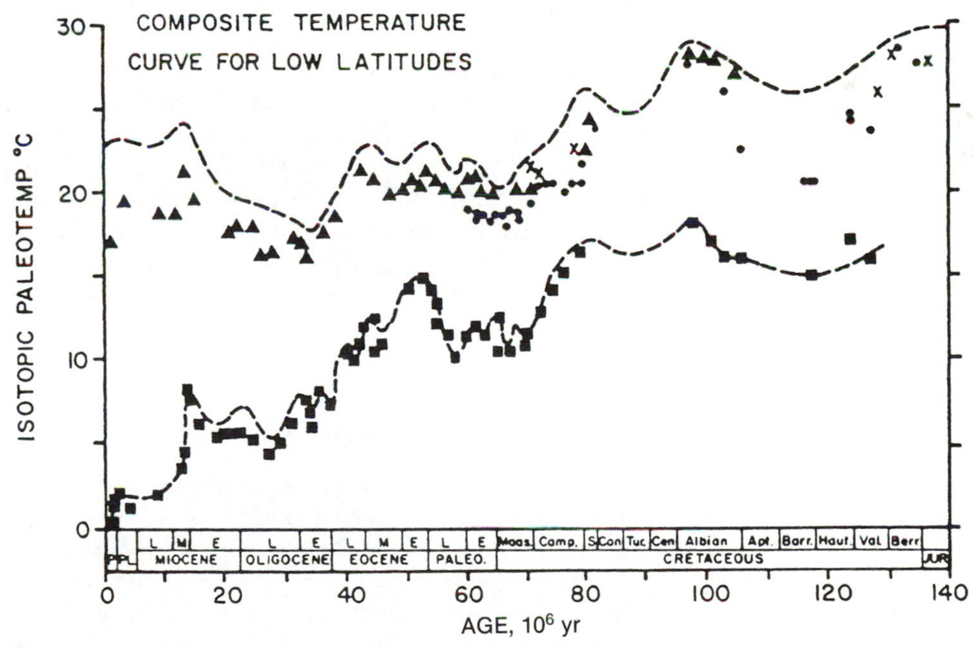

FIGURE 3 Compilation of oxygen isotope paleotemperature data obtained by analysis of benthic and planktonic foraminifera from Deep Sea Drilling Project cores. Bottom curve is drawn through bottom-water data; upper curve is estimate of tropical sea-surface temperatures (Douglas and Woodruff, 1981).

Time-series studies, including both variability analysis and frequency analysis, are possible when continuous global climate signals dominate or are amplified by regional ones. Because of the likelihood of continuous deposition, pelagic sediments from the deep sea, and even older sequences uplifted and exposed on land, are the most promising candidates for time-series study. Well-dated sections provide the most readily interpretable data, but long sections that are only approximately dated may reveal useful repetitions interpretable as having resulted from climate changes. Although the pelagic record is more complete from the mid-Cretaceous and younger seafloor (because of the destruction of ocean crust at subduction zones), ancient sections even as old as Precambrian warrant study. Times in which interesting climatic signals have been observed are the early Pliocene (preceding northern ice caps), the middle Eocene (weak latitudinal temperature gradient), and the late Cretaceous (equable temperature gradient and possibly high carbon dioxide content in the atmosphere).

Pleistocene deep-sea cores show that cyclicity is commonly associated with more than one climate-associated parameter. Phase differences between such parameters contain information, as yet poorly understood, about cause-and-effect chains in the ocean-atmosphere-biosphere-lithosphere system. For example, the coherences and offsets between oxygen isotope cycles, carbon isotope cycles, and carbonate cycles have implications for the sequence of events—such as the buildup and melting of ice, changes in temperature, changes in carbonate sedimentation, and fluctuations in atmospheric carbon dioxide.

Fluctuations of sea level accompany the waxing and waning of continental glaciers. However, nonclimatic factors, such as vertical crustal movements and the rates of seafloor spreading, also cause eustatic sea-level changes. These eustatic changes need to be separated from those caused by climate—in fact they can affect climate. Sea-level fluctuations change the albedo by altering the ratio between exposed land surface and the surface area of inland seas and lakes. Because water on the average is a better absorber of light than is dry land, when sea level is high, increased absorption of solar radiation results. In addition there is an increase in the surface area from which water can enter the atmosphere through evaporation. Hence, humidity rises and there is an increase in the greenhouse effect, which warms the near-surface layers of the atmosphere. Such complex feedback relations as these are still incompletely documented or understood. Are times of high sea level invariably warmer and more equable than those of low sea level? What are the causes of sea-level change? How has sea level changed through geologic time?

A period-by-period and region-by-region compilation of sea-level variation for the continental shelf areas of the world has been initiated by geologists of the petroleum industry employing the techniques of seismic stratigraphy (see Chapter 15). The key to success is exact stratigraphic correlation, using well-dated sections. These data in turn need to be dovetailed with climatic indicators.

EVENT AND EPISODE ANALYSIS

Notable events in Earth history, such as those associated with the termination of the Cretaceous Period, the fragmentation of Gondwana, and widespread volcanism, have left discrete geologic records that will certainly be useful in connection with climatic investigation. Many events may actually occur over longer periods and are more properly referred to as episodes. The analysis of such climatic transitions involves study of conditions prevailing before and after, as well as during, the event. Even though rapid climatic changes are deemed to be caused by single forcing factors, the signals may be overwhelmed by background factors and from competing causes. Careful analysis is required to sort out the various contributing causes of the event. Sea-level change stimulates feedback through a chain of processes that need elucidation. Basic questions regarding many important geologic events are whether they were externally (e.g., extraterrestrially) caused or forced, and how important internal feedback is in the system.

Event analyses of climatic records are successful when gaps in the record are absent or subordinate. A sedimentary sequence whose continuity is established can be characterized statistically by the distribution of events within it. By studying each event in detail, using various climate indicators, it may be possible to distinguish classes of events (Berger *et al.*, 1981).

Times of rapid climate change need to be studied in detail on a worldwide basis if we are to understand the factors that can cause such climatic transitions. At the time of the Cretaceous termination (about 63 Ma), for example, differences in composition of deep-sea sediments (a short-lived increase in iridium concentration) accompanied a marked biological event; this supports the hypothesis of a large meteoritic or cometary impact. Other key times in geologic history that should be elucidated by paleoclimatic investigations are (1) the Albian-Cenomanian (about 100 Ma) transition, (2) the Eocene-Oligocene boundary (38 Ma), (3) the mid-Miocene (15 Ma), (4) the end-of-Miocene (6 Ma), and (5) the mid-Pliocene (3 Ma). A multifaceted approach involving biostratigraphy, magnetostratigraphy, stable isotopes, trace element analysis, and other methods is needed for each of these events and several others.

Some episodes in the distant geologic past may have taken place over millions of years. For example, what happened climatologically at the onset of the Phanerozoic? What caused the appearance of calcareous shells at approximately this time? Was there a change in seawater chemistry or an increase in predation pressure and competition once the first hard shells evolved? In the Precambrian, events may have taken tens of millions of years to rise, culminate, and perhaps to fade—possibly comparable to the long-term cooling in the Cenozoic. Conditions favorable for the formation of some banded-iron formations have not existed for nearly 2 billion years. The origin of intercalations of chert and iron ore in these banded-iron formations is still a subject of research. Perhaps changing climatic and geochemical conditions hold the clue: What were the factors that made deposition of banded-iron formations possible? What environmental condition changed when their deposition ceased?

NEW OPPORTUNITIES IN PALEOCLIMATOLOGY

Climate is determined by two major classes of factors. The first class constitutes the "static setting." This is the arrangement of continents and ocean basins, which can be considered fixed on a short time scale. The second class of factors involves what may be called the "dynamic setting." This includes oceanic and atmospheric circulation, heat transfer, albedo distributions, biosphere patterns, and atmospheric concentration of infrared absorbing gases. A goal of paleoclimatology is to interrelate the static and dynamic settings and determine how they operate together.

New opportunities arise from our present understanding of plate tectonics and continental drift. For times back to about 200 Ma, we know the geography of the static setting and can concentrate on the dynamics of the fluids within it. We now have a set of concepts and methods developed by climatologists working on Pleistocene and modern climates. These concepts and methods can be tested for applicability to pre-Pleistocene climates and promise to reveal much about why long-term climate changes have occurred.

Two examples illustrate the opportunities. The first is the map of the surface of the Earth at the time of the last maximum of glaciation, 18,000 years ago, based on integration of paleoclimatic information from land and from the seafloor (CLIMAP

FIGURE 4 Two long sediment cores from Deep Sea Drilling Project Site 480. Core A was obtained with standard coring procedures, Core B with the recently developed hydraulic piston corer. The layering obvious in the hydraulic piston core will allow detailed climatic analysis heretofore difficult with standard cores.

Project Members, 1976; Denton and Hughes, 1981). Paleotemperature and albedo distributions are presented on a scale fine enough to be useful for numerical modeling of the paleoclimate (Gates, 1976; Manabe and Hahn, 1977).

The second example concerns time-series analysis in which a long climatic record, such as a long deep-sea core (Figure 4), is examined for repetitions and cycles of a climatic signal (Hays *et al.*, 1976; Pisias, 1976). These studies are useful in single stratigraphic sections and offer an escape from many of the frustrations associated with time-slice mapping, where correlations over large distances may be uncertain. Such studies yield insights about the dynamics and variability of climate systems at any one site that cannot be obtained from spatial reconstructions alone. Analysis of Pleistocene climatic records inferred from deep-sea sediments has established relations between orbital parameters and glaciation by identifying astronomical cycles of 19,000 yr, 23,000 yr, 41,000 yr, and perhaps 100,000 yr within the geologic record. A search for these cycles in older records and for other cycles is now possible; they may be present in Mesozoic, Paleozoic, and perhaps even Precambrian sequences. The task is to find long, continuous stratal sequences recording a signal (e.g., thickness variations in limestone-shale couplets) that permits frequency analysis. Even where the time scale is not known accurately, the ratios of different cycles or repetitions may eventually lead to the establishment of such time scales, and the identification of cycles will aid in long-distance correlation.

Existence of both longer and shorter cycles than those orbital cycles so far identified is likely. The analysis of long-period variations in annual layers, if they can be recognized, through geologic time may yield information on the variations of solar radiation incident on the Earth. Unfortunately, there is no simple criterion by which to prove that layering is truly annual. Long-cycle lengths of approximately 250,000 and 450,000 years have been proposed, as well as frequencies of 4-5 m.y., 30 m.y., and 200 m.y. (see Fischer, Chapter 9). Usually such suggestions rest on records that are only two to five times longer than the proposed "cycle," so that they cannot yet be established quantitatively.

THE SYSTEMS APPROACH

Because climate is the result of the air and ocean circulation and their interactions within a geologic setting, a dynamic fluid system is defined, a system that may lend itself to systems analysis. Sediments deposited by these fluids record a composite of global signals and regional or local overprints. It would be useful to attempt to separate the global message from the regional modulations and to investigate the sensitivity of the system. Global effects may be rapid with quick response times or with phase shifts between different components. Regional influences may amplify or dampen or even reverse global signals. Climate extremes and rapid transitions from one state to another are probably associated with times when factors influencing global climate are maximized.

Studies that focus on times of instability and rapid climate transition within an overall stable record may reveal why evolution is apparently punctuated by mass extinctions and relatively rapid speciation, such as those marking the end of the Paleozoic and Mesozoic eras. To what extent were climatic extremes due to variations within the system or to factors external to the system, such as those of extraterrestrial origin? Important questions about extraterrestrial factors are the following: How stable has the solar constant been? Has the Sun's output varied in an oscillatory manner? What influence might large impacts on the Earth (similar to those that produced large craters on the Moon) have had on climate evolution? Have there been times when havoc has been raised in the Earth's surface environment by huge meteorites falling from the sky, as has been suggested for the end of the Cretaceous by Alvarez *et al.* (1980)?

Factors that are not related to the fluid system but are related to the dynamics and mobile Earth include tectonic forces and volcanic activity. Continents have moved about and oceans have opened and widened and closed through geologic time. Mountain ranges rose in many orientations that influenced the air circulation and then were worn away. These marked changes in size and shape and positioning of landmasses, oceans, and mountains have dramatically influenced climates. Ocean gateways between landmasses have opened and closed and have altered the flow of ocean currents and modified the mechanisms of heat transport from low to high latitudes. Ancient ice ages during the Phanerozoic (see Figure 2) may be largely the consequence of times when continents were sited in high latitudes so that air-ocean flow brought heavy and long-lasting snowfall, with the result that ice caps grew upon them. The location of landmasses, the level of the sea, and the presence of ocean gateways in crucial positions are several of the boundary conditions that are responsible for the sensitivity of the fluid system to perturbation.

Geochemical aspects are also intimately tied to climate, mainly through the carbon dioxide content of the atmosphere; increases in carbon dioxide raise the surface temperature. On a long time scale, the average atmospheric carbon dioxide content may be fixed by weathering reactions and carbon storage in soil, rocks, and seawater. However, on a shorter time scale, the atmospheric reservoir is dominated by the much larger adjoining reservoirs of the ocean, the biosphere-soil complex, and the reactive portion of marine organic and calcareous sediments. As the exchange patterns between these reservoirs change, so must the carbon dioxide content of the atmosphere. Rapid increase or decrease of carbon dioxide content that results from temporary buildup or release of carbonate or organic matter may cause, or be associated with, "climate steps" from one climatic state to another (Berger *et al.*, 1981; Revelle, 1981). The existence of steps suggests the existence of inherent instability within the system; this may be tested by a comparison of climatic variability at times long before the event, approaching the event, and after the event.

FINDINGS AND RECOMMENDATIONS

Interpretation of the history of climates on Earth back in geologic time for more than a billion years indicates that the same principles of oceanic and atmospheric circulation have always operated. Over this long interval, however, the composition of the atmosphere and oceans has changed in ways not yet fully determined. Climatic extremes that we are experiencing at present, and that have occurred for the past million years or so, are probably similar to climatic extremes back into the very remote past. Only at rare intervals do the many variables influencing climate so combine that they perturb the system beyond a quasi-steady state, bringing about a transition to a new temperature regime, but one actually not very different from the previous one.

The record of climate and its changes during the last 50 million years reveals a gradual cooling interrupted by a few pertubations. Fairly rapid transitions or steps resulting in a cooler mean surface temperature of a few degrees Celsius have been followed by partial recoveries. Superimposed rhythms on these fluctuations, detectable in the Pleistocene record but so far not discovered in older rocks, are brought about by systematic changes in the Earth's axis and orbit.

Understanding of climate and climatic change, with the eventual goal of arriving at a satisfactory theory of climate, requires an improved data base and more intense efforts in analysis and interpretation. Many disciplines are involved. Toward these ends, we recommend the following:

1. *Geologists, during their investigation of rocks, should continue to strive to glean information bearing on past climates.* Economic geologists, for example, recognize that certain types of mineral deposits were formed under special climatic conditions. Through an understanding of weathering and other climate-controlled processes they may be able to extract climatic information. Even the most ancient sedimentary rocks may reveal some information on climates.

2. Marine geologists and geochemists have a new tool in hand—the hydraulic piston corer—and should obtain many more long sediment cores from the seafloor. Valuable paleoclimatic data, highly resolved with respect to time, can come from well-placed cores, especially from localities where global climatic signals are not masked by local sediment influxes. *We urge the continuation of programs using hydraulic piston cores in any ocean-drilling effort.*

3. *Paleogeographers*, working closely with paleontologists, tectonicists, and stratigraphers, *should prepare paleogeographic maps of the Earth for all practical time intervals of Earth history.* These intervals can be short for time slices in the Tertiary and by necessity must be longer for intervals successively more remote. Such maps will serve as a basis for assembling climatic data and as guides for computer modeling.

4. Meteorologists and climatologists, working at many different scales, should attempt to arrive at explanations for ancient climates. Although satisfactory computer models do not yet exist to explain all aspects of modern climates, such studies for ancient synoptic intervals can nevertheless serve as guides to explain climatic events. Therefore, *models of several types should be applied to the paleogeographic reconstructions of selected times of climatic significance in the geologic past.*

5. *Investigators of sea-level changes should be encouraged to complete as accurately as possible a worldwide eustatic sea-level curve for the geologic past.* Such a curve will aid greatly in correlation from place to place, when the record of advances and retreats of the sea is identified. If other factors remain unchanged, increased flooding should correlate with increased global temperatures because of albedo reduction and increased evaporation. These inferences can be checked independently against the geologic record.

6. *Long stratal sequences displaying variations should be analyzed using the time-series approach.* These sequences may come from long deep-sea cores or from sections exposed on land. Their study may reveal repetitions and cycles in bedding

features, such as thickness or composition, that are related to climate. Among these are orbital and axial phases that influence the amount of radiative energy received from the Sun. With some luck, even ancient Precambrian sedimentary rocks may disclose such information.

REFERENCES

Alvarez, L. W., W. Alvarez, F. Asaro, and H. V. Michel (1980). Extraterrestrial cause for the Cretaceous-Tertiary extinctions: Experiment and theory, *Science 208*, 1095-1108.

Berger, W. H., E. Vincent, and H. R. Thierstein (1981). The deep-sea record: Major steps in Cenozoic ocean evolution, in *Symposium and Results of Deep-Sea Drilling*, R. G. Douglas, J. Warme, and E. L. Winterer, eds., Soc. Econ. Paleontol. Mineral. Spec. Publ. 32.

Berggren, W. A. (1972). Late Pliocene-Pleistocene glaciation, in *Initial Reports of the Deep Sea Drilling Project 12*, U.S. Government Printing Office, Washington, D.C., pp. 953-963.

Beuf, S., B. Biju-Duval, O. DeCharpal, P. Rognon, O. Gariel, and A. Bennacef (1971). *Les Grès du Paléozoique Inférieur au Sahara—Sédimentation et Discontinuitiés, Evolution Structurale d'un Craton*, Inst. Fr. Pétroles-Sci. Tech. Pétroles 18, 464 pp.

Budyko, M. I. (1969). The effect of solar radiation variations on the climate of the Earth, *Tellus 21*, 611-619.

CLIMAP Project Members (1976). The surface of the ice-age Earth, *Science 191*, 1131-1137.

Cloud, P. E., Jr. (1965). Significance of the Gunflint (Precambrian) flora, *Science 148*, 27-35.

Denton, G. H., and T. J. Hughes (1981). *The Last Great Ice Sheets*, Wiley-Interscience, New York, 489 pp.

Douglas, R. G., and F. Woodruff (1981). Deep sea benthic foraminifera, in *The Sea*, Vol. 7, C. Emiliani, ed., Wiley, New York.

Frakes, L. A. (1979). *Climates Throughout Geologic Time*, Elsevier, Amsterdam, 310 pp.

Gates, W. L. (1976). The numerical simulation of ice-age climate with a global general circulation model, *J. Atmos. Sci. 33*, 1844-1873.

Gray, J., and A. J. Boucot, eds. (1979). *Historical Biogeography, Plate Tectonics, and the Changing Environment*, Oregon State U. Press, Corvallis, Ore., 500 pp.

Habicht, J. K. A. (1979). *Paleoclimate, Paleomagnetism, and Continental Drift*, AAPG Studies in Geology No. 9, American Association of Petroleum Geologists, Tulsa, Okla., 31 pp. + 11 maps.

Hambrey, M. J., and W. B. Harland (1981). *Earth's Pre-Pleistocene Glacial Record*, Cambridge U. Press, New York, 1004 pp.

Hansen, J., D. Johnson, A. Lacis, S. Lebedeff, P. Lee, D. Rind, and G. Russell (1981). Climate impact of increasing atmospheric carbon dioxide, *Science 213*, 957.

Hays, J. D., J. Imbrie, and N. J. Shackelton (1976). Variations in the Earth's orbit, pacemaker of the ice ages, *Science 194*, 1121-1132.

Hutchinson, R. W. (1981). Mineral deposits as guides to supracrustal evolution, in *Evolution of the Earth*, R. J. O'Connell and W. S. Fyfe, eds., Geodynamic Series, Geological Society of America and American Geophysical Union, Washington, D.C., pp. 120-140.

Imbrie, J., and J. Z. Imbrie (1980). Modeling the climate response to orbital variations, *Science 207*, 943-953.

Imbrie, J., and K. P. Imbrie (1979). *Ice Ages—Solving the Mystery*, Enslow Publishers, Short Hills, N.J., 224 pp.

James, H. L. (1966). Chemistry of the iron-rich sedimentary rocks, *U.S. Geol. Surv. Prof. Pap. 440.*

Manabe, S., and D. G. Hahn (1977). Simulation of the tropical climate of an ice age, *J. Geophys. Res. 82*, 3389-3911.

McElhinny, M. W., and D. A. Valencio, eds. (1981). *Paleoreconstruction of the Continents*, Geodynamic Series, Geological Society of America and American Geophysical Union, Washington, D.C., 200 pp.

NRC Climate Board (1982). *CO_2 and Climate: A Second Assessment*, National Academy Press, Washington, D.C., 72 pp.

NRC Climate Research Board (1979). *Carbon Dioxide and Climate: A Scientific Assessment*, National Academy of Sciences, Washington, D.C., 22 pp.

NRC Geophysics Study Committee (1977). *Energy and Climate*, National Academy of Sciences, Washington, D.C., 158 pp.

NRC U.S. Committee for the Global Atmospheric Research Program (1975). *Understanding Climatic Change*, National Academy of Sciences, Washington, D.C., 239 pp.

Pisias, N. G. (1976). Late Quaternary variations in sedimentation rate in the Panama Basin and the identification of orbital frequencies in carbonate and opal deposition rates, *Geol. Soc. Am. Mem. 145*, 375-391.

Revelle, R. R. (1981). Introduction to *The Sea*, Vol. 7, C. Emiliani, ed., Wiley, New York.

Rubey, W. W. (1951). The geologic history of sea water, *Geol. Soc. Am. Bull.* *62*, 1111-1147.

Schneider, S. H., and R. E. Dickinson (1974). Climate modeling, *Rev. Geophys. Space Phys.* *12*, 447-493.

Seyfert, C. K., and L. A. Sirkin (1979). *Earth History and Plate Tectonics: An Introduction to Historical Geology*, Harper & Row, New York.

Shackleton, N. H., and N. D. Opdyke (1977). Oxygen isotope and paleomagnetic evidence for early northern hemisphere glaciation, *Nature* *270*, 216-219.

U.S. Congress (1978). *National Climate Program Act*, Public Law 95-367, 95th Congress, September 17, 1978.

Ziegler, A. M., C. R. Scotese, W. S. McKerrow, M. E. Johnson, and R. K. Bambach (1979). Paleozoic paleogeography, *Ann. Rev. Earth Planet. Sci.* *7*, 473–502.

BACKGROUND

The Role of Prediction in Paleoclimatology

1

JOHN IMBRIE
Brown University

INTRODUCTION

In the early, descriptive phase of any historical science, such as paleoclimatology, the primary task is to provide a narrative of past events. As facts accumulate, investigation enters an interpretive phase in which attempts are made to explain why these events occurred. Initially, these explanations are likely to be *ad hoc* in nature, verbal in form, and therefore difficult to pin down and test. Gradually, as data accumulate and as physical insights into causal mechanisms improve, it may be possible to transform such qualitative explanations into a numerical model of the natural system in which specific responses (system output) can be estimated from a knowledge of specific forcing functions (system input). The model can then be used to make *hindcasts*, i.e., to make predictions of system behavior over some past interval of time for which the forcing functions are known. The principal value of such a model is that the concepts on which the model is based can be tested by comparing hindcasts with observations. The test is particularly powerful if comparisons are made over an interval of time that was not used to develop the model.

In some cases it is possible to use a quantitative model to make *forecasts*, i.e., to make predictions of system behavior over some future interval of time. In order to use a model in this forecast mode, one must either be able to predict what values the forcing function will assume in the future or be investigating a system in which the behavior at any given time is strongly conditioned by previous values of a forcing function.

For many years, the science of paleoclimatology has been in a descriptive phase. But as other chapters in this volume make clear, the field has already begun to move into an interpretive phase. So far, however, most interpretations are formulated as verbal narratives rather than as quantitative predictions. The aim of this chapter is to point out the value of building paleoclimatic models that are capable of making quantitative predictions and to foster the development of such models by reviewing some of the types of quantitative, predictive models that have been used in climatology.

The first section of this chapter is a brief review of the nature of the climate system. The second section considers some general properties of quantitative, deterministic models. In the last section, six examples of quantitative climatic predictions are cited.

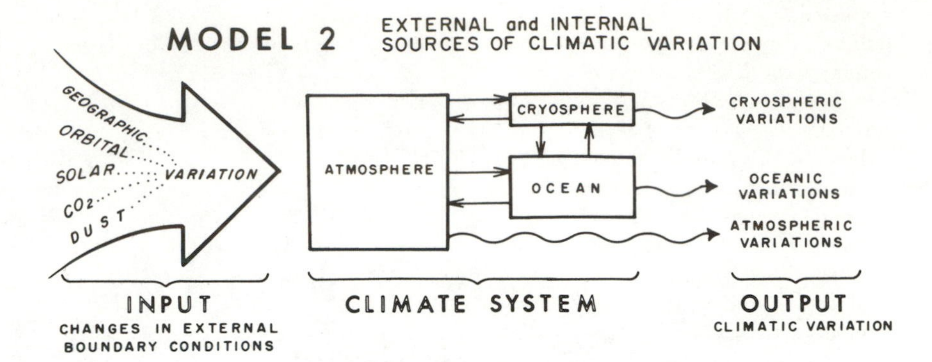

FIGURE 1.1 A simple model of the climate system illustrating internal and external sources of climatic variation.

THE CLIMATE SYSTEM

In order to formulate a predictive climate model it is necessary to have in mind some clear definition of the climate system. A conventional and useful definition (Figure 1.1) considers the climate system to consist of four interacting components: the atmosphere, the ocean, the cryosphere, and the surface of the land, including the terrestrial biota (NRC Panel on Climatic Variation, U.S. Committee for GARP, 1975). For simplicity, the last of these components is not shown in Figure 1.1, and the other components are not subdivided. The system boundary conditions are defined as the global geography, the geometry of the Earth's orbit, the output of the Sun, and the concentrations of carbon dioxide and dust in the atmosphere. Two of these variables—orbital geometry and solar output—are clearly external to the climate system by any definition, because the values they assume are not influenced by the climatic state. However, some aspects of global geography—notably sea level and the elevation of the crust in high latitudes—are not completely independent of climate, so that the listing of geography as an external boundary condition makes sense only if the term is defined as the location of continents and oceans. Finally, we should note that the atmospheric concentrations of carbon dioxide and dust depend to some extent on the climatic state and are therefore not, strictly speaking, external variables. They are often placed in that category for the convenience of modelers, however.

Based on the definitions discussed above and illustrated in Figure 1.1, two quite different sources of climatic variation can be recognized (Mitchell, 1976). One of these sources is a change in external boundary conditions. Any such change will force the system as a whole to move toward a new equilibrium state and will therefore give rise to variations in the cryosphere, ocean, and atmosphere. This kind of climatic variation is described as being *externally forced*. Another type of climatic variation originates within the climatic system itself owing to interactions among the components of the system. These variations are *internally forced* and will occur even if all the external boundary conditions are fixed.

PROPERTIES OF DETERMINISTIC CLIMATE MODELS

From the above comments about the structure of the climate system, it is clear that the task of making quantitative predictive models is a challenging one (see Schneider and Dickinson, 1974, and Leith, 1978, for useful reviews). In general, there are two types of predictive models, *deterministic* and *stochastic* (Mitchell, 1976). Deterministic models view changes in one or more components (or subcomponents) of the climate system as forced by (1) changes in other components or (2) changes in external boundary conditions. Moreover, such models assume that the actual climatic narrative—the time-dependent behavior of the system component being modeled—can be predicted from a knowledge of changes in other internal components or from a knowledge of changes in external boundary conditions. Stochastic models, on the other hand, assume that random variations of internal origin make it impossible to predict the time-dependent behavior of the system. Instead, they make statistical predictions concerning the amount of variability to be expected over particular frequency bands (Hasselmann, 1976; Frankignoul and Hasselmann, 1977).

This chapter will discuss only deterministic models. Before proceeding, we examine some general elements of modeling strategy (Table 1.1). For example, in certain deterministic models the forcing function is identified as a change in one or more *external* boundary conditions (e.g., a change in global geography or orbital geometry). In other models, however, a change in one component of the climate system (e.g., atmospheric temperature) is viewed as a response to a change in another component (e.g., the sea-surface temperature or the extent of ice sheets), without taking feedback relationships

TABLE 1.1 Some Properties of Deterministic Climate Models

Category	Possible Modeling Strategies
Nature of forcing	External or quasi-external
Type of response	Equilibrium or time-dependent
Prediction mode	Hindcast or forecast
Mathematical form	A wide range from simple, highly parameterized models to complex statistical-dynamical models

between the prescribed variable (the predictor) and the calculated response (the predictand) into account. This type of forcing will be referred to as *quasi-external* forcing.

Another distinction to be made between modeling strategies has to do with the type of response. In many cases the response calculated is an *equilibrium* response to some specified change in an external or quasi-external forcing function. In this type of model, attention is focused on the new equilibrium value rather than on the time-dependent nature of the change to the new equilibrium. Such a formulation is reasonable provided that the characteristic time scale of changes in the forcing function is much longer than the characteristic time scale of the response. However, when these two time scales are of the same order of magnitude, so that the response never has a chance to approach equilibrium values, it is desirable to use a *time-dependent* model in which the response is explicitly calculated as a function of time. Such time-dependent models are sometimes referred to as *differential models* (Imbrie and Imbrie, 1980).

As previously discussed, predictions can be made in either the *hindcast* or *forecast* mode, depending on whether the changes modeled have actually occurred.

Finally, we should note that a wide variety of mathematical procedures are used in climate modeling. For many purposes it is useful to consider this variety as ranging across a wide range of complexity, from simple statistical models at one end of the scale to complex, statistical-dynamical models at the other. The simplest models make little demands on our knowledge of the underlying physics. The investigator is content to assume that the response being modeled is linearly related to one or more input variables, and his modeling strategy is to use regression methods to extract the constants of proportionality from a given set of data. As knowledge of the underlying mechanisms becomes more secure, the investigator may attempt to capture certain elements of the physical processes in a small number of differential equations, leaving the balance to be treated as statistical parameterizations. Such models of intermediate complexity predict the behavior of only a small number of climatic variables. At the other end of the range of complexity, more of the underlying physics is explicitly captured in differential equations, and fewer processes are treated statistically. Examples include the general-circulation models of the atmosphere, which take a global array of seasonal input parameters and calculate the seasonal response in three dimensions (Gates, Chapter 2 in this volume, 1976a, 1976b; Manabe and Hahn, 1977).

EXAMPLES

Examples to illustrate some of the modeling strategies already in use are given in Table 1.2. Donn and Shaw (1977) predict the atmospheric temperature field as a response to changing global geography since the Triassic (Figure 1.2). Imbrie and Imbrie (1980) hindcast late Pleistocene glacial history and forecast the glacial history of the next 100,000 years. Their model is shown in Figures 1.3 and 1.4. Gates (1976a, 1976b) and Manabe and Hahn (1977) have used estimates of ice-sheet distribution and sea-surface temperatures at the last glacial maximum, 18,000 years ago, as a basis for calculating the equilibrium response of the global atmosphere. Namias (1980) uses information on Pacific sea-surface temperature, and other information, to forecast weather patterns over the United States one season in advance. Barnett (1981) uses information on the Pacific wind field to hindcast sea-surface temperature off Peru, 12 months in advance (Figure 1.5).

TABLE 1.2 Examples of Climate Prediction Using Quantitative, Deterministic Models

Type of Forcing	Predictor (Characteristic Time Scale in Years[a])	Predictand (Characteristic Time Scale in Years[a])	Reference	Prediction Mode	Type of Model
External	Global geography (10^7)	Atmospheric temperature field (10^{-1})	Donn and Shaw (1977)	Hindcast	Complex equilibrium
	Orbital geometry (10^4)	Global ice volume (10^4)	Imbrie and Imbrie (1980)	Hindcast, forecast	Simple time-dependent
Quasi-external	Global SST field Ice sheets (10^4)	Global atmospheric fields of T, SST, winds (10^{-1})	Gates (1976b); Manabe and Hahn (1977) CLIMAP Project Members (1976; in press)	Hindcast	Complex equilibrium
Quasi-external	Pacific SST (10^{-1})	U.S. weather (10^{-1})	Namias (1980)	Forecast	Simple time-dependent
	Pacific wind field (10^{-1})	SST off Peru (10^{-1})	Barnett (1981)	Hindcast	Simple time-dependent

[a]T, temperature; SST, sea-surface temperature.

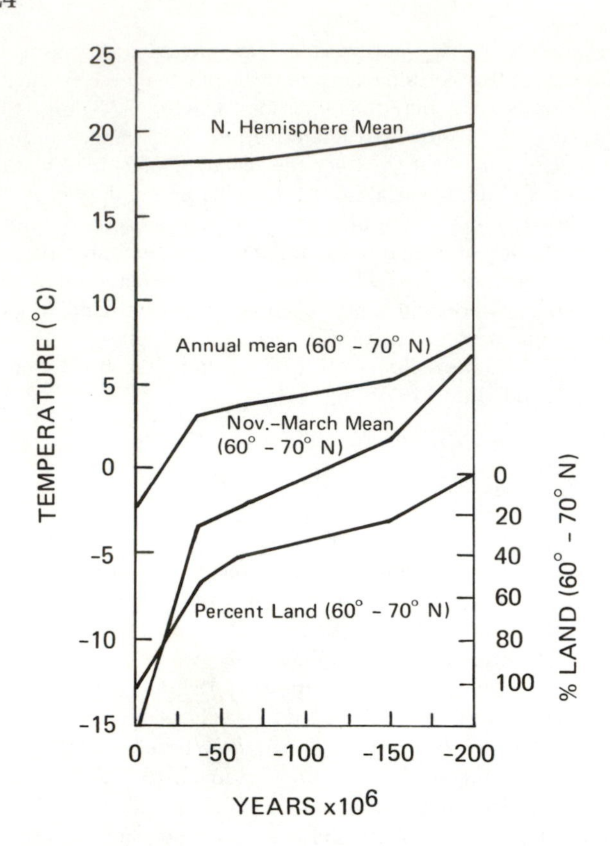

FIGURE 1.2 Change in mean temperature for the northern hemisphere (upper curve) and 60-70° N latitudinal zone (middle curves) computed for the past 200 m.y. Lower point for present on hemispheric curve is based on actual unaveraged grid-point data. Other curves are based on zonal means for land and water, respectively. Lower curve shows percentage of land in 60-70° N latitudinal zone (from Donn and Shaw, 1977, with permission of the Geological Society of America).

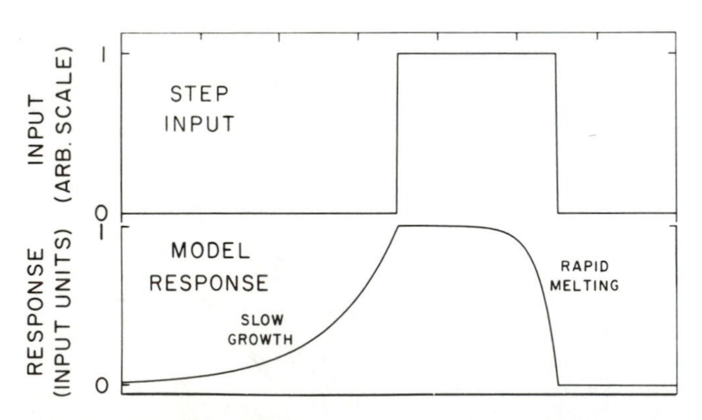

FIGURE 1.3 Response characteristics of a simple model for predicting global ice volume as a function of changes in orbital geometry. The response function has a mean time constant of 17,000 years and a ratio of 4:1 between the time constants of glacial growth and melting. Lower curve shows the system response to an arbitrary step input (Imbrie and Imbrie, 1980, copyright 1980 by the American Association for the Advancement of Science).

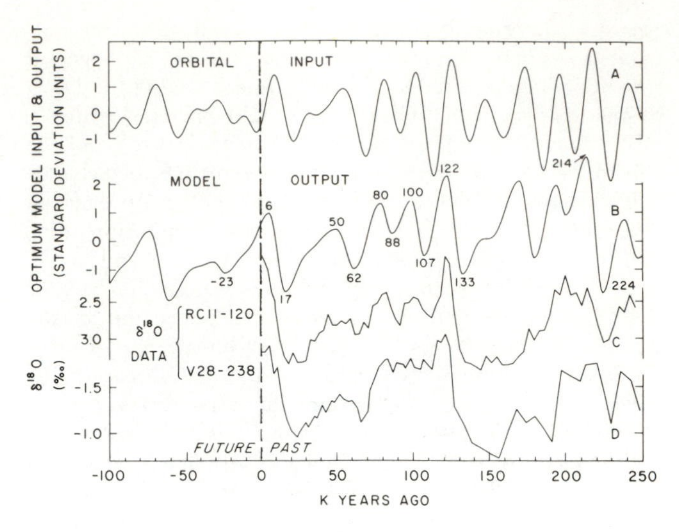

FIGURE 1.4 Hindcasts and forecasts of global ice volume derived from the response model shown in Figure 1.3. A, Input to the model. This curve represents changes in the incoming radiation during July at 65° N. B, Model output. C, Oxygen isotope curve for deep-sea core RC11-120 from the southern Indian Ocean. D, Oxygen isotope curve for deep-sea core V28-238 from the Pacific Ocean. Curves C and D are taken as estimates of changes in the global volume of land ice (Imbrie and Imbrie, 1980, copyright 1980 by the American Association for the Advancement of Science).

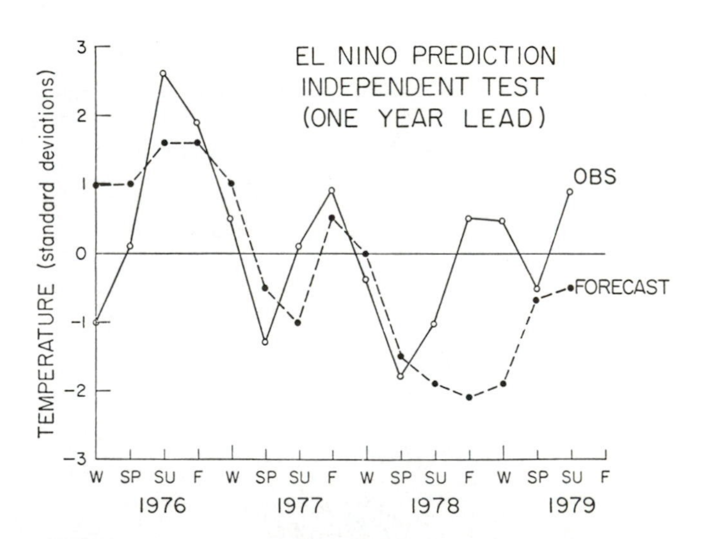

FIGURE 1.5 Observed and predicted values of sea-surface temperature off Peru (Barnett, 1981). The model uses variations in the Pacific wind field as input and is calibrated for the period 1950-1975. Thus Barnett considers the predictions as forecasts, whereas in this chapter they are defined as hindcasts. Unpublished work by T. P. Barnett, Scripps Institution of Oceanography, shows that the predictions are considerably improved if the assumption of stationarity of the seasonal signals is removed.

CONCLUSIONS

The examples of climatic hindcasting and forecasting reviewed in this paper are cited not because they solve fundamental problems in paleoclimatology, but because they illustrate a powerful method of attack on those problems: the formulation and testing of numerical, predictive models. So long as our explanations for past climates remain essentially qualitative in form, causal theories will be difficult to evaluate. It will, for example, be difficult to distinguish causal relationships from mere correlations and perhaps impossible to assess the relative importance of different causes that act together to produce a given effect. These problems are particularly acute in pre-Pleistocene climatology. Here it is often tempting to explain a given climatic change in terms of a correlated change in some particular aspect of the geometry of continents and oceans—ignoring other aspects of that geometry and implicitly denying the possibility that changes in atmospheric chemistry, in solar output, or in other climatic boundary conditions may be involved.

In the rapidly developing and challenging field of paleoclimatology, there is clearly an urgent need to test our ideas about the underlying physical mechanisms by developing quantitative, predictive models. As such models are substituted for the evasive verbal narratives that now serve as explanations of past climates, the science of paleoclimatology will come of age.

REFERENCES

Barnett, T. P. (1981) On the predictability of ocean/atmosphere fluctuations in the tropical Pacific, *J. Phys. Oceanogr. 11*.

CLIMAP Project Members (1976). The surface of the ice-age Earth, *Science 191*, 1131-1137.

CLIMAP Project Members (in press). Seasonal reconstructions of the Earth's surface at the last glacial maximum, *Geol. Soc. Am. Map and Chart Series*, No. 36.

Donn, W. L., and D. M. Shaw (1977). Model of climate evolution based on continental drift and polar wandering, *Geol. Soc. Am. Bull. 88*, 390-396.

Frankignoul, C., and K. Hasselmann (1977). Stochastic climate models, part II. Application to sea-surface temperature anomalies and thermocline variability, *Tellus 29*, 284-305.

Gates, W. L. (1976a). Modeling the ice-age climate, *Science 191*, 1138-1144.

Gates, W. L. (1976b) The numerical simulation of ice-age climate with a global general circulation model, *J. Atmos. Sci. 33*, 1844-1873.

Hasselmann, K. (1976). Stochastic climate models, part I. Theory, *Tellus 28*, 473-485.

Imbrie, J., and J. Z. Imbrie (1980). Modeling the climatic response to orbital variations, *Science 207*, 943-953.

Leith, C. E. (1978). Predictability of climate, *Nature 276*, 352-355.

Manabe, S., and D. G. Hahn (1977). Simulation of the tropical climate of an ice age, *J. Geophys. Res. 82*, 3889-3911.

Mitchell, J. M. (1976). An overview of climatic variability and its causal mechanisms, *Quat. Res. 6*, 481-493.

Namais, J. (1980). The art and science of long-range forecasting, *EOS Trans. Am. Geophys. Union 61*, 449-450.

NRC Panel on Climatic Variation, U.S. Committee for GARP (1975). *Understanding Climatic Change: A Program for Action*, National Academy of Sciences, Washington, D.C., 239 pp.

Schneider, S. H., and R. E. Dickinson (1974). Climate modeling, *Rev. Geophys. Space Phys. 12*, 447-493.

Paleoclimatic Modeling—A Review with Reference to Problems and Prospects for the Pre-Pleistocene

2

W. LAWRENCE GATES
Oregon State University

INTRODUCTION

While problems of paleoclimate have long held the attention of geologists, biologists, and climatologists, they have only recently been regarded as a subject to which mathematical modeling could be applied successfully. From painstakingly assembled evidence of climatic changes in the geologic past at scattered locations around the world, the volume of paleoclimatic data has caused a veritable flood during recent years as new techniques of dating and new sources of information have been developed. The unifying key to this evidence of global pre-Pleistocene climates is the theory of plate tectonics, through which the relative global geometry of the continents and oceans may be at least tentatively reconstructed. These reconstructions provide the large-scale boundary conditions for climate, whereas the accompanying biological and geophysical data provide verifying evidence for climatic change; what is missing is the reconstruction of the paleoclimate itself, so that the global distribution and evolution of the climate over geologic time may be seen in parallel to the large-scale changes of the Earth's surface and the evolution of the Earth's flora and fauna. This reconstruction of the Earth's climatic history is one of the great unsolved problems of geophysics and poses a particularly exciting challenge to the climate modeler.

In addition to its unifying role in paleoclimatic research, the modeling of the structure and evolution of climate over geologic time presents a unique opportunity to test the performance of climate models under the widest possible variety of boundary conditions. The comparison of modeled paleoclimates with the paleoclimatic evidence is a valuable supplement to the models' calibration, which would otherwise be confined to modern climates, and thereby increases confidence in the models' use in the estimation of possible future climates. The systematic evaluation of model-simulated paleoclimates may also be expected to serve as a guide in the identification of the accuracy needed in paleoclimatic data and in the assembly and analysis of new data in regions of particular climatic significance. In view of the uncertainty that surrounds the global distribution of many of the pre-Pleistocene boundary conditions required by climate models, disagreement between the modeled and "observed" paleoclimate may be at least partly due to this source; experimentation with alternate but plausible distributions of mountains and shallow seas, for example, may improve the accuracy of both the simulated climate and the geologic reconstruction. Once a best fit is obtained for a particular geologic epoch, further experimentation may permit the testing of hypotheses of paleoclimatic change and thereby provide the elements of a consistent physical synthesis of pre-Pleistocene climates.

The purpose of this chapter is to review and evaluate climate modeling as a tool for paleoclimatic research. To this end, I shall first present an overview of the techniques and present status of climate modeling, followed by a brief review of past applications of climate models to paleoclimatology. With this background, I shall then consider the development of a research strategy for modeling pre-Pleistocene climates in particular. More comprehensive reviews of climate modeling have been given by Schneider and Dickinson (1974) and Saltzman (1978), and current model performance has been reviewed by the World Meteorological Organization (1979).

THE CLIMATE SYSTEM AND CLIMATE MODELING

Underlying much of the resurgence of interest in the problem of climate that has occurred in the past decade or so is the increasingly clear demonstration that numerical models are capable of simulating many important aspects of the global climate with considerable accuracy. This has prompted the use of such models to study the climatic impacts of a variety of factors, such as increased CO_2 concentration in the atmosphere and changes in the character of the Earth's surface. Even though climate models have not yet been applied extensively to the prediction of the time-dependent evolution of future climate in terms of specific seasons or specific years, experimentation with climate models has provided new insight into the many factors that can influence climate and its changes.

The Physical Basis of Climate Modeling

The Earth's climate is the result of the interaction among a wide variety of physical processes, many of which are familiar as the ingredients of daily weather. On the longer time scales of climate, however, it is not only the atmosphere that is involved but also the behavior of the world's oceans and ice masses and the nature of the land surface and the associated biomass. These elements are usually identified as components of the *climate system*, which thus formally includes the atmosphere, hydrosphere, cryosphere, surface lithosphere, and global biomass (NRC Panel on Climatic Variation, U.S. Committee for GARP, 1975). These components of the complete climate system are shown schematically in Figure 2.1, together with some of the principal interactions among them.

The *atmosphere* is of course the most prominent and familiar component of the climate system and is central to the common perception of climate itself. In addition to the average of statistical distribution of the wind, pressure, temperature, and humidity, the atmospheric climate includes the distribution of cloudiness and precipitation, the distribution of shortwave and long-wave radiation, and the atmospheric composition and aerosol concentration. On the global scale the distribution of many of these atmospheric properties is closely related to the surface properties of the *ocean*, including the ocean's configuration relative to the continents. The oceanic climate itself includes the statistical distribution of ocean temperature, salinity, and current, together with the distribution of dissolved gases and suspended matter. Of particular importance to the atmosphere is the distribution of the sea-surface temperature, which largely controls the vertical flux of heat and moisture through the atmospheric boundary layer.

In addition to the properties of the atmosphere and ocean, the global distribution of *ice and snow* is an important compo-

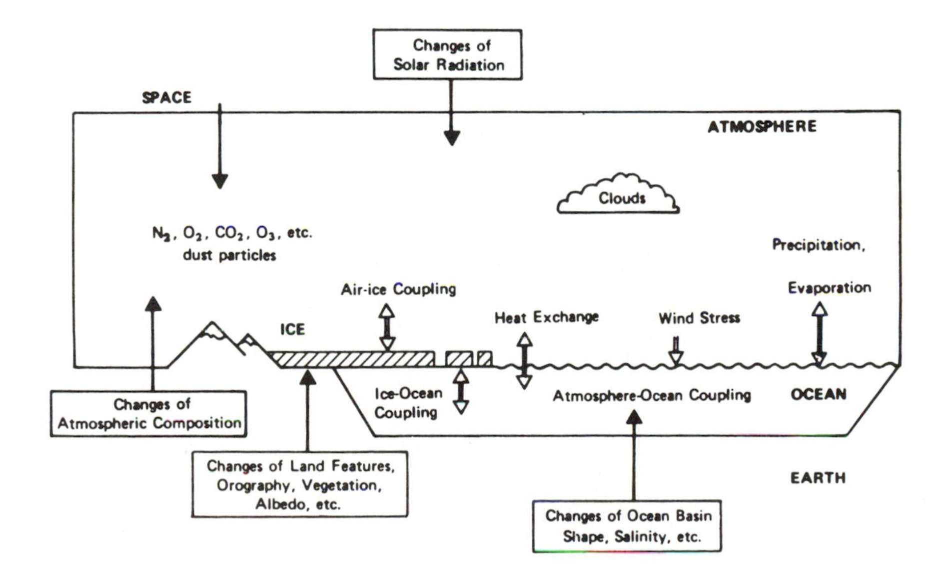

FIGURE 2.1 Schematic representation of the climate system and the principal interactions between its components (from NRC Panel on Climatic Variation, U.S. Committee for GARP, 1975).

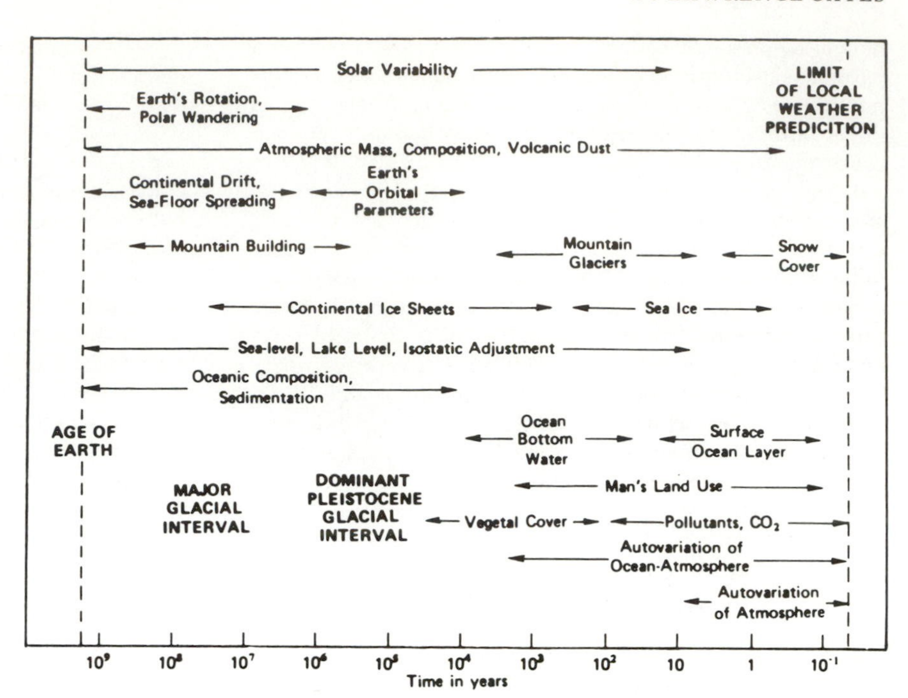

FIGURE 2.2 The characteristic time scales over which possible internal and external causes of climatic change take place (from NRC Panel on Climatic Variation, U.S. Committee for GARP, 1975).

nent of the climate system. The climate of the world's ice masses includes the statistics of their areal extent and thickness, and their temperature and motion, whether for the continental ice sheets, mountain glaciers, sea or lake ice, or surface snow cover. These latter elements respond relatively rapidly to atmospheric and oceanic conditions, while the former and more massive elements change much more slowly. In this respect the climate of the cryosphere is like that of the *land surface* and *associated biomass*, both of which contain elements of widely differing response times to the prevailing atmospheric (and oceanic) conditions. The climate of the Earth's land surface and biomass in turn includes the statistics of the temperature and water content of the surface soil, the surface's albedo or reflective characteristics for solar radiation, and the surface vegetation or land use, all of which have important effects on surface evapotranspiration and surface heat and moisture fluxes.

Taken together, these climatic statistics or characteristics constitute the total climate system, which also formally includes the variability (and higher-order statistics) of the various climatic elements in addition to the distribution of the means. Defined in this way the climate of the Earth has not yet been completely determined, for large portions of the global atmosphere and ocean remain unobserved on a systematic basis, and the analysis necessary for an adequate description of even the atmosphere's variability has not yet been completed. The determination of climate is further complicated by the fact that the statistics (or climate) of the complete climatic system can in principle change in response to a wide variety of influences originating outside the system, as well as in response to influences from within the system itself. As illustrated in Figure 2.1, a distinction is therefore made between *internal* and *external* causes of climate change, with the latter presumably being independent of the climate of the (internal) system.

In view of many processes that serve to relate one portion of the climate system to another, and the disparate response times of the system's atmospheric, oceanic, cryospheric, land surface, and biomass components to these internal processes as well as to external ones, the history of the Earth's climate can be expected to show variations over a wide range of time scales. As illustrated in Figure 2.2, these variations extend from the age of the atmosphere itself to the shortest-term climatic variation (about 1 month) that can be identified. Even if the external factors causing climate change were concentrated in well-defined frequencies, the high degree of coupling or interaction among internal processes in the climate system would be sufficient to create a virtually continuous spectrum of climatic variation. It is precisely these nonlinear feedback processes, the most prominent of which are illustrated in Figures 2.1 and 2.2, that make the analysis of climate and climate change a difficult if not impossible task without the assistance of models. (As we shall see, however, even with models there remain significant uncertainties, although the process is now at least a systematic one.)

Although a truly comprehensive climate model would include the interactions among all components of the climate system, usually only a portion of the complete system is modeled at any one time, with the remaining components held fixed (or ignored). This tactic is due both to our limited observational understanding of many of the interactions involving the nonatmospheric portions of the system and to the difficulty of coping with the vastly different time scales over which different portions of the climate system respond (see Figure 2.2).

The fastest-responding component of the climate system is of course the atmosphere, which effectively adjusts to new external conditions on time scales of the order of months. This relatively rapid response is determined by the characteristic growth rate of synoptic-scale baroclinic disturbances (the familiar transient highs and lows of mid-latitudes) and their associated transports of heat and momentum and by the more rapid adjustments that occur through smaller-scale convective and turbulent processes. The response of the ocean occurs over time scales ranging from the order of months in the surface waters in response to atmospheric changes to the order of centuries in the case of the deeper ocean. These time scales are determined by the vertical mixing of the upper ocean by wind and convection and by the relatively high thermal and mechanical inertia of the bulk of the oceanic water mass. Except for the synoptic-scale and seasonal variation of sea ice and surface snow cover as a result of surface oceanic and atmospheric heat fluxes, the response to the cryosphere may be even slower than that of the oceans; mountain glaciers typically respond to changes in atmospheric conditions and snow accumulation over time scales of centuries, while the more massive continental ice sheets respond over thousands of years. These time scales are determined by the relatively high viscosity, latent heat, and albedo of the ice and by the large volumes of ice involved. Finally, the response of the land surface and associated biomass occurs on time scales from months in the case of seasonal vegetative cover to centuries and beyond in the case of changes in surface-soil properties by erosion and desertification. These and other aspects of the physical basis of climate and climate modeling are reviewed in more detail elsewhere (World Meteorological Organization, 1975).

Faced with this complexity of space and time scales, it has been found convenient to introduce a scientific definition of climate. We may define climate, or more precisely a climatic state, as the statistics of one or more components of the climatic system over a specified domain and specified time interval, including the mean, variance, and other moments (NRC Panel on Climatic Variation, U.S. Committee for GARP, 1975). We may therefore speak, for example, of the atmospheric climate of Washington or of North America during July or of the climate of the global atmosphere-ocean system over a year, decade, or century. This definition recognizes that the climate varies significantly in both space and time and avoids the sometimes troublesome questions of homogeneity and statistical equilibrium. Climatic change may then be defined as the difference between climatic states of the same kind, and we may speak, for example, of the interannual climatic change between two Januaries or of the climatic change between two decades. When practical applications of climate are considered, these definitions are often further restricted to apply to only the atmosphere near the surface and to the more readily observed variables, as in the traditional classifications of climate in terms of the average seasonal regimes of surface-air temperature and precipitation. There are therefore many types of climate and climatic change, the most important aspect of which is the identification of the space and time scales involved for that portion of the climate system being considered.

The Formulation and Structure of Climate Models

For the present purposes, climate models may be considered to be physical-mathematical representations of the structure and variation of the large-scale climate. Most of these focus on the atmosphere, although there are some that consider aspects of the coupled atmosphere-ocean-ice system. For the most part, the continental ice sheets, which play such a large part in the geologic history of the Earth's surface, have not been fully incorporated into the interactive climate system nor have the longer-term changes of the land surface and biomass been adequately modeled.

The dynamical basis of climate models rests on the fundamental equations that presumably govern the system's motion, mass, and energy, together with whatever additional physical information is necessary to describe the system and its boundary conditions. For the atmosphere, these relations may be written in the form of the equations describing the conservation of horizontal momentum, the conservation of heat, and the conservation of mass:

$$\frac{\partial u}{\partial t} = -\mathbf{v}_H \cdot \nabla u - \omega\frac{\partial u}{\partial p} + \frac{uv \tan\phi}{a} + 2v\Omega\sin\phi - \frac{1}{a\cos\phi}\frac{\partial\Phi}{\partial\lambda} + F_\lambda, \quad (2.1)$$

$$\frac{\partial v}{\partial t} = -\mathbf{v}_H \cdot \nabla v - \omega\frac{\partial v}{\partial p} - \frac{u^2 \tan\phi}{a} - 2u\Omega\sin\phi - \frac{1}{a}\frac{\partial\Phi}{\partial\phi} + F_\phi, \quad (2.2)$$

$$\frac{\partial T}{\partial t} = -\mathbf{v}_H \cdot \nabla T - \omega\frac{\partial T}{\partial p} - \frac{\alpha\omega}{c_p} + \frac{Q}{c_p}, \quad (2.3)$$

$$\frac{\partial q}{\partial t} = -\mathbf{v}_H \cdot \nabla q - \omega\frac{\partial q}{\partial p} + (E - C), \quad (2.4)$$

$$\frac{\partial\omega}{\partial p} + \nabla \cdot \mathbf{v}_H = 0, \quad (2.5)$$

where a is the radius of an assumed spherical Earth with latitude ϕ, longitude λ, and rotation rate Ω, and ∇ is a gradient operator on an isobaric surface. Here the dependent variables are the eastward (u) and northward (v) components of the horizontal wind $\mathbf{v}_H$, the vertical mass flux ω ($= dp/dt$ in the isobaric coordinate system in which the pressure p plays the role of the vertical independent variable), the geopotential Φ of an isobaric surface, the temperature T, and the atmospheric specific humidity q. Together with the hydrostatic equation $\partial\Phi/\partial p = -\alpha$ and the atmospheric equation of state $p\alpha = RT$, where α is the specific volume and R the gas constant, Eqs. (2.1)-(2.5) constitute a closed dynamical system in terms of the variables u, v, ω, T, q, and α once the frictional forces F_λ and F_ϕ, the diabatic heating Q, the net evaporation E minus the net condensation C, and the necessary boundary conditions are specified. For most applications, the Earth's geophysical constants (Ω, a, and gravity), the Earth-Sun geometry, the atmospheric composition, and the distribution and elevation of the continents relative to the oceans are assumed given.

Equations (2.1)-(2.4) describe the local variation of the velocity, temperature, and moisture with time (t) as a result of the horizontal and vertical advection accompanying the large-scale field of motion over the spherical Earth, the Coriolis forces induced by the Earth's rotation, the pressure forces, and the local frictional, heating, and moisture source/sink effects; Eq. (2.5), on the other hand, serves to determine the vertical motion in terms of the isobaric divergence of the horizontal velocity and is a diagnostic relation in contrast to the prognostic Eqs. (2.1)-(2.4). These equations, often with further approximations, are the basis of the numerical weather-prediction methods now used routinely by the National Weather Service. Similar equations exist for the variation of the velocity, temperature, and salinity of the ocean and form the basis of operational models for predicting the large-scale oceanic circulation and sea-surface temperature. The amount of water in the surface soil and the depth of snow on the surface are sometimes also treated as dependent variables through equations describing their net accumulation as a function of the local precipitation, evaporation, and surface runoff and/or melting.

For the purposes of climate, however, some of the most important physical processes are not those displayed explicitly in these equations (all of which can in principle be determined with reasonable accuracy) but rather those that are contained implicitly in the frictional and heating terms. Included in the symbol F, for example, are all the smaller-scale convective and turbulent transfers of momentum within the atmosphere and between the atmosphere and the underlying surface, whether it is ocean, land, or ice; in the case of the ocean, this includes the wind stress at the sea surface and the ocean's frictional drag against the bottom and lateral boundaries of the basin. These frictional effects serve as a brake on the intensity of the circulation, which is otherwise sustained by the kinetic energy converted from the total potential and internal energy of the system, the ultimate source of which is the Sun's radiation. The symbol Q contains not only this net heating from insolation but also the heating from the transfer of long-wave radiation within the atmosphere and between the atmosphere, the clouds, and the Earth's surface. Also included in Q is the latent heating accompanying the processes of condensation and evaporation and the transfers of sensible heat by conduction and convection.

Many of these momentum and heat fluxes occur in the boundary layer near the Earth's surface and in the convective-scale motions associated with clouds in the case of the atmosphere. Generally, these fluxes are not explicitly resolved in climate models, in spite of the fact that they contain much of the model's physics, and their successful parameterization or representation in terms of the large-scale variables is an important aspect of modeling. Such parameterization usually rests heavily on empirical information and affords an opportunity to adjust or tune the model to some extent to observed (present) conditions. On the time scales of climate such model parameterization is of particular importance, because in the long run it is the accumulation of small imbalances in friction and heating that may cause the climate to change as the system seeks a new statistical equilibrium. The net incoming and outgoing radiation at the top of the atmosphere are in only approximate balance over periods of a year or longer, and it is possible for the deeper oceans and the ice sheets, for example, to store or release large amounts of heat effectively over long periods of time by slow but progressive changes in their temperature and volume.

The remaining aspect of the formulation of a climate model is the specification of the boundary conditions necessary for its solution. In the case of the atmosphere, these conditions consist of the incoming solar radiation at the top of the (model) atmosphere and the elevation and albedo of the Earth's surface. Without the ocean and ice included as interactive components, it is also necessary to specify the sea-surface temperature and the distribution of surface ice as boundary conditions for the atmospheric model as sketched in Figure 2.3. The boundary conditions for an oceanic model are the net fluxes of momentum, heat, and moisture at the ocean surface (which would automatically be determined by a coupled atmosphere) and the geometry of the ocean basin. Similarly, for models of the ice sheets and land surface the necessary boundary conditions

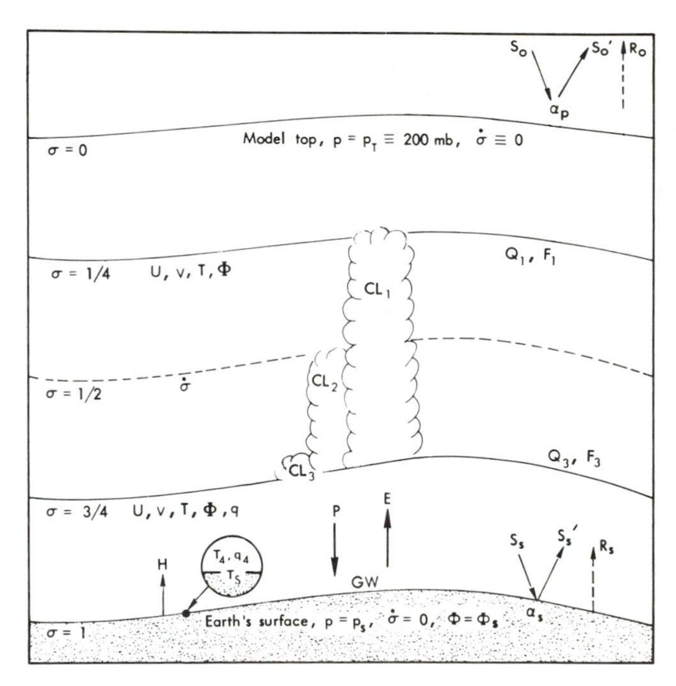

FIGURE 2.3 Schematic representation of the structure and principal variables of an atmospheric GCM. Symbols are as defined in the text, except for σ, which is a scaled vertical coordinate $= (p - p_T)(p_s - p_T)^{-1}$, surface sensible heat flux H, ground wetness GW, solar radiation S, terrestrial radiation R, cloudiness CL, and albedo α (from Schlesinger and Gates, 1979).

are the net fluxes of water and heat with the overlying atmosphere and the geometry of the Earth's surface.

Because the atmosphere is the central conduit (and the smallest reservoir) of the climate system's energy and contains most of the high-frequency variations, it is the resolution of the atmospheric portion of a climate model that is usually the most demanding. An exception to this occurs when the atmosphere is itself treated in a parametric or statistical fashion, in which case the necessary resolution is determined by the highest-frequency component of the climate system that is explicitly modeled. The climate models of lowest spatial resolution are those that effectively represent the entire atmosphere as a single point; those of highest spatial resolution are capable of resolving features of several-hundred-kilometer horizontal dimension and a few kilometer vertical dimension. Because the computational effort involved in solving a model generally varies as the cube of the spatial resolution, the higher-resolution models generally proceed relatively slowly on even the fastest of modern computers, whereas models of less spatial resolution proceed hundreds or thousands of times faster. This computational disparity effectively restricts the more detailed climate models to the simulation of a few years or decades. The more highly parameterized (and therefore less-detailed) models are able to simulate effectively the behavior of at least some aspects of the climate over much longer periods of time. As we shall see, the problems of paleoclimate require an effective solution to this computational dilemma.

In general, atmospheric and oceanic models (and hence climate models) are conveniently classified as either general-circulation models (GCMs) or statistical-dynamical models (SDMs), depending on whether synoptic-scale motions are resolved. In the atmosphere these motions are the familiar cyclones and anticyclones of mid-latitude weather, which occur on characteristic scales of the order of 10^3 km, whereas the corresponding motions in the ocean (the so-called mesoscale oceanic eddies) occur on scales of the order of 10^2 km. Most of the kinetic energy in both fluids is associated with these transient disturbances, and they represent the dominant mode of instability whereby potential and internal energy is transformed into kinetic energy; these motions also play an important role in the maintenance of the global balances of angular momentum, heat, and total energy. GCMs are therefore essentially weather models that have been designed to run on climatic time scales and from which the statistics of the climate are extracted by averaging the solutions over the desired intervals of time (and possibly space as well). SDMs, on the other hand, may be regarded as more directly addressing the climatic averages themselves, although not without a possibly serious loss of information. Depending on the essential space and time scales of the problem being considered, both modeling approaches may be useful; for the problem of pre-Pleistocene climates in particular, a blending of the two approaches is indicated, as will be discussed below.

As already noted, climate models are invariably solved with the aid of high-speed computers. This requires the design of appropriate numerical algorithms to replace the physical differential equations [such as Eqs. (2.1)-(2.5)], as well as the adoption of resolution intervals in both space and time; in a well-designed numerical scheme the solution of the approximating equations should be reasonably close to the (generally unknown) solution of the basic differential equations themselves. Although other techniques are available, most GCMs are solved with the aid of either finite-difference or spectral approximations to the models' physical equations, in which case the maximum allowable time step is limited to the order of 1 h in order to ensure the solution's computational stability. Moreover, it is also generally desirable to design the numerical solution of a GCM in such a way that the basic integral invariants of the flow (such as the total energy, total angular momentum, and total square of the vorticity) will be at least approximately conserved in the absence of sources and sinks.

When these conditions are met, the numerical solution of an atmospheric GCM proceeds on modern computers at speeds between roughly 10 and 1000 times the speed of the evolving climate itself, depending primarily on the number of levels in the vertical and the amount of sophistication included in the parameterization of the friction, heating, and moisture source terms. The numerical solution of SDMs, on the other hand, may be carried out without the restrictions of computational stability, which so severely slow a GCM's solution because they do not explicitly resolve the atmospheric (or oceanic) synoptic-scale motions. It is therefore not uncommon for a time-dependent SDM to advance in steps of several years, although special attention is sometimes given to seasonal changes. SDMs also lend themselves readily to the examination of equilibrium or time-independent solutions, such as those that are built around the requirement of an exact heat balance at the Earth's surface. By employing different degrees of spatial averaging, SDMs may be developed in zero-, one-, two-, or three-dimensional versions (Saltzman, 1978); a zero-dimensional model is one in which averaging is performed both horizontally and vertically over the climate system, a one-dimensional model has only vertical or latitudinal resolution, and two-dimensional models are usually developed in either the meridional or horizontal planes. Zonally averaged atmospheric models in particular have been widely used to study the latitudinal dependence of climate, although they necessarily obscure the longitudinal structure of climate features such as those due to the relative spacing of oceans and continents and require parameterization of the effects of synoptic-scale disturbances.

When examining the output of a climate model, whether in the form of an extended numerical integration of a GCM or the equilibrium numerical solution of an SDM, it is usually assumed that the statistical properties of the solution (i.e., the climate) depend only on the specification of the model's internal physical properties and on the specified boundary conditions. In this deterministic view of climate, the initial conditions from which a particular integration is started are viewed as unimportant in the determination of the statistical properties of the equilibrium climatic state; when the more slowly responding components of the climatic system are included, however, more time will generally be required to achieve statistical equilibrium. As reasonable as this sounds, it is known that the climate given by at least some relatively simple models is not unique and that more than one stable solution exists for identical boundary conditions. This property has its origin in the model's parameterization of nonlinear feedback effects, and whether it is also present in more general climate models has not yet been satisfactorily determined. The solutions of certain simplified climate systems under invariant boundary conditions are also known to switch in a seemingly irregular and unpredictable manner between two seemingly distinct states; whether such behavior is (or has been) exhibited by the Earth's climate is not known. Lest these considerations cast doubt on the validity of paleoclimatic reconstruction, it should be noted that the Earth's climate has necessarily followed only one climatic path during geologic time even if multiple paths were available and that this uniqueness should therefore in principle be reproducible.

Characteristic Model Performance

To give a clear idea of the nature and accuracy of the solutions to be expected from climate models, we first consider the characteristic performance of current atmospheric GCMs. The distribution of the average sea-level pressure simulated for the month of July in a recent integration of a two-level atmospheric GCM is shown in Figure 2.4, along with the observed climatological distribution (Schlesinger and Gates, 1979). While a number of local simulation errors may be noted (the most prominent of which is the model's evident failure to simulate correctly the position of the quasi-stationary high-pressure cells over the mid-latitude oceans in the summer hemisphere), the model has satisfactorily portrayed most of the large-scale features of the pressure field. A similar level of skill is present in the corresponding solutions from other GCMs and for other months of the year when the insolation and the sea-surface temperature are assigned their observed seasonal variations. Figure 2.4 also illustrates the typical resolution of the continents and oceans achieved by global GCMs, which in this case is that given by a 4° latitude and 5° longitude grid.

The global distribution of total precipitation simulated during July in the same model integration just considered is shown in Figure 2.5, together with the climatological average July precipitation. As was the case with sea-level pressure, the model can be seen to have successfully simulated the large-scale distribution of precipitation, including the band of maximum rainfall near the equator and the low precipitation in the subtropical desert and semiarid regions. Although there are errors of 100 percent or more in the estimate of local precipitation in some areas, the coherence in the simulated precipitation pattern and its systematic shift with the seasons (not shown here) lends confidence to the overall realism of this and other similarly performing GCMs. This performance is especially

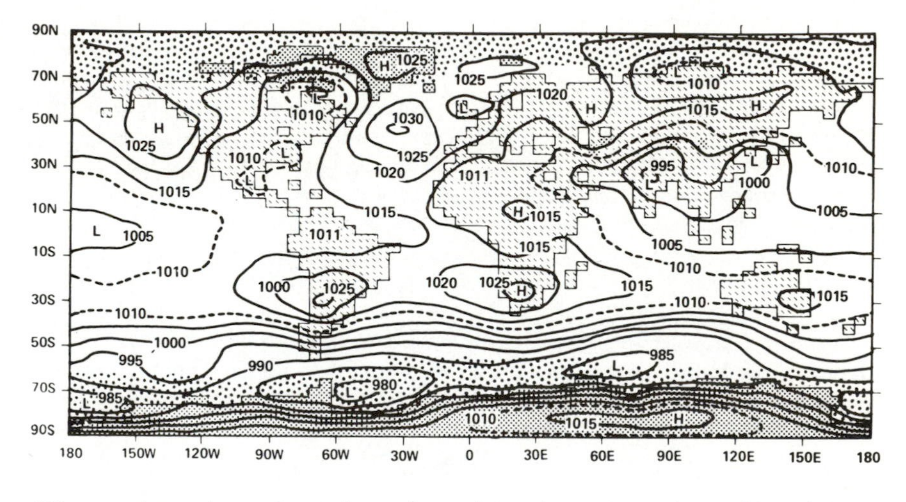

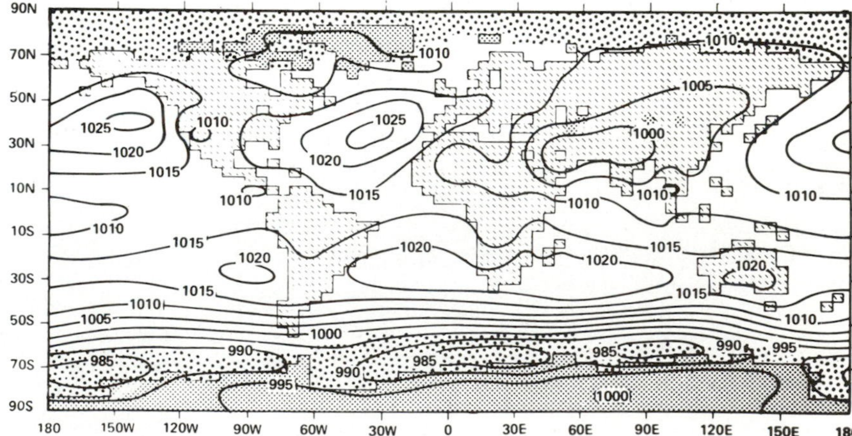

FIGURE 2.4 The distribution of July sea-level pressure (in millibars) as simulated (above) and as observed (below). Here stippling over the ocean denotes the assigned locations of sea ice (from Schlesinger and Gates, 1979).

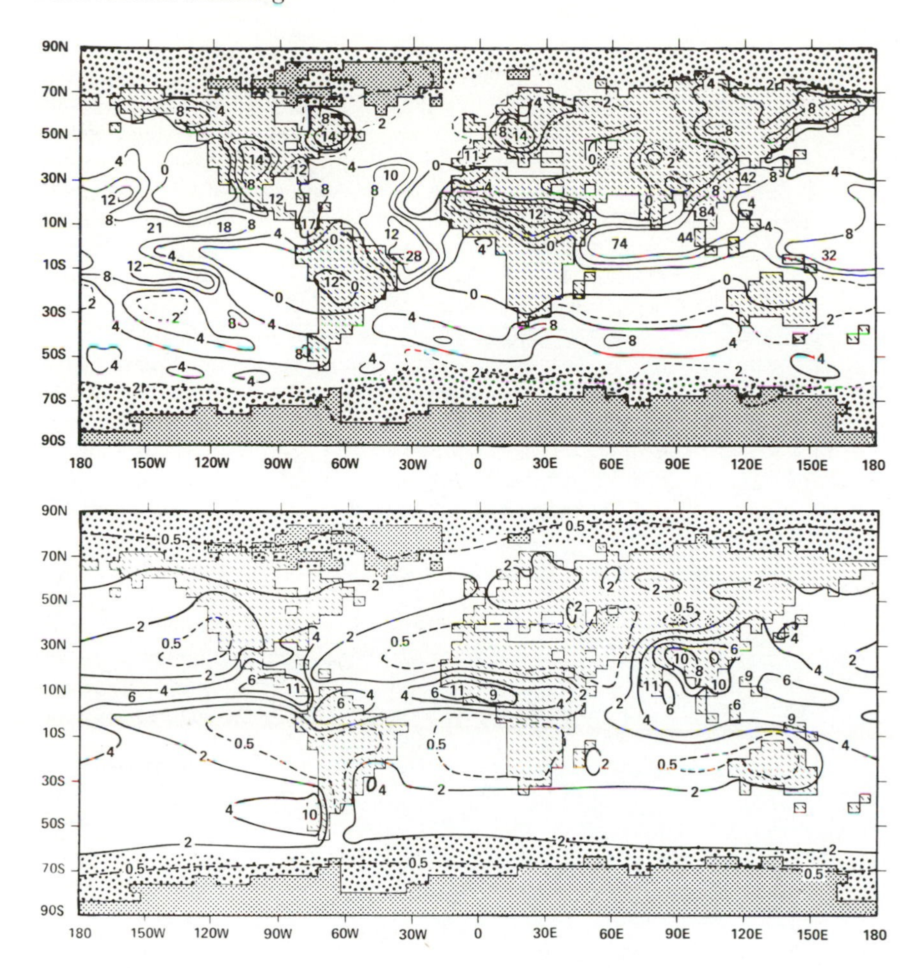

FIGURE 2.5 Same as Figure 2.4 except for July precipitation (in millimeters per day).

noteworthy in the case of precipitation, of which approximately half of that in mid-latitudes and nearly all of that in low latitudes is simulated to occur in association with convective-scale motions; the effective parameterization of this convective rainfall is one of the more difficult and sensitive aspects of climate modeling.

The characteristic performance of atmospheric SDMs may be illustrated by the zonally averaged distribution of mean air temperature simulated by a simple energy-balance mode, as shown in Figure 2.6, along with the observed average distribution (North and Coakley, 1979). This model (and most other SDMs with latitudinal resolution) has obviously reproduced the climatological meridional surface-temperature variation, especially in middle and higher latitudes, with an accuracy equal to that achieved by GCMs. As is usually the case with the more highly parameterized models, however, the extent to which this agreement with observation has been caused by the use of climatological data in the tuning or adjustment of the model's overall heat balance needs to be determined. In spite of this proviso (or perhaps because of it), SDMs have generally shown an acceptable level of accuracy in simulating other elements of the climate, including the pressure and zonal winds (not shown here), and have proven useful in the depiction of at least a first-order solution for those aspects of the climate that are least sensitive to parameterization and spatial averaging.

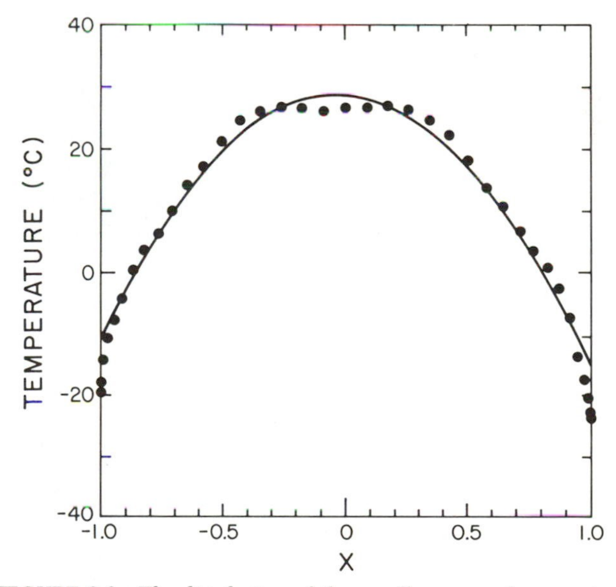

FIGURE 2.6 The distribution of the zonally averaged mean atmospheric temperature (in degrees Celsius) during March, April, and May as simulated by a simple climate model (curve) and as observed (dots). The abscissa X is the sine of the latitude (from North and Coakley, 1979).

More detailed information on the performance of current atmospheric GCMs and SDMs has been given by the World Meteorological Organization (1979).

In parallel with atmospheric GCMs, general circulation models have also been developed for the global ocean. The characteristic performance of these oceanic GCMs is illustrated in Figure 2.7 by the distribution of the annual average sea-surface temperature simulated by a six-level oceanic model in response to atmospheric wind stress and heat flux at the surface (Bryan *et al.*, 1975). In comparison with the observed average distribution, which is also shown, the model has successfully reproduced the large-scale features of the sea-surface temperature distribution even though there are local errors of several degrees Celsius. The success of this simulation, as well as that of similar oceanic GCMs, is partly the result of the advection of temperature by the large-scale current systems, which the model successfully simulates (not shown here) and partly due to the model's depiction of the local vertical mixing of heat. This performance is of particular importance to the atmosphere in view of the close correspondence observed between the variations of sea-surface temperature and rainfall in the lower latitudes—a result of the surface temperature's control of surface evaporation and the subsequent convective condensation. There is therefore an expectation that when these mechanisms are allowed to interact freely in a coupled ocean-atmosphere GCM, there will be an improvement in at least the tropical simulation of both the sea-surface temperature and precipitation; the fact that this has not occurred to a marked degree in the few integrations of coupled GCMs that have been made so far is testimony to the difficulty of properly parameterizing the turbulent fluxes in the back-to-back atmospheric and oceanic surface-boundary layers and to the lack of an adequate treatment of the disparate response times of the large-scale atmospheric and oceanic circulation to these mutually determined surface fluxes.

In addition to the atmosphere and ocean, models have also been successfully developed for the cryosphere. Among these are numerical models of the growth and movement of sea ice,

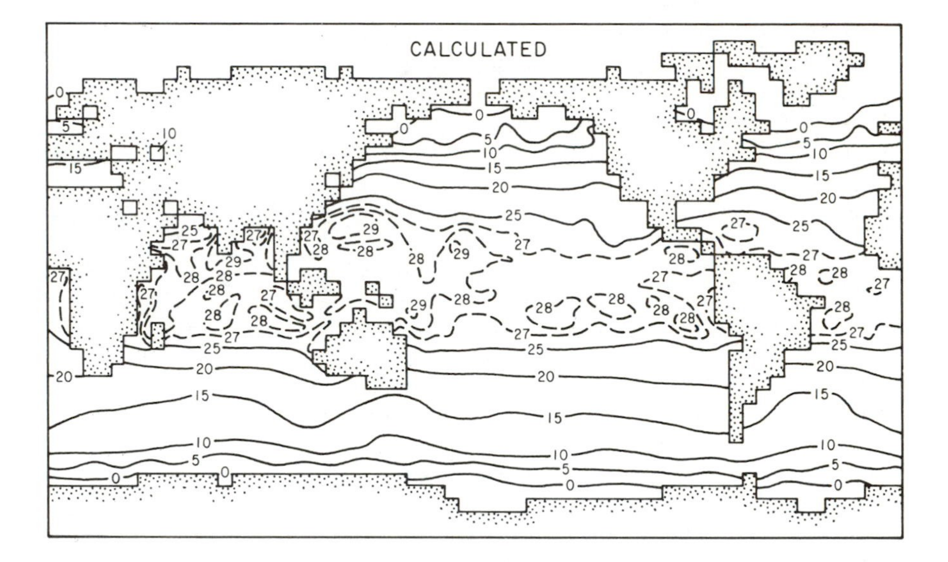

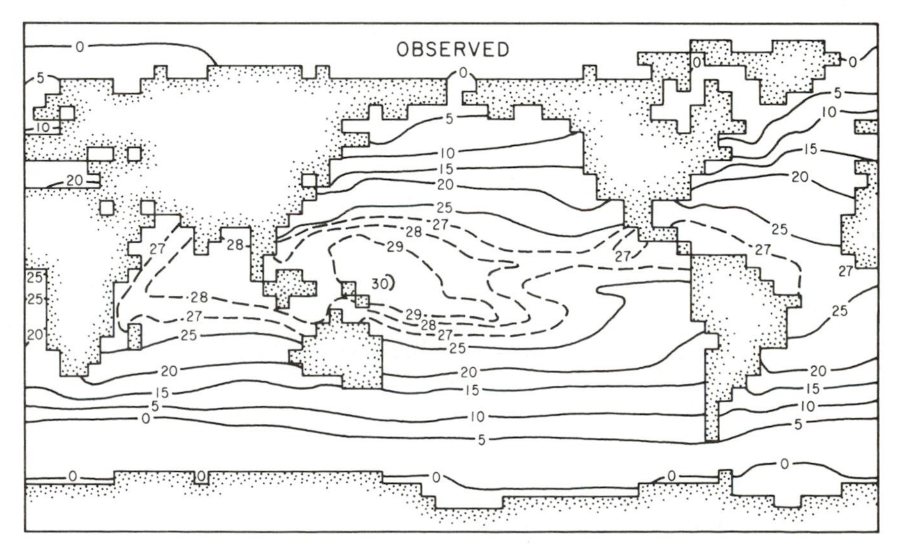

FIGURE 2.7 The distribution of annually averaged sea-surface temperature (in degrees Celsius) as simulated by an oceanic GCM (above) and as observed (below) (from Bryan *et al.*, 1975; with permission of the American Meteorological Society).

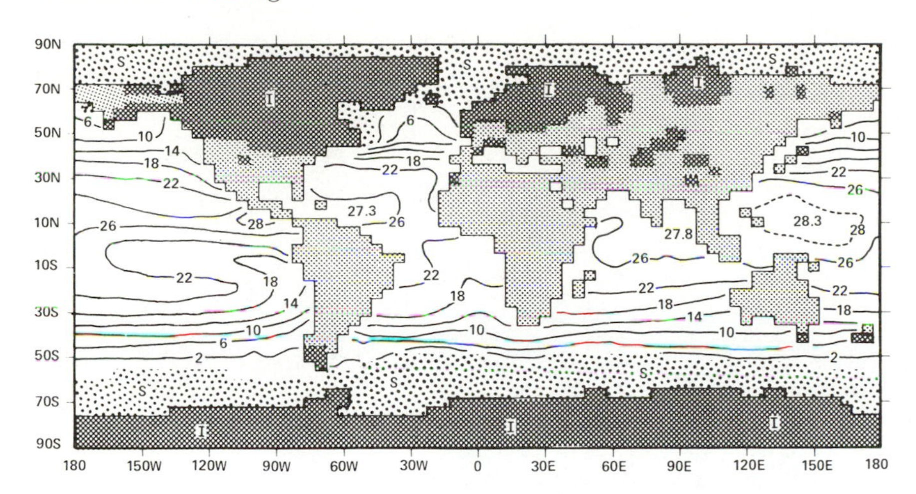

FIGURE 2.8 The distribution of sea-surface temperature (in degrees Celsius), sea ice (*S*, denoted over the oceans by stippling) and land ice (*I*, denoted over the land by cross-hatching). Here the continental outlines correspond to a sea-level lowering of 85 m as seen on a 4° latitude and 5° longitude grid (from Gates, 1976a; copyright 1976 by the American Association for the Advancement of Science).

which are usually integrated as part of the solution of an oceanic model. The seasonal variations of sea ice in the Arctic and Antarctic are reasonably well simulated by the simpler of such models that consider only the local vertical fluxes of heat above and beneath the ice. Of at least equal importance to the overall modeling of the climatic system are the models of the ice sheets (and to a lesser extent, models of mountain glaciers). In response to a prescribed net surface accumulation of mass and to a prescribed net surface-heat flux (which are the most important elements of the local atmospheric climate as far as the ice sheet is concerned), dynamical ice-sheet models that consider both the sheet's internal temperature and velocity have successfully simulated the development of ice sheets on continental dimensions. Whereas these models may be regarded as ice-sheet GCMs, their dynamics are somewhat simpler than those of their atmospheric and oceanic counterparts because of the high viscosity of the ice. The integration of such models can therefore be carried out relatively easily. In solutions extending over hundreds and thousands of years the friction with the underlying ground and the possibility that the ice may melt at the bottom may exert a major influence on the behavior of an ice sheet as it either approaches or retreats from an equilibrium configuration (Budd, 1969). It might also be noted that an ice-sheet GCM has not yet been solved in concert with an atmospheric or oceanic GCM, so the long-term effects of their mutual interaction cannot be adequately assessed; this interaction, however, is likely to be an important element in the history of the Earth's glaciation.

PALEOCLIMATIC MODELING

On the basis of the background and review of climate modeling given above, we are now in a position to consider the specific problem of modeling the paleoclimate, with particular reference to the pre-Pleistocene. After reviewing the problem of paleoclimatic data assembly and the present status of paleoclimatic numerical experiments, we shall consider a specific strategy for pre-Pleistocene climate modeling.

Paleoclimatic Data Assembly and Model Testing

The relative scarcity of suitable paleoclimatic records and the difficulty that usually accompanies their extraction dictates that careful consideration be given to all paleoclimatic data. These proxy data are typically found at scattered locations over the globe and can be interpreted in terms of only a limited number of climatic variables. As shown elsewhere in this volume, such paleoclimatic "data of opportunity" are also generally less abundant as one proceeds to older geologic periods.

From the viewpoint of climate modeling, paleoclimatic data serve two purposes: first, they are necessary for the realistic specification of the boundary conditions at the Earth's surface that are required by climate models, and, second, they are necessary for the verification of at least portions of the models' simulated climate. As has been noted previously, the most demanding surface-boundary conditions are those required by atmospheric GCMs; these consist of the location and elevation of the land surface relative to a possibly changed sea level, the sea-surface temperature, the distribution (and elevation) of ice sheets and sea ice, and the surface albedo, which serves to distinguish land-surface types and vegetative cover. Ideally these data should be provided at every point of the model's global grid and be changed with the seasons if appropriate for the model's time integration. No paleoclimatic data set fully meets these requirements, although that assembled by the CLIMAP Project Members (1976) for the Wisconsin ice age of about 18,000 yr. B.P. (years before present) shown in Figure 2.8 comes closest. For earlier epochs atmospheric climate modelers will have to proceed on the basis of less precise knowledge of conditions at the Earth's surface and will have to develop the grid-point boundary-condition data sets required by GCMs with the help of interpolation schemes and the most physically consistent estimates that can be made.

In the paleoclimatic application of oceanic GCMs, the required boundary conditions consist of the location and geometry of the global ocean basins and the water mass and heat fluxes at the ocean surface. From the growing body of data col-

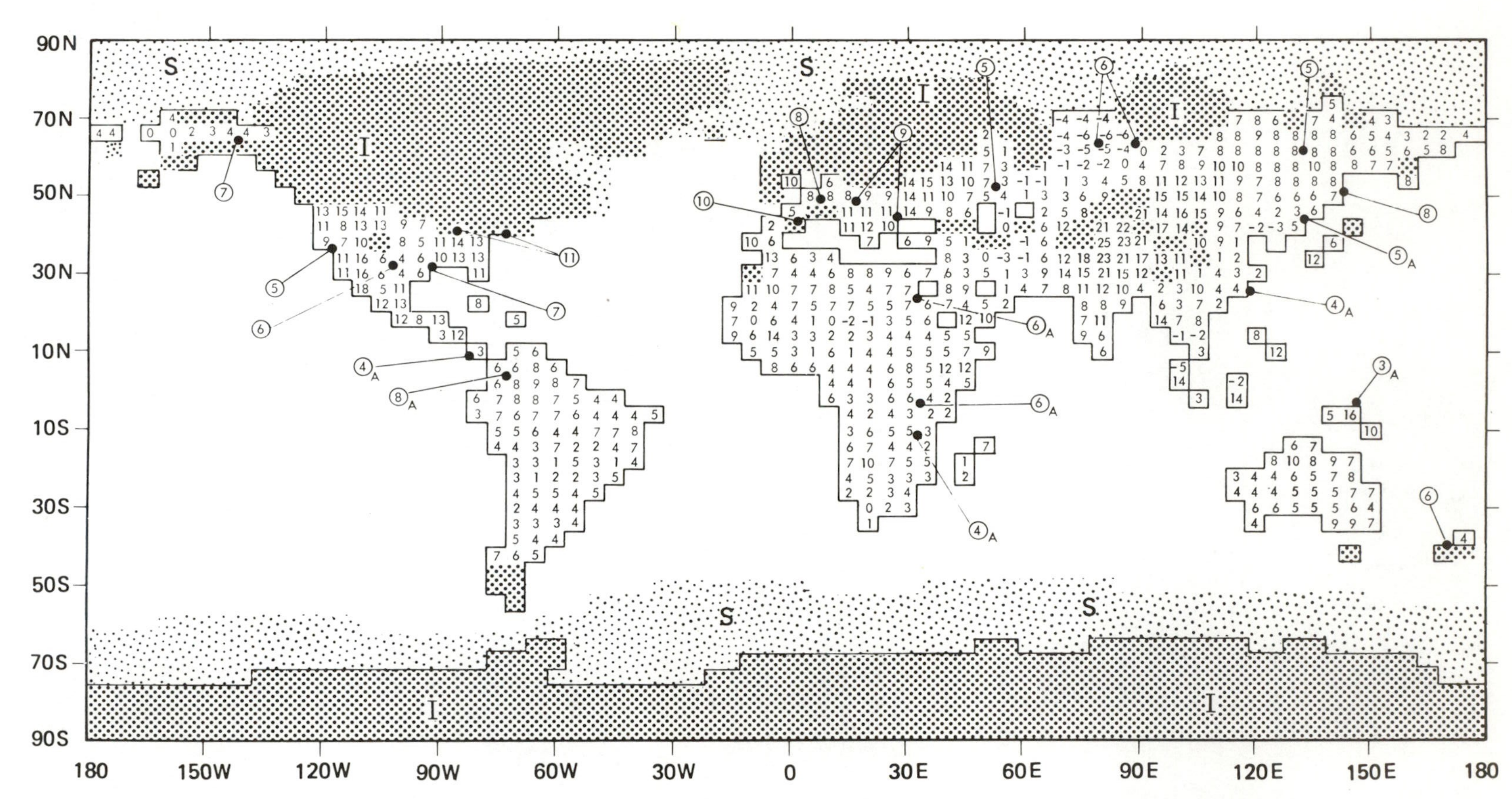

FIGURE 2.9 The distribution of the difference (present minus ice age) of July surface air temperature (in degress Celsius) over ice-free continental areas as simulated by an atmospheric GCM for the Wisconsin ice age (18,000 yr B.P.). Here the stippled areas over the ocean denote the locations of ice-age sea ice (*S*), and the more heavily shaded areas denote the locations of ice-age land ice (*I*). The encircled data are the "observed" changes as assembled by CLIMAP from pollen analyses and periglacial evidence, with the subscript A denoting observed annual (rather than July) changes (from Gates, 1976a; copyright 1976 by the American Association for the Advancement of Science).

lected in association with studies of plate tectonics, the history of the shape, depth, and global position of the major ocean basins is now emerging from which it may be possible to extract the information required by models (Smith *et al.*, 1973; Smith and Briden, 1977; Scotese *et al.*, 1979; Ziegler *et al.*, 1979). Of particular importance for the large-scale oceanic circulation, and hence for the distribution of global climate, is the presence (or absence) of zonal passages and shallow seas and the geographic locations of intermediate and bottom water formation. In the simulation of a paleo-ocean, it may be expected that the surface layers will show a relatively rapid adjustment to the imposed surface fluxes and that the response of deeper water will require many years. The details of the surface fluxes are probably not critical for the study of the ocean's spin-up characteristics, although these fluxes should be made compatible (or actually be coupled) with an atmospheric model if the oceanic climate is to achieve a realistic equilibrium state. A similar remark applies to the application of ice-sheet models, which as previously noted have not yet been integrated in concert with their atmospheric and oceanic counterparts.

After a climate model has been integrated under suitable paleoclimatic conditions, there remains the question of verifying its results against whatever paleoclimatic data are available (and have not already been used as boundary conditions for the model). Using an atmospheric GCM as an example, the time history of a climate simulation will generally yield the global distribution of a wide variety of surface climatic variables (in addition to variables in the free atmosphere itself). These include the surface-air temperature, pressure and wind velocity, surface evaporation and sensible heat flux, net surface fluxes of long-wave and shortwave radiation, precipitation in the form of either rain or snow, degree of ground wetness, local snow depth (if any), and cloudiness. If the model simulation extends over seasonal or annual time scales, then the corresponding seasonal and interannual variability of such variables is automatically furnished as well. Because the detail of the simulated data far exceeds that which is ever likely to be available for even relatively recent geologic periods, model verification will generally consist of the comparison with limited "observed" proxy data at selected locations. If such comparison shows reasonable agreement, and especially if agreement occurs for more than one variable and in more than one location, then the model's results may be accepted provisionally. The importance of such verification for multiple variables at multiple sites is an especially important factor in our appraisal of the realism of the simulated climate, observed climate, or both as the elements of the simulated climate at least are guaranteed to be physically consistent with one another and in their spatial distribution. While it is of course true that a model is never better than the data on which it rests, a locally poor verification of a particular simulated climatic variable may be due to the low quality of the verifying proxy data rather than to a failure of the model itself.

Although these two uses of paleoclimatic data—for boundary conditions and for verification—are formally independent from the modeler's viewpoint, they are often closely related in practical paleoclimatic work. For example, the derivation of one proxy data set may depend on another set, as in the calibration of paleobotanical evidence in terms of either temperature and rainfall or the implications that the specification of the land-surface albedo has in terms of surface vegetation. Furthermore, a proxy data set that serves as a boundary condition for one model may serve as verification data for another, more general model, as in the case of the sea-surface temperature inferred from the analysis of oceanic sediments. In the CLIMAP effort, most of the available proxy data were for the specification of model surface-boundary conditions, leaving relatively little independent data for verification. It is hoped that this allocation will be reversed as paleoclimatic data assembly proceeds in parallel with climatic reconstructions with progressively more comprehensive models.

Paleoclimatic Model Experiments

As a preview of the results that may be expected from future model simulations of the pre-Pleistocene, it is useful to review in more detail some of the results that have been obtained for the Wisconsin (Pleistocene) ice age. The first application of a comprehensive atmospheric model to the simulation of ice-age climate was that of Alyea (1972), who used a hemispheric geostrophic model in conjunction with idealized surface-boundary conditions representing an ice age. He found, as might have been anticipated, a generally cooler and drier climate along with indications of slight shifts in the pattern of the zonal wind circulation. Generally similar results were subsequently found by Williams *et al.* (1974), who used a global GCM but with surface-boundary conditions drawn from diverse estimates that schematically represented conditions during the Wisconsin ice age. Only a qualitative verification of these model experiments was attempted, and the degree of model dependence in the results has not been determined satisfactorily.

By far the most comprehensive paleoclimatic data assembly yet achieved is that of the CLIMAP Project Members (1976). These data, representing conditions at the height of the Wisconsin ice age about 18,000 yr B.P., include the global distribution of sea-surface temperature and sea ice (inferred from the fossil record in deep-sea sediments), the extent and thickness of ice sheets (reconstructed from glacial evidence and ice-sheet budgets), the albedo of the land surfaces (inferred from botantical evidence and soil characteristics), and a specification of global sea level relative to the present (inferred from coral-reef data and ice-volume estimates). These data have been used by Gates (1976a, 1976b) and Manabe and Hahn (1977) in the reconstruction of the July ice-age climate with global GCMs. The results of these experiments (in which seasonal effects were omitted) are represented in Figures 2.9 and 2.10, which show the simulated change of July surface-air temperature (with respect to the model's simulation for present-day conditions) and the simulated ice-age July sea-level pressure, respectively. In Figure 2.9 the model is seen to have simulated an ice-age cooling of 10°C or more in extensive ice-free areas over the continents, even though the average specified cooling of the ice-age oceans relative to modern July was only about 2°C. In comparison with the limited verification data also shown in Figure 2.9, the model's simulated cooling is considered reasonably accurate.

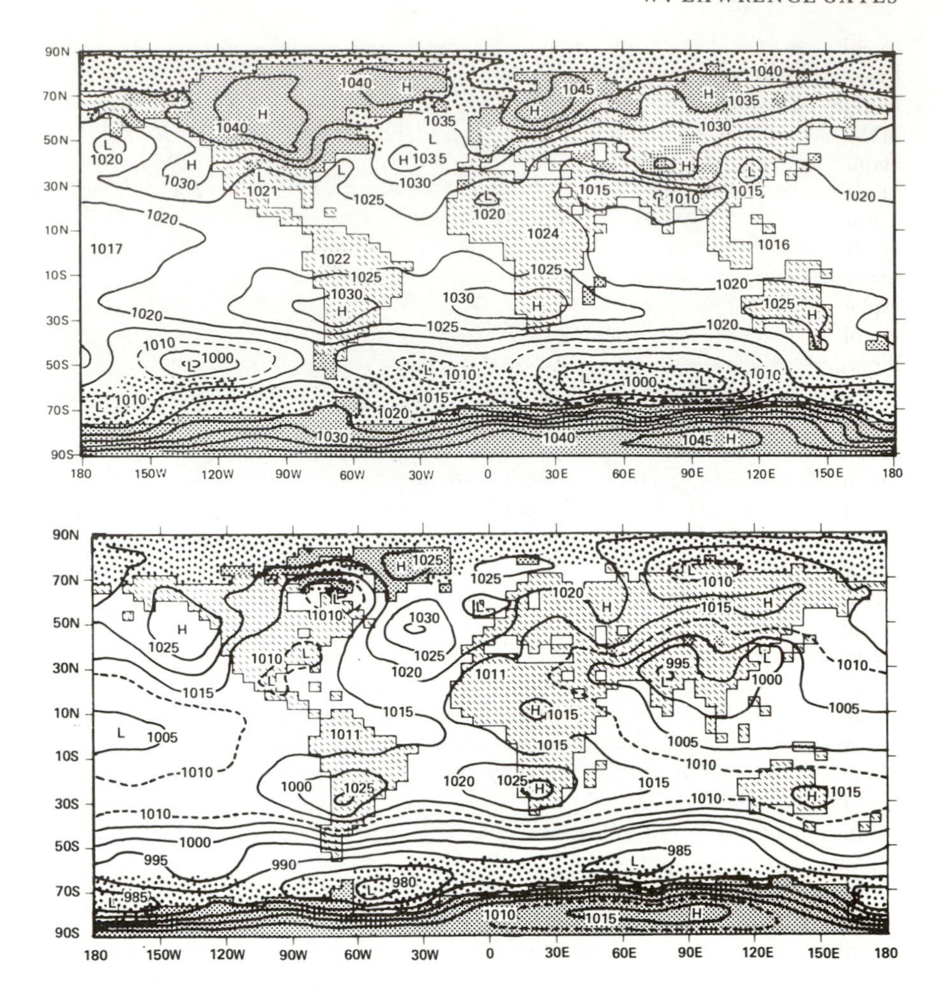

FIGURE 2.10 The distribution of July sea-level pressure (in millibars) as simulated for Wisconsin ice-age conditions (above) and for modern conditions (below); see also Figure 2.8 (from Gates, 1976b; with permission of the American Meteorological Society).

From the simulated July sea-level pressure shown in Figure 2.10, the major ice-age change in the surface circulation is seen to be that associated with the semipermanent anticyclones found over the major ice sheets in North America and Europe; this pattern implies a drastic change from the present characteristic July climate over the eastern and southern portions of these continents. Although the models of both Gates (1976a, 1976b) and of Manabe and Hahn (1977) show the July ice-age climate in general to be slightly drier than at present, less confidence can be placed in the models' simulated changes of local precipitation (not shown here) for reasons discussed earlier. These results have been generally confirmed in new GCM experiments using the revised global data sets for both January and July recently assembled by CLIMAP (Peterson *et al.*, 1979).

The performance of SDMs in paleoclimatic reconstruction is illustrated in Figure 2.11 by the results of Saltzman and Vernekar (1975) for glacial maximum conditions. Here the simulated ice-age changes of zonally averaged temperature, wind, and the evaporation-precipitation difference are generally similar to the zonal average of the results given by GCMs. Other SDMs have been used to simulate glacial-interglacial cycles in response to either variations of the Earth's orbital parameters (Pollard, 1978) or to internal feedback between the ice albedo and temperature (Held and Saurez, 1974). A preliminary attempt to simulate the climatic effects of changed continental positions has also been made by Donn and Shaw (1975). Such experiments serve to illustrate the sensitivity of the climate in such models (and perhaps in all models) to the relative absorption of heat by the atmosphere, ocean, and ice. Without additional paleoclimatic data it is difficult to verify such models adequately even for the Pleistocene, although they may nevertheless prove useful in conjunction with other models in the exploration of earlier climates.

A Strategy for Modeling Pre-Pleistocene Climate

In view of the complexity of the climate system and the effort involved in achieving even a modest degree of success for modern and recent Holocene climates, it is not realistic to expect a rapid breakthrough in the simulation of pre-Pleistocene climate; it will in all likelihood require several decades of work before the paleoclimatic data and modeling techniques necessary for satisfactory results will be in hand. And even if virtually unlimited computing resources were available (which is certainly not the case), it would be utterly unrealistic to expect to simulate with GCMs the day-by-day evolution of climate over pre-Pleistocene time scales. What is needed is a modeling strategy that combines the valuable features of both GCMs and SDMs in an economical and effective fashion while guiding the acquisition of new proxy data and the development of new modeling techniques.

Such a strategy in fact is suggested by a basic characteristic of the climate system itself, namely, the widely differing response times of the system's major atmospheric, oceanic, cryospheric, land surface, and biomass components. The behavior of the more inertial portions of the system suggests that these components effectively respond to the average or integral of the climatic information from the more rapidly varying components and that they therefore require such information only at relatively widely separated times. Hence, in the solution of an atmospheric model (and specifically an atmospheric GCM) it is permissible to neglect the variation of all other components of the system (except perhaps the oceanic surface layer and associated sea ice) and to integrate only over that time interval sufficient to generate a statistically representative atmospheric climate; this interval is certainly longer than a season but probably less than a decade. For an oceanic GCM (and perhaps also for the surface biomass) whose response time is intermediate between that of the ice sheets and the atmosphere, this strategy suggests that integration needs to be performed in response to a periodically updated atmospheric climate only over that time interval sufficient to generate a representative oceanic climate, with the land surface and ice sheets held fixed as boundary conditions; this interval is probably of the order of a few centuries. Finally, for the ice sheets themselves (and perhaps also for the land surface), this in turn suggests that models describing their variation need to be integrated in response to a periodically updated climate only over that time interval sufficient to generate a representative ice-sheet climate.

As summarized in Figure 2.12, this strategy results in a series of snapshots of the climate of the various components of the climate system, with each snapshot exposed only long enough to acquire representative information for use by another component and with the interval between snapshots selected in accord with each component's natural response time. Although there are many conceivable variations of such a modeling strategy, its basic philosophy appears to offer a way in which long climatic time scales can be addressed without unnecessary computation. In this scheme, a climate-system component is linked to any slower-reacting components by boundary conditions and to any faster-reacting components by their climatic statistics; the common practice of using an atmospheric model with all other variables held fixed may be recognized as a part of this procedure. If the complete strategy is implemented, the evolution of the slow components will periodically update the boundary conditions for the faster components, whereas the transfer of climatic statistics in the other directions affords an opportunity for the development and use of appropriate parameterizations or entire SDMs.

Such a snapshot strategy for modeling paleoclimate may be initiated at any particular point in geologic time (or extended into the future, for that matter). Although we do not yet know the behavior of the climate system under different arrange-

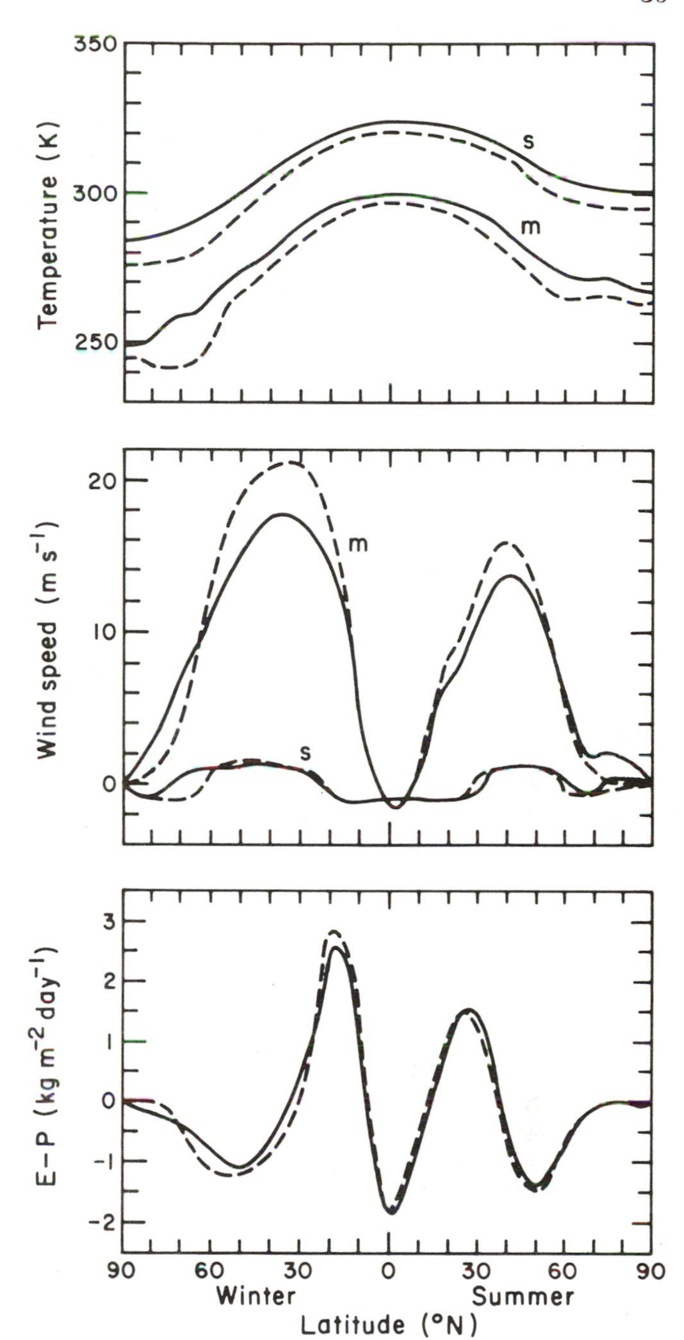

FIGURE 2.11 The distribution of zonally averaged potential temperature (top panel), zonal wind (middle panel), and evaporation-precipitation difference (bottom panel) as simulated by an atmospheric SDM for modern conditions (full line) and for 18,000 yr B.P. glacial conditions (dashed line). In the top and middle panels the letters *s* and *m* refer to surface values and mean atmospheric values, respectively (from Saltzman and Vernekar, 1975).

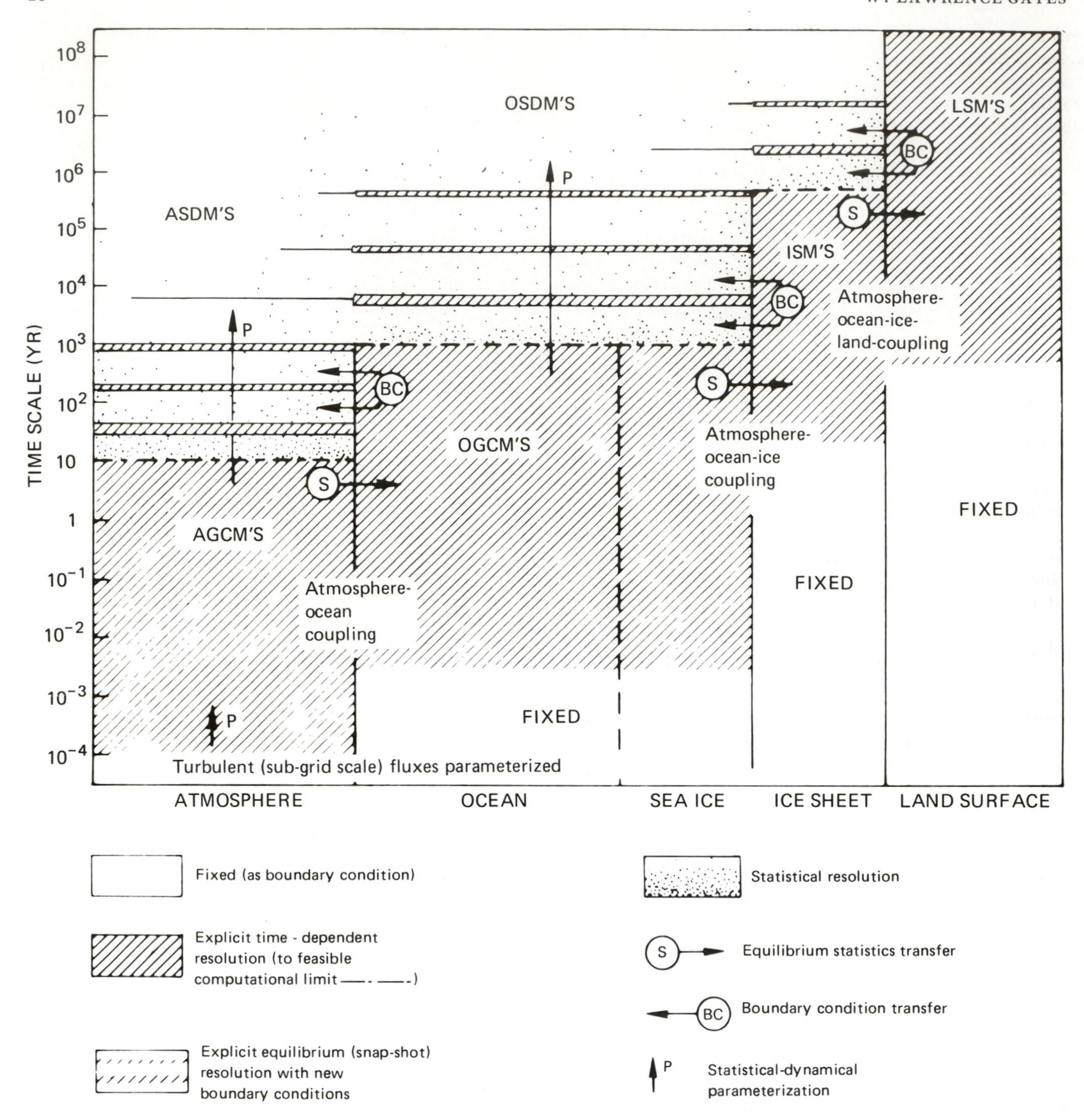

FIGURE 2.12 Schematic summary of a snapshot strategy for integrating the components of the coupled climate system. Here the symbol *S* denotes the transfer of equilibrium climatic statistics from one component to a more slowly varying component, *BC* denotes the provision of boundary-condition information in the reverse direction, and *P* denotes the possibility of parameterization (from Gates, 1975).

ments of the continents and oceans, there are certain times when the climate should be more interesting to study than at other times; these include in particular those periods when the continents were away from the poles and when the global ocean was free of meridional obstructions in low latitudes (Frakes, 1979). This suggests that the late Cambrian and early Devonian might be interesting times at which to attempt a reconstruction of the climate, perhaps starting with a prescribed ocean and then progressively relaxing this constraint with the strategy described above. Other considerations, not the least of which is the likely availability of appropriate paleoclimatic data, suggest that simulating the climate at some time during the Carboniferous, Permian, and Jurrasic would also be of considerable interest.

SUMMARY AND CONCLUDING REMARKS

From this review it is seen that the successful application of climate modeling to the pre-Pleistocene will depend primarily on (1) the assembly of the paleoclimatic data necessary for the specification of the model boundary conditions and for model verification and (2) the development of a modeling strategy for treating the disparate characteristic response times of the interacting components of the climatic system. It is particularly important that the geometry and relative positions of the oceans and continents be specified as accurately as possible for those geologic periods chosen for study and that data that may be used for verification of the model-simulated climate be assembled for as many climatic elements and locations as possible. These considerations are perhaps more important than the absolute accuracy of the individual data or their spatial density, in view of the inevitable questions concerning the representativeness of local data. Finally, it may be remarked that in the assembly of paleoclimatic proxy data for either model boundary conditions or verification particular attention should be given to the rapidly accumulating body of quantitative information from paleomagnetic, tectonic, and isotopic evidence, in addition to the use of the more qualitative climatic information from biogeographic sources.

Although there is always room for improvement, available models of the atmosphere and ocean are clearly capable of simulating the global structure of the currently observed seasonal climates with acceptable accuracy, and the limited application that has been made of atmospheric models to the simulation of the Wisconsin ice-age climate has given encouraging results. The most serious shortcomings of present climate models (aside from the errors associated with inadequate resolution) are believed to occur as a result of the improper parameterization of the surface-boundary layer and of the unresolved convective-scale motions. For successful application to the simulation of paleoclimates, and especially those of the pre-Pleistocene, there is the additional need for an effective and economical method of treating the widely differing response rates of the atmosphere, ocean, ice sheets, and land surface.

One such method is a snapshot strategy, whereby interacting components of the climate system are intermittently coupled to provide the statistics of the forcing functions, boundary conditions, or both needed for a sustained integration. Such a strategy can serve as the key organizing element of a comprehensive paleoclimatic research program (NRC Panel on Climatic Variation, U.S. Committee for GARP, 1975; NRC Committee on Geology and Climate, 1978), wherein there is an active interplay between the identification, assembly, and analysis of paleoclimatic data on the one hand and the design, application, and evaluation of a hierarchy of climate models on the other. Whereas such a program calls for the dedication of considerable resources over several decades, and will require an extraordinary degree of collaboration among Earth, atmospheric, and oceanic scientists, it is probably the most effective way to advance the scientific study of pre-Pleistocene climates.

REFERENCES

Alyea, F. F. (1972). Numerical simulation of an ice-age paleoclimate, *Atmos. Sci. Paper No. 193*, Colorado State U., Ft. Collins, Colorado, 134 pp.

Bryan, K., S. Manabe, and R. C. Pacanowski (1975). A global ocean-atmosphere climate model, Part II. The oceanic circulation, *J. Phys. Oceanogr. 5*, 30-46.

Budd, W. F. (1969). The dynamics of ice masses, *ANARE Science Report, Ser. A (IV), No. 108*, Melbourne.

CLIMAP Project Members (1976). The surface of the ice-age Earth, *Science 191*, 1131-1137.

Donn, W. L., and D. Shaw (1975). The evolution of climate, Proc. WMO/IAMAP Symp. Long-Term Climatic Fluctuations (18-23 August 1975, Norwich), *WMO No. 421*, World Meteorological Organization, Geneva, pp. 53-62.

Frakes, L. A. (1979). *Climates Throughout Geologic Time*, Elsevier, Amsterdam, 310 pp.

Gates, W. L. (1975). Numerical modelling of climate change: A review of problems and prospects, Proc. WMO/IAMAP Symp. Long-Term Climatic Fluctuations (18-23 August 1975, Norwich), *WMO No. 421*, World Meteorological Organization, Geneva, pp. 343-354.

Gates, W. L. (1976a). Modeling the ice-age climate, *Science 191*, 1138-1144.

Gates, W. L. (1976b). The numerical simulation of ice-age climate with a global general circulation model, *J. Atmos. Sci. 33*, 1844-1873.

Held, I. M., and M. J. Saurez (1974). Simple albedo feedback models of the icecaps, *Tellus 26*, 613-629.

Manabe, S., and D. G. Hahn (1977). Simulation of the tropical climate of an ice age, *J. Geophys. Res. 82*, 3889-3911.

NRC Committee on Geology and Climate (1978). *Geological Perspectives on Climatic Change*, National Academy of Sciences, Washington, D.C., 46 pp.

NRC Panel on Climatic Variation, U.S. Committee for GARP (1975). *Understanding Climatic Change—A Program for Action*, National Academy of Sciences, Washington, D.C., 239 pp.

North, G. R., and J. A. Coakley (1979). Simple seasonal climate models, Report of the JOC Study Conference on Climate Models: Performance Intercomparison and Sensitivity Studies, W. L. Gates, ed., *GARP Publications Series No. 22, Vol. II*, World Meteorological Organization, Geneva, pp. 715-727.

Peterson, G. M., T. Webb III, J. E. Kutzbach, T. van der Hammen, T. A. Wijmstra, and F. A. Street (1979). The continental record of environmental conditions at 18,000 yr B.P.: An initial evaluation, *Quat. Res. 12*, 47-82.

Pollard, D. (1978). An investigation of the astronomical theory of the ice ages using a simple climate-ice sheet model, *Nature 272*, 233-235.

Saltzman, B. (1978). A survey of statistical-dynamical models of the terrestrial climate, *Adv. Geophys. 20*, 183-304.

Saltzman, B., and A. D. Vernekar (1975). A solution for the northern hemisphere climatic zonation during a glacial maximum, *Quat. Res. 5*, 307-320.

Schlesinger, M. E., and W. L. Gates (1979). Numerical simulation of the January and July global climate with the OSU two-level atmospheric general circulation model, Report No. 9, Climatic Research Institute, Oregon State U., Corvallis, 102 pp.

Schneider, S. H., and R. E. Dickinson (1974). Climate modeling, *Rev. Geophys. Space Phys. 12*, 447-493.

Scotese, C. R., R. K. Bambach, C. Barton, R. van der Voo, and A. M. Ziegler (1979). Paleozoic base maps, *J. Geol. 87*, 217-277.

Smith, A. G., and J. C. Briden (1977). Mesozoic and Cenozoic paleocontinental maps, Cambridge U., Cambridge, England.

Smith, A. G., J. C. Briden, and G. E. Drewry (1973). Phanerozoic world maps, Organisms and Continents Through Time, *Spec. Pap. Palaeontol. No. 12*, Paleontol. Assoc., London, 42 pp.

Williams, J., R. G. Barry, and W. M. Washington (1974). Simulation of the atmospheric circulation using the NCAR global circulation model with ice age boundary conditions, *J. Appl. Meteorol. 13*, 305-317.

World Meteorological Organization (1975). The Physical Basis of Climate and Climate Modelling, *GARP Publications Series No. 16*, World Meteorological Organization, Geneva, 265 pp.

World Meteorological Organization (1979). Report of the JOC Study Conference on Climate Models: Performance, Intercomparison and Sensitivity Studies, W. L. Gates, ed., *GARP Publication Series No. 22*, World Meteorological Organization, Geneva, Vols. 1 and 2, 1049 pp.

Ziegler, A. M., C. R. Scotese, W. S. McKerrow, M. E. Johnson, and R. K. Bambach (1979). Paleozoic paleogeography, *Ann. Rev. Earth Planet. Sci. 7*, 473-502.

Climate Steps in Ocean History—Lessons from the Pleistocene

3

WOLFGANG H. BERGER
Scripps Institution of Oceanography and *Universität Kiel*

INTRODUCTION

There is evidence from the study of deep-sea sediments that the climatic (and therefore geochemical and evolutionary) history of the ocean is characterized by a series of transitions from one state to another. The general impression of instability, which arises from contemplating a history full of transitions, has been captured in the phrase "commotion in the ocean" (Berggren and Hollister, 1977). Prime examples (see Table 3.1) of major transitions are the Cretaceous termination (Thierstein, Chapter 8), the Eocene-Oligocene boundary event (Benson, 1975; Kennett and Shackleton, 1976), the mid-Miocene oxygen isotope shift (Savin, 1977), the 6 Ma (million years ago) carbon isotope shift (Keigwin, 1979; Vincent *et al.*, 1980), and the 3 Ma northern glaciation onset (Berggren, 1972; Shackleton and Opdyke, 1977). Also there are many less spectacular events, some of which apparently are of a recurring kind and belong to cyclic or quasi-cyclic phenomena (see Fischer and Arthur, 1977; Haq *et al.*, 1977; van Andel *et al.*, 1977; Berger, 1979; Arthur, 1979).

There are two fundamentally different ways to approach the discussion of climate steps recorded in sediments. One is to take each event as a unique occurrence that calls for a unique explanation. For example, the 3 Ma northern glaciation event might be viewed as a result of closing the Isthmus of Panama, with the deflection of previously westward traveling Caribbean waters into the Gulf Stream resulting in increased moisture supply in high latitudes, and hence increased snowfall. Alternatively, the same event might be seen as an inevitable consequence of a general cooling trend with strong positive feedback setting in, from albedo increase, once a snow-cover lasts for a certain part of the year. The first approach emphasizes a cause that is external to the climate-producing system, the second focuses on positive feedback mechanisms within the system. The second approach can be applied to an entire class of events, as well as to the amplification of cyclic signals within the period in question. It need not, of course, exclude the search for prime causes, both for the general trend onto which the event is grafted and for the exact timing of the event.

This chapter summarizes some concepts in connection with event analysis (Berger *et al.*, 1977, 1981; Thierstein and Berger, 1978; Vincent *et al.*, 1980). The basic proposition is to separate the external causes (e.g., irradiation changes, opening or closing of basin connections) from the internal sources of insta-

TABLE 3.1 Examples of Fast Climatic Transitions in the Cenozoic

1. Terminations of the late Pleistocene (0.8 Ma)
   Effect : Rapid deglaciations, warming of mid-latitudes, pCO_2 change.
   Refs. : Emiliani, 1955, 1966, 1978; Broecker and van Donk, 1970; Dansgaard *et al.*, 1971; Shackleton and Opdyke, 1973; Berger and Killingley, 1977; Berger *et al.*, 1977; Rooth *et al.*, 1978; Delmas *et al.*, 1980.
2. 1-Ma Event
   Effect : Onset of large climatic fluctuations after period of quiescence.
   Refs. : Shackleton and Opdyke, 1976; van Donk, 1976; Vincent and Berger, 1982.
3. 3-Ma Event
   Effect : Onset of Pleistocene-type climatic fluctuations.
   Refs. : Berggren, 1972; Shackleton and Opdyke, 1977; Keigwin, 1978, 1979.
4. Messinian Salinity Crisis (~5.5 Ma)
   Effect : Isolation of Mediterranean through regression, strong cooling.
   Refs. : Hsü *et al.*, 1973, 1977; van Couvering *et al.*, 1976; Adams *et al.*, 1977.
5. 6-Ma Carbon Shift
   Effect : Isotope ratios of the ocean's carbon shift to lighter values, presumably owing to organic carbon input from regression and erosion. pCO_2 change(?).
   Refs. : Bender and Keigwin, 1979; Keigwin, 1979; Vincent *et al.*, 1980; Berger *et al.*, 1981.
6. Mid-Miocene Oxygen Shift(?) (~15 Ma)
   Effect : Isotope ratios of the ocean's oxygen shift to heavier values, presumably owing to Antarctic ice buildup.
   Refs. : Douglas and Savin, 1975; Savin *et al.*, 1975; Shackleton and Kennett, 1975; Savin, 1977.
7. Mid-Oligocene Oxygen Shift(?) (~30 Ma)
   Effect : First occurrence of rather heavy oxygen isotope values in deep-sea benthic foraminifera, presumably owing to polar bottom-water formation.
   Refs. : Savin, 1977; Arthur, 1979.
8. Eocene Termination Event (~38 Ma)
   Effect : Cooling in high and low latitudes, expansion of polar highs. Significant changes in deep-sea benthic foraminifera and of high-latitude planktonic foraminifera. Rapid drop of the carbonate compensation depth.
   Refs. : van Andel and Moore, 1974; Benson, 1975; Kennett and Shackleton, 1976; Savin, 1977; Burchardt, 1978.

bility. These sources of instability are of two kinds. The first are regular amplification mechanisms, such as albedo feedback, whose strength is more or less proportional to the excursion from the steady-state condition. The second are strong interferences from "transient reservoirs" of geochemical disequilibrium, which are rather unpredictable. An important feature of the concept of transient reservoirs is that an exchange of "disequilibrium energy" between transient reservoirs of different kinds can lead to oscillations of the type observed in mechanical and electrical systems.

THE TASKS OF PALEOCLIMATOLOGY

The need for step analysis as a means to advance the science of paleoclimatology must be framed within the entire scope of this discipline. The task of paleoclimatology is to record and explain the climatic trends and events that have occurred throughout the Earth's history. In order to develop models for sequences of climatic states, we must study periods of some duration, with adequate sampling sets of states and their transitions. The Pleistocene, especially the late Pleistocene, which includes the largest known climatic fluctuations, is such a set. One central task is to analyze the Pleistocene record in a way that provides analogies for the understanding of more ancient climates. In doing so, it is useful to focus on the systematic aspects of climate (Kominz and Pisias, 1979; Imbrie and Imbrie, 1980) rather than on the physical aspects of climate, which tend to be poorly constrained for this time period.

The Pleistocene is itself part of a long-term climatic trend—that of an overall cooling since the early Tertiary. It also contains a trend within it—that of ever-increasing amplitudes of climatic excursions (see e.g., Shackleton and Opdyke, 1976). Both the onset of northern glaciation and the trend within the Pleistocene are reasonable to ascribe to an increase in positive feedback within the system, the obvious candidate being an increase in the role of snow cover in the heat budget of the Earth's surface. Likewise, instability evidently grows with growing ice caps. Rapid large transgressions become possible through the storage of continental ice masses; such transgressions can cause rapid changes in albedo (water is dark; land is bright; see Table 3.2).

TABLE 3.2 Albedo Values of Ocean and Land Surfaces[a,b]

Annual global average:	14
Ocean:	
low latitudes:	4-7
mid-latitudes:	4-19
high latitudes:	6-50
Great lakes:	
min. (summer):	6
max. (winter):	55
Land:	
desert:	20-30
grasslands, coniferous, and deciduous forests:	15
wetlands:	10
tropical rain forests:	7
snow-covered land:	35-82
Antarctic Continental Ice Cap:	85
pack ice in water:	40-55

[a]Reflectivity in percent of incident light, during noon.
[b]Source: Compilation of Hummel and Reck, 1978.

The buildup of continental ice causes overall regression and may, itself, be a consequence of regression (Hamilton, 1968)—a prime example of positive feedback. Regression, besides leading to increased albedo, may create other sources of instability in addition to snow and ice, as we shall see.

The most striking feature of the Pleistocene record—the climate cycles driven by the Milankovitch mechanism—reveals the activity of strong positive feedback (Kukla, 1975; Suarez and Held, 1979). Parts of the pre-Pleistocene record show cycles also, but generally of a much smaller amplitude. We must assume that before the Ice Age the lack of strong albedo feedback allowed negative feedback to hold sway in dampening climatic fluctuations. Negative feedback, of course, ultimately provides the climatic stability that allowed life to exist on Earth for billions of years. However, negative feedback typically has substantial time lags, which presumably are of the same order as the leads and lags between various climatic indices, as well as the climatic response times. In the Pleistocene such lags are of the order of 5000-10,000 years (Moore *et al.*, 1977; Imbrie and Imbrie, 1980). Thus, when strong positive feedback exists, large climatic excursions can develop before the system brakes itself.

To study the various ways in which sea-level fluctuations produce (and interact with) trends, cycles, and events and the roles of positive and negative feedback in amplifying and dampening climatic input functions would appear to be the most challenging tasks of paleoclimatology. We are still at the beginning of meeting this challenge. Correlations of various climatic, geochemical, and evolutionary signals with sea-level fluctuations have been suggested (Fischer and Arthur, 1977), but the linking mechanisms remain obscure. Quantitative analysis of pre-Pleistocene climatic trends has been attempted (Donn and Shaw, 1977) but without consideration of feedback. The most advanced studies are those modeling the climatic conditions of the last glacial (Gates, 1976a, 1976b; Manabe and Hahn, 1977) and those that extract frequencies and phase shifts from Pleistocene deep-sea records (Hays *et al.*, 1976; Pisias, 1976; Moore *et al.*, 1977). The modeling of climatic conditions is not the same as the modeling of climatic change. The extraction of frequencies is of prime importance for finding the input functions, but it leaves open the question of internal feedback. Thus, the most intriguing mystery of the ice ages—how climate can change so fast—remains unresolved.

Phase shifts between different climate-related signals are potentially revealing as far as cause-and-effect chains, much as one would expect from the analysis of time segments of rapid change. However, (1) shifts may differ between various types of climate excursions, so that the result of a bulk analysis extending over a long period may be misleading; and (2) shifts between signals may be produced artificially through mixing processes in the record as a result of changes in the concentration of the signal carriers (Hutson, 1980).

There can be little doubt that the various tasks of paleoclimatology would be greatly facilitated if we had a detailed record of a number of climatic steps and some idea about the processes associated with them. Before going any further, however, we must ask whether suitable steps exist at all.

THE REALITY OF STEPS

The reality of rapid climatic change was first demonstrated by Emiliani (1955) through oxygen isotope stratigraphy of long continuous deep-sea records. These records provide the strongest support for the Milankovitch mechanism of Pleistocene climatic fluctuations (Figure 3.1). Geomagnetic dating of an isotope stratigraphy in the western equatorial Pacific (Shackleton and Opdyke, 1973) established the chronology that allowed the correct identification of the periodicities involved (Hays *et al.*, 1976; Pisias, 1976). The isotope fluctuations, of course, are not a direct representation of an irradiation curve but the result of a convolution of radiation input with climatic feedback mechanisms involving ocean, atmosphere, snow cover, ice cover, and vegetation. When studying the curves generated by Emiliani (1955, 1966) and by Shackleton and Opdyke (1973, 1976) we note a striking phenomenon, important especially for the survival of higher organisms. The fluctuations never go beyond a certain maximum value on either side of the range (see Figure 3.1). Obviously, there is a limit to warming: radiation of heat into space increases approximately as the fourth power of the Earth's surface temperature (the Stefan-Boltzmann law). The rapid rise of backradiation with increas-

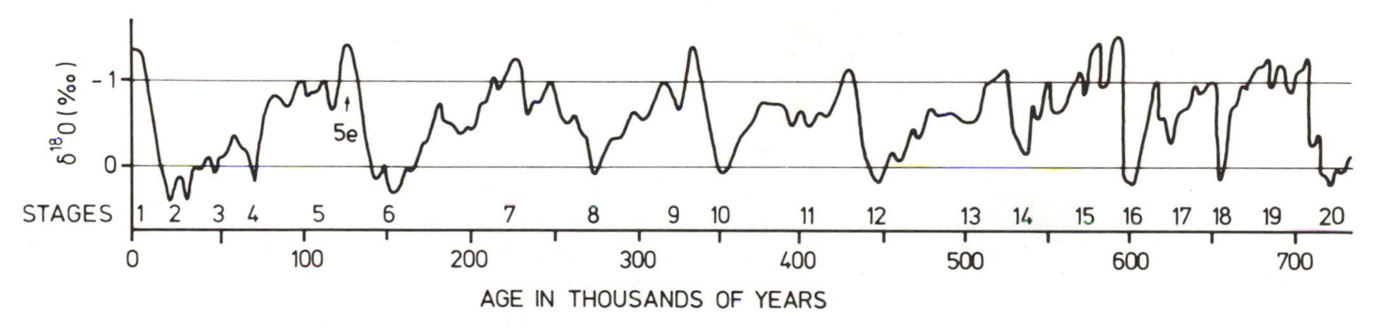

FIGURE 3.1 Composite $\delta^{18}O$ curve of Emiliani (1978), showing sawtooth pattern and well-defined limits of $\delta^{18}O$ maxima and minima. The rapid change after most $\delta^{18}O$ maxima suggests attainment of a critical setting (interpreted as a buildup of sufficiently large transient reservoirs) and a "runaway" effect once melting reaches some critical rate (interpreted as reservoir collapse).

ing temperature constitutes efficient negative feedback. But what negative feedback is preventing continued cooling? Why does not the Earth cover itself with ice? Once ice spreads, albedo increases. Thus, less and less of the incoming radiation is retained for heating the Earth's surface—ice becomes the stable phase over larger and larger areas. Yet, the Earth does not become white but remains blue and brown.

One type of negative feedback that has been invoked to prevent total glaciation is the decrease of moisture transport that accompanies a general lowering of temperatures (cold air cannot hold much water) and a covering up of the sea surface in high latitudes with pack ice (Emiliani, 1978). Also, migration of the polar front toward mid-latitudes prevents moist tropical air from reaching the original centers of glacial growth. The glaciers in high latitudes would "starve" under such conditions. But would they disappear? And would glaciers not continue to grow in mid-latitudes?

The question of negative feedback, which prevents the Earth from icing over, is open. One important factor, probably, is the removal of positive albedo feedback from falling sea level. Once the sea level falls to the shelf edge, a further drop does not result in much decrease of ocean surface; thus the albedo stabilizes.

We now turn to the third important feature of the oxygen isotope record. Glacial maxima are followed, almost inevitably, by glacial minima, and the transitions are extremely rapid. The phenomenon was emphasized by Broecker and van Donk (1970), who called the transitions "terminations" and gave them numbers. They also coined the term "sawtooth" pattern to characterize the alternations between rapid deglaciations and more gradual (but fluctuating) buildup of ice.

The smoothing activity of bioturbation on the seafloor is such that a change of the rapidity suggested by the deep-sea record of Termination I or Termination II (11,000 and 127,000 yr ago, respectively) is extremely difficult to envisage. Both an instantaneous flip-over from one climatic state to the next (Peng *et al.*, 1977) and an overshoot phenomenon (Berger *et al.*, 1977; Berger, 1978) have been suggested to account for this difficulty. If bioturbation worked then as it does today, almost any physically reasonable transition should be more gentle than that observed.

One possibility is that we are looking at a hiatus in sedimentation. A gap in the recording would juxtapose different stages in history. If we entertain this notion of a gap, we are then faced with the necessity of providing a short event, an impulse, that produces nondeposition or erosion at the correct time. Thus the question of rapid change within the system would return, having merely been shifted from paleoclimatology to geochemistry. It is true that cessation (or great reduction) of bioturbation also would help; the problem might then be shifted to deep-sea biology. However, such shifting of responsibility does not come to grips with the central problem: that the system is changing rapidly and that this has consequences for the circulation of the ocean and atmosphere as well as for their chemical composition, and hence for climate and evolution.

What is the importance of the deglaciation event, other than demonstrating the existence of rapid climatic change, or climate steps? Can we learn something about climate steps in general, even though physical mechanisms may vary widely?

We may assume that a system in rapid transition cannot in any way be thought of as being intermediate between the previous and the subsequent state. The best support for this proposition again comes from the study of Pleistocene deep-sea sediments, this time from a region where high sedimentation rates allow a detailed look at Termination I, namely, the Gulf of Mexico (Kennett and Shackleton, 1975; Emiliani *et al.*, 1975). The oxygen isotope records of planktonic foraminifera show a marked excursion toward light values at the end of a rapid (but apparently pulsating) rise from the glacial maximum (Figure 3.2). The curves, in fact, look much like the standard amplitude-versus-time response plots in the textbooks of systems analysis, familiar to students of mechanical and electrical engineering. In principle, such plots describe the response of a system to a step input. The steepness of the transition from one state to the next is a measure for the sensitivity of the system, as is the amount of overshoot (Figure 3.3). Essentially we see here the result of the competition between negative feedback, which slows the transition, and positive feedback, which accelerates it and builds up the overshoot. The overshoot is characterized by being short-lived and by preceding another period that mimics the original state, i.e., a kind of undershoot. To produce this rebound, positive feedback is active in reverse. More oscillations can then follow, depending on the strength (or weakness) of the dampening processes in the system.

Let us, for the sake of argument, accept the proposed analogy between the oxygen isotope record of the Gulf of Mexico and a two-phase step response (Figure 3.4). Have we thereby done any more than introduce some terms from systems analysis to the description of a set of phenomena, without a net increase in knowledge or understanding?

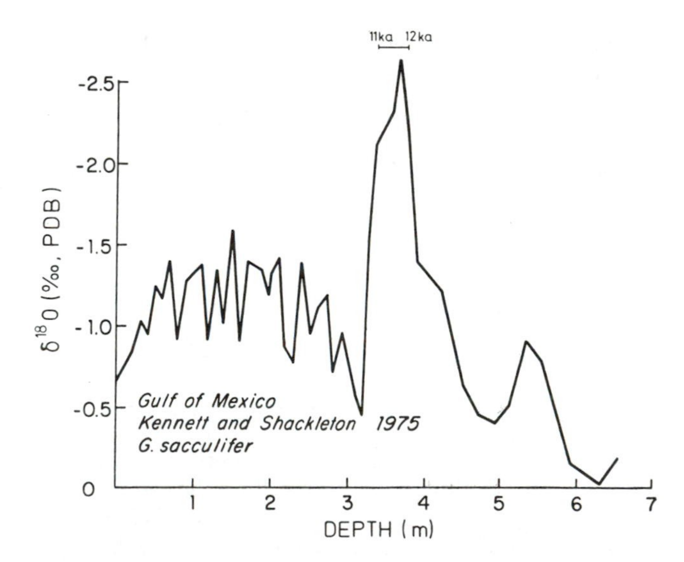

FIGURE 3.2 Oxygen isotope record of the last glaciation in the Gulf of Mexico by Kennett and Shackleton (1975), as dated by correlation with a similar (^{14}C-controlled) record of Emiliani *et al.* (1975).

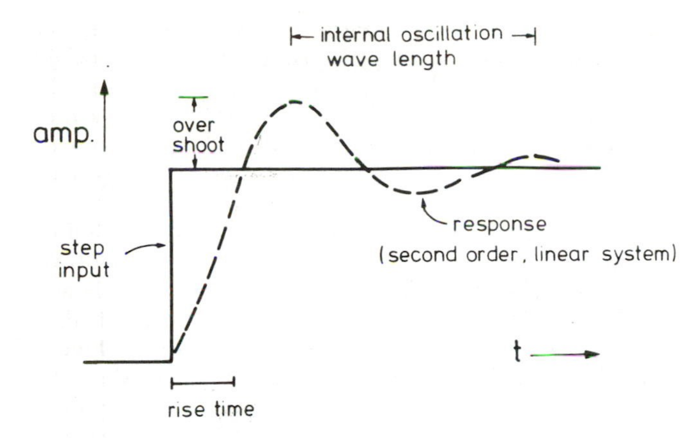

FIGURE 3.3 Sketch of three types of response to a step input in a two-dimensional system.

On the contrary, I suggest that we have now changed the mode of attack on the problem. Recall that the Gulf of Mexico isotope anomalies were originally interpreted by their discoverers as indicating an unusual influx of meltwater from the disintegrating Laurentian Ice Sheet (Kennett and Shackleton, 1975; Emiliani *et al.*, 1975). According to these authors the low $^{18}O/^{16}O$ ratio in the meltwater produced the anomaly, after mixing with the seawater of the Gulf. Ensuing discussions accepted the explanation (which appeared reasonable) but questioned Emiliani's ^{14}C date of 11,500 yr ago because it did not agree with evidence on land, regarding the course and the timing of meltwater flow. In fact, these discussions may have missed the point. The meltwater influx could have been at a maximum well *before* the isotope anomaly. The rapid change in the isotope signal that precedes the anomaly is as much witness to the rapid introduction of meltwater as is the anomaly itself. We cannot dismiss the possibility that the difference in $\delta^{18}O$ between the upper waters of the Gulf and the global ocean was just as great during this period, which corresponds to the "ramp" of the signal, as during the anomaly itself. Thus, an exclusive focus on the anomaly is not the right approach: the anomaly and the ramp belong together, just as the step-function analogy would suggest. Evidently, the anomaly might be expected at the end of the rapid change, whether or not the flux of the Mississippi River increased at anomaly time. Thus, a purely formal consideration—seeing the record as a step response—changes the argument considerably.

THE SEARCH FOR POSITIVE FEEDBACK

The questions that arise within the step-function analogy are quite general: What physical processes limit the rate of transition from glacial to postglacial values? Is the gradient of $\delta^{18}O$ versus time steeper in the Gulf than outside of it? Is there an overshoot phenomenon outside the Gulf also? If so, can we identify positive feedback mechanisms that could produce it? What is the nature of the "rebound" following the anomaly? If it has the structure of a pendulum swing, why is the period near 2000 yr? And which are the reservoirs of disequilibrium that pass the equivalent of kinetic and potential energy between them? We cannot answer these questions at present. However, we might usefully consider where to begin the search.

In the present case, naturally, the most obvious candidates for physical processes providing positive feedback are those having to do with the melting of ice. The melting takes heat, and there is a limit to the rate at which heat can be transported to the site of melting. Incidentally, if much of the heat comes in the form of rain, the runoff will be a mixture of rain and meltwater and its oxygen isotope composition will be somewhere between that of rain and glacial ice. To get positive feedback, we must ask that the melting, once started at some minimum rate, enhance further melting. For example, increased local absorption of radiation by exhumed debris on top of glacial ice could help. The effect would seem insufficient, however, because it can be removed quickly by snow cover. Continuing vigorous heat transfer from the tropics would appear necessary.

How can strong initial melting change the heat budget of the entire system in favor of its continuation? Meltwater does mainly two things: it raises sea level, and it decreases the salinity of the ocean. A sea-level rise decreases the albedo: as the ocean surface expands, absorption increases and more of the Sun's radiation is used to heat the Earth's surface. A rise in sea level can also, presumably, destabilize those parts of the glacial ice that rest on shelf and can be floated (J. T. Andrews, University of Colorado, personal communication, 1979). Conceivably such floating could favor the occurrence of ice surges of the type envisaged by Wilson and others (Wilson, 1964, 1969; Hollin, 1972; Flohn, 1974), which would accelerate the rise of sea level. (In the ice-surge hypothesis, the focus is on the attendant increase in albedo and cooling, however.)

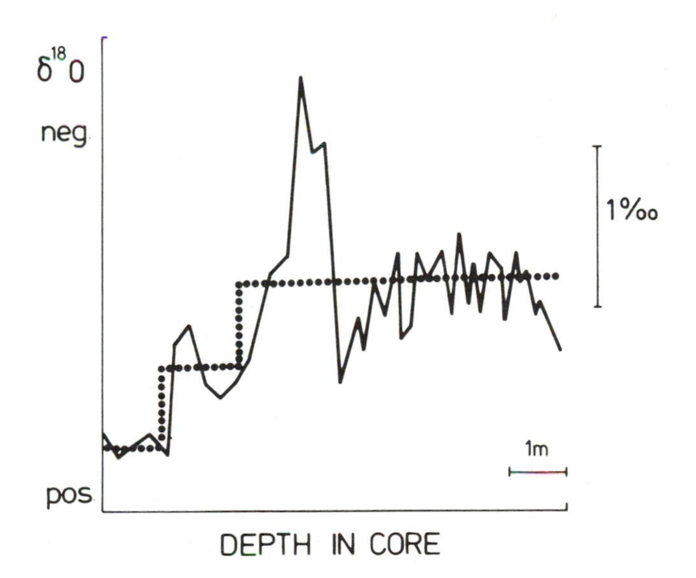

FIGURE 3.4 The Gulf of Mexico record (Figure 3.2) interpreted as a response of a two-dimensional system to a couplet of step inputs.

Assimilation of the freshwater into the ocean takes time. The time constant of mixing is nonnegligible, being of the order of 1000 yr in the present ocean. The rate depends on two factors: the strength of the mixing drive and the vertical stability of the water column. Mixing is both wind- and density-driven and is therefore closely tied to temperature gradients and salinity patterns. Stability depends on the density profile, and is likewise tied to temperature and salinity distributions. The relationships are such that an increase in the forces responsible for mixing, which are derived largely from the planetary temperature gradient, also leads to an increase in the opposing stability—mixing rate is resistant to change. However, the introduction of meltwater can change this situation, as it is much lighter than seawater and tends to float. A low-density surface layer does not prevent mixing by current shear and eddies through wind, but it can potentially interfere with deep- and bottom-water production.

In theory, there is some range of the rate of introduction of meltwater where there is essentially no effect on mixing. At the upper end of this range there must be a critical point, where the rate of influx begins to exceed the ability of the ocean to assimilate the added freshwater at the given rate of mixing. At this point the mixing rate must slow. The questions are, what *is* the critical input rate, and was it ever reached during deglaciation? The first question can be attacked by modeling; the second must be read from the record, for example, through comparison of stable isotope stratigraphies of deep-sea benthic and planktonic foraminifera.

Let us assume that the critical influx is indeed reached, and mixing slows. At this point the system changes its mode of operating. Because mixing has now slowed, the value for the critical rate of influx will begin to fall. The system develops a strong positive feedback by building up stable stratification, with a halocline at the bottom of the wind-mixed layer in the broad sense, say, between 500 and 1000 m. Furthermore, the ocean will now "remember" this condition for some time, in fact, for about 1000 yr or so. The value for the critical rate of influx will be lowered during this entire period, rising but slowly to its original level.

The point of the discussion is this: if early in deglaciation there is a strong pulse of meltwater influx, stable stratification can be maintained through a series of lesser pulses later. There is in fact evidence that sea level rose in pulses (Fairbridge, 1961; Mörner, 1975), but the matter is still under discussion.

Meltwater influx, then, can conceivably set into motion a strong nonlinear positive feedback mechanism, which goes beyond albedo decrease from ocean-area expansion and which involves the development of a low-salinity layer of the type suggested by Worthington (1968). But how can this potential for feedback be translated into an energy budget favorable for melting?

These questions call for some rigorous modeling; at present, we can only guess what a low-salinity lid on the ocean would do to the climate. Presumably, a lack of communication with cold deep waters would allow low-latitude waters to heat up considerably, enhancing the meridional temperature gradient and hence the heat transport to higher latitudes. Also, the temporary decoupling of three fourths of the ocean mass from the heat budget should increase climatic instability: the inertia of the system is decreased. If true, this would favor delivery of meltwater in pulses, thus maintaining instability.

While seeking strong positive feedback, we discovered a source of instability that can, once activitated, develop feedback for instability itself. Within the unstable system, small changes in input (e.g., from the Sun's radiation) can be translated into larger climatic fluctuations. We have here one way to produce the rapid changes in climate that characterize the transition from glacial to postglacial time.

Is the development of instability typical for fast climate transitions? Are there nonglacial climate-changing mechanisms analogous to the melting of ice? If yes, analogous in which sense?

THE PHENOMENON OF RESERVOIR COLLAPSE

Glacial ice may be seen as a transient reservoir of water, outside the main ocean basin, which, when reunited with the ocean, suddenly raises sea level with all the attendant effects on climate through a decrease in albedo and an increase of supply of moisture to the atmosphere. We can view the melting of ice caps as a "reservoir collapse" that feeds on itself once destruction proceeds at a minimum rate. Glacial ice is a reservoir of *freshwater*, hence the potential for additional complications. Can we envisage other types of transient reservoirs? Perhaps so. Several other possibilities are shown in Figure 3.5. The most obvious transient reservoirs are adjunct ocean basins and marginal seas.

A reservoir of glacial ice is analogous to a reservoir of water in an isolated basin. If the basin can run dry, as the Mediterranean did at 5 Ma (Hsü *et al.*, 1977), the analogy is almost perfect. We can produce, through alternating emptying and filling of such a basin, rapid transgressions and regressions. The volume of the Mediterranean Sea, for example, would have allowed almost instantaneous changes in sea level of near 10 m in the Messinian, whereas that of the South Atlantic might have allowed changes of about 50 m in the Aptian (Berger and Winterer, 1974). If such changes occurred, we should see them both in the deep-sea record and on the shelves. We also should expect substantial evidence for climatic instability. On the one hand, emptying a marginal sea can affect albedo over a large area—including not only the basin itself but also the hinterland around it, which depends on moisture from the basin to maintain its vegetation. On the other hand, rapid transgressions and regressions have their own global effects on albedo and moisture distribution.

The potential for the existence of isolated basins is quite large since the breakup of Pangaea. Salt deposits at ocean margins, e.g., around the Atlantic (Emery, 1977), suggest that large isolated basins existed at various times in the Mesozoic and in the late Paleozoic. If such basins did indeed exist, we should see the evidence in the global correlation of fast sea-level fluctuations during certain periods.

Isolated basins can provide separate transient reservoirs of water, but they can also provide reservoirs of water of low or

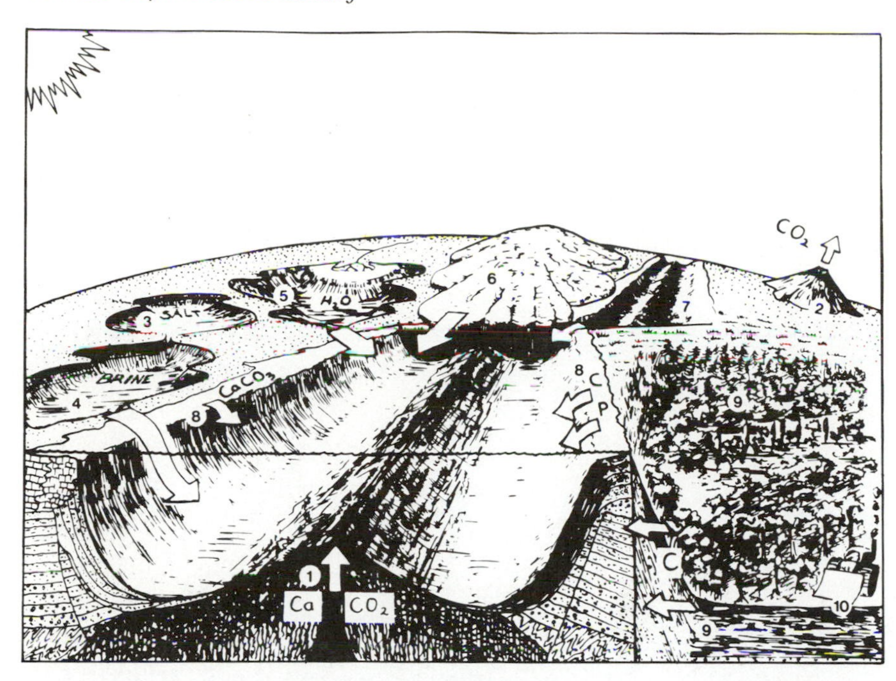

FIGURE 3.5 Diagrammatic representation of sources of instability. External: Sun and mantle processes (1, 2). (Feedback from ice growth and decay on mantle processes cannot be excluded, however.) Internal: transient reservoirs of geochemical energy; 3 to 5, marginal seas with unusually saline waters or freshwaters; 6, continental ice masses; 7, adjunct ocean basins with salinity deviations; 8, easily erodable shelf carbonates, carbon deposits, and phosphatic sediments; 9, "soil" carbon accessible to fast erosion; 10, man's activity (industrial CO_2, deforestation, accelerated erosion). The transient reservoirs 3 to 7 both influence and respond to sea level variation, the reservoirs 8 and 9 collapse during regression. Source 10 may eventually respond to negative feedback from climatic change.

high salinity. Injection of such waters, presumably occurring repeatedly during periods of critical isolation, could have had a profound influence on the history of evolution of climate and life (Gartner and Keany, 1978; Thierstein and Berger, 1978; Berger and Thierstein, 1979). More generally, the existence of semi-isolated reservoirs of brackish or supersaline water is a source of instability whose scale is tied to the size of the reservoir, the degree of deviation of salinity from the global average, and the potential flux exchange.

There are, even today, two large reservoirs that are almost isolated and that could become much more so with a drop of sea level of between 100 and 200 m: the Mediterranean Sea and the Arctic Ocean. The Mediterranean is anomalously salty and plays an important role in the deep circulation of the Atlantic Ocean. The history of the Mediterranean outflow may be closely tied to that of North Atlantic bottom-water production (Reid, 1979). Deep circulation in the Mediterranean apparently reversed its direction in the earliest Holocene because of an increased supply of freshwater (Kullenberg, 1952; Williams *et al.*, 1978), which removed one source of heavy deep water in the North Atlantic. The effects (if any) on the deep circulation have not been modeled; I suspect they were substantial. The Arctic Ocean has unusually low salinities in its surface waters. Its connection with the world ocean is somewhat tenuous; at least one geologist suggests that it was severed entirely during glaciation (M. Vigdorchik, INSTAAR, personal communication, 1978). If, as expected, the connection between the Arctic and Pacific Oceans was cut off during glacials, and that between the Arctic and Atlantic Oceans was greatly reduced, the Arctic Ocean might have collected brackish water throughout. Thus, when sea level first rose, low-salinity water from the Arctic Ocean could have helped to start off the deglaciation feedback chain postulated earlier.

The development of transient-water reservoirs and hence of a potential for strong climatic instability is not necessarily restricted to the time since the breakup of Pangaea. In earlier times back-arc basins might have provided transient reservoirs under favorable conditions. Again there is a recent analog: the Pleistocene record of the Japan Sea contains layers with brackish-water diatoms, suggesting substantial isolation (Burckle and Akiba, 1978).

Although the occurrence of transient-water reservoirs is a fact, the effect of such reservoirs on climate is virtually unstudied. The reservoirs are part of the hydrological cycle; they build up energy within this cycle, which might be released all at once, with quasi-catastrophic consequences. Are there other geochemical cycles for which this might also be true?

TRANSIENT CARBON RESERVOIRS

The carbon cycle is another obvious candidate for the presence of transient reservoirs. The phosphorus cycle also is a likely choice (Arthur and Jenkyns, 1980), as is the sulfur cycle (Holser, 1977). That the carbon cycle is intimately connected to climatic change is obvious from an overall parallelism of carbon isotope fluctuations with oxygen isotope fluctuations in the deep-sea record on various scales (Broecker, 1973; Berger, 1977b; Fischer and Arthur, 1977; Shackleton, 1977).

Various mechanisms have been proposed through which the linkage could be achieved. Long-term fluctuations in the ratio of ^{13}C to ^{12}C in carbonates were ascribed by Tappan (1968) to variations in the accumulation rate of organic carbon. This idea has since been elaborated on in several guises. Rapid fluctuations in $\delta^{13}C$ in the Pleistocene were related by Shackleton (1977) to the buildup and destruction of tropical rain forests,

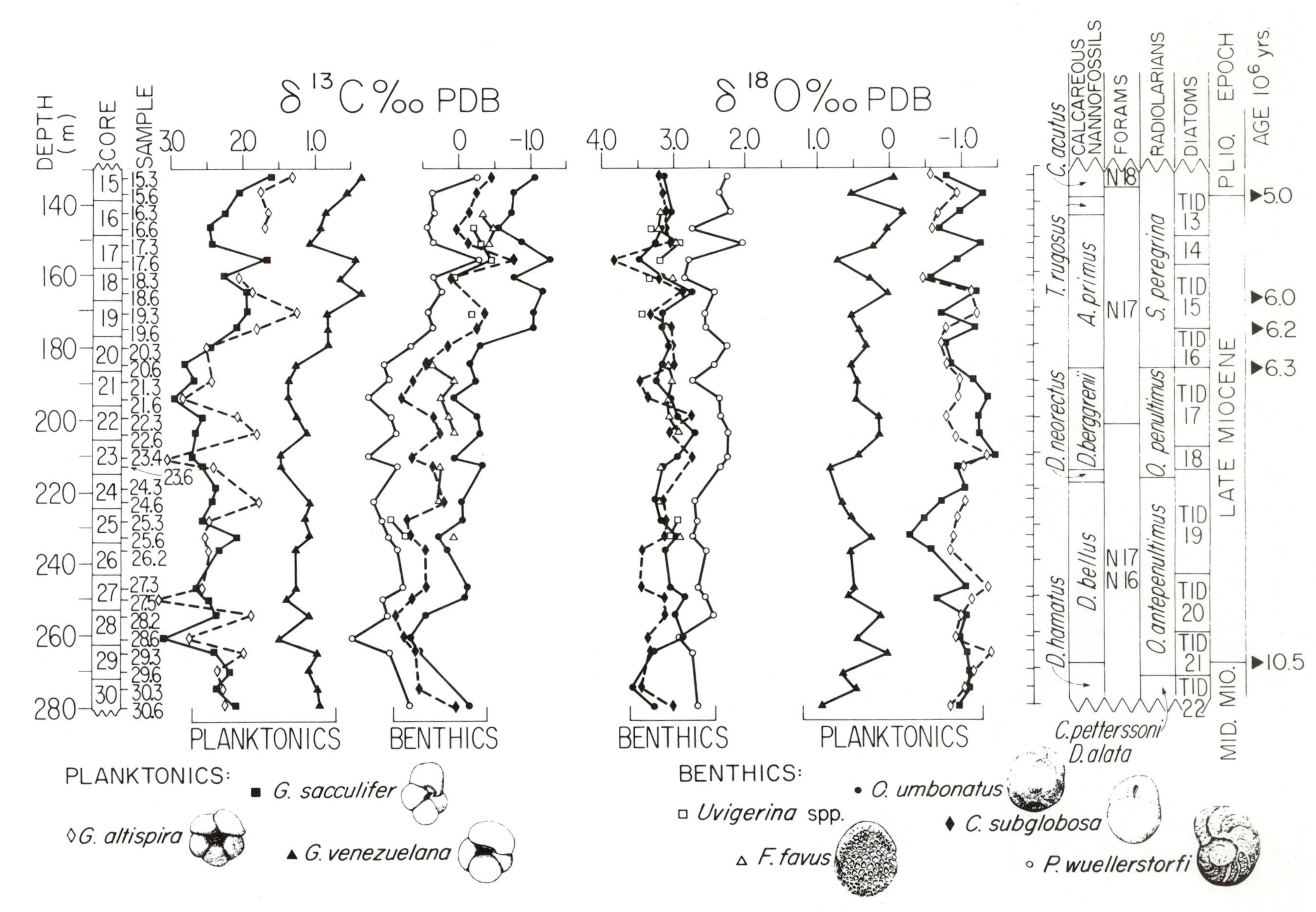

FIGURE 3.6 Upper Miocene isotopic stratigraphies, DSDP Site 238, tropical central Indian Ocean, from Vincent *et al.* (1980). Note the carbon isotope "shift" of approximately 0.8 ‰ toward lighter values upward in the section between samples 19-6 and 20-6. Note the possibility of an "overshoot" and a "rebound," and the (phase-shifted?) fluctuations in the $\delta^{18}O$ signal.

which represent a substantial part of the biosphere. We can extend this idea to include the carbon of extratropical forests, marshes and swamps, peat and "soil carbon" in shelf and coastal deposits (all readily available for decay or erosion) in a pool of "transient carbon"; short-term variations in the size of this reservoir should have effects on the chemistry of the ocean and on the CO_2 content of the atmosphere (Figure 3.5). Direct evidence for changes in pCO_2, from glacial to postglacial time, was recently reported from ice cores (Delmas *et al.*, 1980).

There is little question that the size of the transient carbon reservoir must have varied considerably during the Pleistocene. Apparently, the last glacial maximum was dry, deserts being widespread and the tropical rain forests being greatly reduced (Sarnthein, 1978). In contrast, the early Holocene was a wet period over large areas; the reversal of the deep circulation in the Mediterranean is part of this phenomenon. During the transition from the dry glacial to the wet Climatic Optimum, the transient carbon reservoir was affected both by the buildup of the biosphere (essentially forests) and the erosion of soil carbon. The buildup extracts carbon from the ocean-atmosphere system; the erosion delivers carbon to it. If these processes fluctuate, with a phase shift near 180°, there may be a remarkable potential for introducing climatic instability via variation of the CO_2 content of the atmosphere. That the carbon chemistry of the ocean underwent major changes during deglaciation is clear from large vertical excursions of carbonate preservation levels on the seafloor, during a short period (Berger, 1977a). If stable stratification developed during deglaciation, the exchange of CO_2 between ocean and atmosphere must have been severely affected (Worthington, 1968).

How do these observations and speculations bear on pre-Pleistocene climates? Evidently, the waxing and waning of ice caps is a sufficient but not a necessary condition for producing fluctuations in the transient carbon reservoir. Any mechanism producing fluctuations in sea-level and dry-wet cycles will do.

Of special interest is the possibility that the size of the transient carbon reservoir can be greatly increased if the deep ocean provides temporary carbon storage through changes in mixing time and oxygenation of the deep sea. Reactive organic carbon can accumulate on a poorly oxygenated seafloor and can be redelivered to the system on improvement of aeration. Opportunities for instability and for rapid climatic change might arise from the presence of such a marine transient carbon reservoir. For example, small radiation input cycles, leading to slight fluctuations in the oxygenation of various parts of the ocean via pulsating production of deep and bottom waters, could then translate into a pulsating supply of CO_2 from the ocean to the atmosphere. The existence of oxygenation cycles in Mesozoic and early Tertiary deep-sea sediments is of interest in this connection (Dean *et al.*, 1978).

OVERSHOOT AND REBOUND

Earlier, when discussing the nature of steps, we have seen that a step input can produce an overshoot and a rebound toward the original condition. In the case of deglaciation, the period known as "Alleröd" apparently was an overshoot and the "Younger Dryas" was a rebound. Having identified two components of the climate system that can store disequilibrium—the hydrosphere and the active carbon sphere—it is now in principle possible to construct a two-dimensional oscillating system. In such a system the disequilibrium is passed back and forth from one compartment to the other; the rate of transfer determines the periodicity of the oscillation.

In our deglaciation example, stable stratification (a "meltwater lid") might lead to CO_2 buildup in the deep sea (Worthington, 1968), which leads to CO_2 loss in the atmosphere and hence cooling. Release of the deep-stored CO_2 after mixing could then increase the CO_2 content of the atmosphere and produce warming. The CO_2-induced cooling and warming, of course, would feed back into meltwater pulsing. An oscillation period of 2000 to 3000 yr would seem reasonable, in view of a 1000-yr mixing time for the normal ocean. Incidentally, the CO_2 fluctuation hypothesis agrees well with the CO_2-concentration record reported from Antarctic ice (Delmas *et al.*, 1980).

In a recent study of oxygen and carbon isotope variations in the late Miocene focusing on the Magnetic Epoch-6 Carbon Shift, Vincent *et al.* (1980) presented evidence for a step followed by increased fluctuations in climatic signals (Figure 3.6). One is tempted to identify an overshoot and a rebound in the carbon isotope stratigraphy, following the step at 6.2 Ma. The wavelength of such an oscillation would be several hundred thousand years. However, it is difficult in this instance to separate possible oscillations from a general increase in instability that was presumably introduced by the factor causing the step. If regression and the isolation of the Mediterranean was a crucial factor in producing the signal, this creation of a transient reservoir would likewise be expected to increase climatic instability. In any case, one would like to see a closer spacing of samples, because the typical frequencies of climatic fluctuations—whether they be oscillations or not—should contain clues to the responsiveness of the transient reservoirs involved. Unfortunately, the quality of the cores traditionally recovered by the *Glomar Challenger* sets severe limits for stratigraphic resolution. The new method of hydraulic piston coring removes this obstacle and offers a better definition of climatic fluctuations; this tool should be used to full advantage.

SUMMARY AND CONCLUSIONS

The deep-sea record provides a number of stratigraphic intervals showing a rapid transition from one climatic-geochemical state to another. These intervals provide an opportunity to study the dynamics of the ocean-atmosphere system on a scale from 10^3 to 10^6 yr. Step-function analysis provides useful concepts for the study of such transitions, as can be readily demonstrated using the last "termination" event in the Pleistocene record. There is an intriguing possibility that a system in transition oscillates, because of the passing of geochemical "disequilibrium energy" from one transient reservoir to another, analogous to mechanical systems (potential versus kinetic energy) and electric circuits (magnetic versus electric fields). In a glaciated world, the likely candidates for transient reservoirs are ice caps and temporary carbon pools in forests and soils and on shallow seafloors. Geologic periods associated with strong positive feedback in the system (variable snow-cover) and with a large transient reservoir potential (ice caps, semi-isolated basins, reactive terrestrial and marine soil carbon and biocarbon) are characterized by climatic instability. The breakup of Pangaea and the subsequent dispersion of continents and creation of semi-isolated ocean basins must have produced constellations of instability at various stages in the evolution of present-day geography. One type of evidence for such constellations is in the salt deposits of continental margins.

From an operational point of view, system analysis suggests an approach that considers the following questions when studying climate steps: (1) Climate step or hiatus? (2) Strongly or weakly damped? (3) Overshoot and rebound present? (4) External cause (astronomy, tectonics)? (5) Nature of internal feedback? (6) Likelihood of transient reservoir collapse? (7) Nature of disequilibrium oscillations? Even tentative answers to such questions should lead to fruitful working hypotheses regarding climatic change over geologic time spans. The concept of interacting transient reservoirs suggests that climatic systems are not strictly deterministic, that is, they are "almost-intransitive" (Lorenz, 1968) even over long periods of time.

ACKNOWLEDGMENTS

My interest in climate steps developed in the course of studies done in collaboration with R. F. Johnson, J. S. Killingley, H. R. Thierstein, and E. S. Vincent. Hannes Vogler (on hearing the "meltwater spike" story several years ago) first suggested to me to apply impulse-function concepts from systems analysis. W. L. Gates kindly criticized an earlier draft of the manuscript. The work was funded by the National Science Foundation (Oceanography Section) and by the Office of Naval Research (Marine Geology and Geophysics).

REFERENCES

Adams, C. G., R. H. Benson, R. B. Kidd, W. B. F. Ryan, and R. C. Wright (1977). The Messinian salinity crisis and evidence of late Miocene eustatic changes in the world ocean, *Nature 269*, 383-386.

Arthur, M. A. (1979). Paleoceanographic events—recognition, resolution, and reconsideration, *Rev. Geophys. Space Phys. 17*, 1474-1494.

Arthur, M. A., and H. Jenkyns (1980). Significance of rock phosphate and other geochemical sediments in the Cretaceous and Miocene oceans (abs.), *26ᵉ Congrès Géologique International Résumés*, 1357.

Bender, M. L., and L. D. Keigwin, Jr. (1979). Speculations about the upper Miocene change in abyssal Pacific dissolved bicarbonate ^{13}C, *Earth Planet. Sci. Lett. 45*, 383-393.

Benson, R. H. (1975). The origin of the psychrosphere as recorded in changes of deep-sea ostracod assemblages, *Lethaia 8*, 69-83.

Berger, W. H. (1977a). Deep-sea carbonate and the deglaciation preservation spike in pteropods and foraminifera, *Nature 269*, 301-304.

Berger, W. H. (1977b). Carbon dioxide excursions and the deep-sea record: Aspects of the problem, in *The Fate of Fossil Fuel CO_2 in the Oceans*, N. R. Anderson and A. Malahoff, eds., Plenum, New York, pp. 505-542.

Berger, W. H. (1978). Oxygen-18 stratigraphy in deep-sea sediments: Additional evidence for the deglacial meltwater effect, *Deep Sea Res. 25*, 473-480.

Berger, W. H. (1979). Impact of deep-sea drilling on paleoceanography, in *Deep Drilling Results in the Atlantic Ocean: Continental Margins and Paleoenvironment*, M. Talwani, W. Hay, and W. F. B. Ryan, eds., Maurice Ewing Series 3, American Geophysical Union, Washington, D.C., pp. 297-314.

Berger, W. H., and J. S. Killingley (1977). Glacial-Holocene transition in deep-sea carbonates: Selective dissolution and the stable isotope signal, *Science 197*, 563-566.

Berger, W. H., and H. R. Thierstein (1979). On Phanerozoic mass extinctions, *Naturwissenschaften 66*, 46-47.

Berger, W. H., and E. L. Winterer (1974). Plate stratigraphy and the fluctuating carbonate line, in *Pelagic Sediments on Land and under the Sea*, K. J. Hsü and H. Jenkyns, eds., Spec. Publ. Int. Assoc. Sedimentol. 1, pp. 11-48.

Berger, W. H., R. F. Johnson, and J. S. Killingley (1977). "Unmixing" of the deep-sea record and the deglacial meltwater spike, *Nature 269*, 661-663.

Berger, W. H., E. Vincent, and H. R. Thierstein (1981). The deep-sea record: Major steps in Cenozoic ocean evolution, in *Symposium and Results of Deep-Sea Drilling*, R. G. Douglas, J. Warme, and E. L. Winterer, eds., Soc. Econ. Paleontol. Mineral. Spec. Publ. 30.

Berggren, W. A. (1972). Late Pliocene-Pleistocene glaciation, in *Initial Reports of the Deep Sea Drilling Project 12*, U.S. Government Printing Office, Washington, D.C., pp. 953-963.

Berggren, W. A., and C. D. Hollister (1977). Plate tectonics and paleocirculation—commotion in the ocean, *Tectonophysics 38*, 11-48.

Broecker, W. S. (1973). Factors controlling CO_2 content in the oceans and atmosphere, in *Carbon and the Biosphere*, G. M. Woodwell and E. V. Pecan, eds., AEC Symposium 30, pp. 32-50.

Broecker, W. S., and J. van Donk (1970). Insolation changes, ice volumes, and the O^{18} record in deep-sea cores, *Rev. Geophys. Space Phys. 8*, 169-198.

Burchardt, B. (1978). Oxygen isotope paleotemperatures from the Tertiary period in the North Sea area, *Nature 275*, 121-123.

Burckle, L. H., and F. Akiba (1978). Implications of Late Neogene fresh water sediment in the Sea of Japan, *Geology 6*, 123-127.

Dansgaard, W., S. J. Johnson, H. B. Clausen, and C. C. Langway (1971). Climatic record revealed by the Camp Century ice core, in *Late Cenozoic Glacial Ages*, K. K. Turekian, ed., Yale U. Press, New Haven, Conn., pp. 37-56.

Dean, W. E., J. V. Gardner, L. F. Jansa, P. Cepek, and E. Seibold (1978). Cyclic sedimentation along the continental margin of Northwest Africa, in *Initial Reports of the Deep Sea Drilling Project 41*, U.S. Government Printing Office, Washington, D.C., pp. 965-989.

Delmas, R. J., J. M. Ascencio, and M. Legrand (1980). Polar ice evidence that atmospheric CO_2 20,000 yr BP was 50% of present, *Nature 284*, 155-157.

Donn, W. L., and D. M. Shaw (1977). Model of climate evolution based on continental drift and polar wandering, *Geol. Soc. Am. Bull. 88*, 390-396.

Douglas, R. G., and S. M. Savin (1975). Oxygen and carbon isotope analyses of Tertiary and Cretaceous microfossils from the Shatsky Rise and other sites in the North Pacific Ocean, in *Initial Reports of the Deep Sea Drilling Project 32*, U.S. Government Printing Office, Washington, D.C., pp. 509-520.

Emery, K. O. (1977). Structure and stratigraphy of divergent continental margins, in *Continuing Education Course Note Series No. 5*, Am. Assoc. Petrol. Geol., pp. B1-B20.

Emiliani, C. (1955). Pleistocene temperatures, *J. Geol. 63*, 538-578.

Emiliani, C. (1966). Paleotemperature analysis of Caribbean cores

P6304-8 and P6304-9 and a generalized temperature curve for the past 425,000 years, *J. Geol. 74*, 109-126.

Emiliani, C. (1978). The cause of the ice ages, *Earth Planet. Sci. Lett. 37*, 349-352.

Emiliani, C., S. Gartner, B. Lidz, K. Eldridge, D. K. Elvey, T. C. Huang, J. J. Stipp, and M. F. Swanson (1975). Paleoclimatological analysis of Late Quaternary cores from the Northeastern Gulf of Mexico, *Science 189*, 1083-1088.

Fairbridge, R. W. (1961). Eustatic changes in sea-level, in *Physics and Chemistry of the Earth, Vol. 4.*, L. H. Ahrens *et al.*, eds., Pergamon, New York, pp. 99-185.

Fischer, A. G., and M. A. Arthur (1977). Secular variations in the pelagic realm, *Soc. Econ. Paleontol. Mineral. Spec. Publ. 25*, 19-50.

Flohn, H. (1974) Background of a geophysical model of the initiation of the next glaciation, *Quat. Res. 4*, 385-404.

Gartner, S., and J. Keany (1978). The terminal Cretaceous event: A geologic problem with an oceanographic solution, *Geology 6*, 708-712.

Gates, W. L. (1976a). Modeling the ice-age climate, *Science 191*, 1138-1144.

Gates, W. L. (1976b). The numerical simulation of ice-age climate with a global general circulation model, *J. Atmos. Sci. 33*, 1844–1873.

Hamilton, W. (1968). Cenozoic climatic change and its cause, *Meteorol. Monogr. 8*, 128-133.

Haq, B. U., I. Premoli-Silva, and G. P. Lohmann (1977). Calcareous plankton paleobiogeographic evidence for major climatic fluctuation in the early Cenozoic Atlantic Ocean, *J. Geophys. Res. 82*, 3861-3876.

Hays, J. D., J. Imbrie, and N. J. Shackleton (1976). Variations in the Earth's orbit, pacemaker of the ice ages, *Science 194*, 1121-1132.

Hollin, J. T. (1972). Interglacial climate and Antarctic ice surges, *Quat. Res. 2*, 401-408.

Holser, W. T. (1977). Castastrophic chemical events in the history of the ocean, *Nature 267*, 403-408.

Hsü, K. J., M. B. Cita, and W. B. F. Ryan (1973). The origin of the Mediterranean evaporites, in *Initial Reports of the Deep Sea Drilling Project 13*, U.S. Government Printing Office, Washington, D.C., pp. 1203-1231.

Hsü, K. J., L. Montadert, D. Bernoulli, M. B. Cita, A. Erickson, R. E. Garrison, R. B. Kidd, F. Melières, C. Müeller, and R. Wright (1977). History of the Mediterrranean salinity crisis, *Nature 267*, 399-403.

Hummel, J., and R. A. Reck (1978). A global surface albedo model, *Res. Publ., General Motors, GMR 2607*, 47 pp.

Hutson, W. H. (1980). Bioturbation of deep-sea sediments: Oxygen isotopes and stratigraphic uncertainty, *Geology 8*, 127-130.

Imbrie, J., and J. Z. Imbrie (1980). Modeling the climatic response to orbital variations, *Science 207*, 943-953.

Keigwin, L. D., Jr. (1978). Pliocene closing of the Isthmus of Panama based on biostratigraphic evidence from nearby Pacific Ocean and Caribbean Sea cores, *Geology 6*, 630-634.

Keigwin, L. D., Jr. (1979). Late Cenozoic stable isotope stratigraphy and paleoceanography of DSDP sites from the east equatorial and north central Pacific Ocean, *Earth Planet. Sci. Lett. 45*, 361-382.

Kennett, J. P., and N. J. Shackleton (1975). Laurentide ice sheet meltwater recorded in Gulf of Mexico deep-sea cores, *Science 188*, 147-150.

Kennett, J. P., and N. J. Shackleton (1976). Oxygen isotopic evidence for the development of the psychrosphere 38 m.y. ago, *Nature 260*, 513-515.

Kominz, M. A., and N. G. Pisias (1979). Pleistocene climate: Deterministic or stochastic? *Science 204*, 171-173.

Kukla, G. J. (1975). Missing link between Milankovitch and climate, *Nature 253*, 600-603.

Kullenberg, B. (1952). On the salinity of the water contained in marine sediments, *Medd. Oceanogr. Inst. Göteborg 21*, 1-38.

Lorenz, E. N. (1968). Climatic determinism, *Meteorol. Monogr. 8*, 1-3.

Manabe, S., and D. G. Hahn (1977). Simulation of the tropical climate of an Ice Age, *J. Geophys. Res. 82*, 3889-3911.

Moore, T. C., N. G. Pisias, and G. R. Heath (1977). Climate changes and lags in Pacific carbonate preservation, sea surface temperature, and global ice volume, in *The Fate of Fossil Fuel CO_2 in the Oceans*, N. R. Anderson and A. Malahoff, eds., Plenum, New York, pp. 145-165.

Mörner, N. A. (1975). Eustatic amplitude variations and world glacial changes, *Geology 3*, 109-110.

Peng, T. M., W. S. Broecker, G. Kipphut, and N. Shackleton (1977). Benthic mixing in deep sea cores as determined by ^{14}C dating and its implications regarding climate stratigraphy and the fate of fossil fuel CO_2, in *The Fate of Fossil Fuel CO_2 in the Oceans*, N. R. Anderson and A. Malahoff, eds., Plenum, New York, pp. 355-373.

Pisias, N. G. (1976). Late Quaternary variations in sedimentation rate in the Panama Basin and the identification of orbital frequencies in carbonate and opal deposition rates, *Geol. Soc. Am. Mem. 145*, 375-391.

Reid, J. L. (1979). On the contribution of the Mediterrranean Sea outflow to the Norwegian-Greenland Sea, *Deep Sea Res. 26A*, 1199-1223.

Rooth, C. G. H., C. Emiliani, and H. W. Poor (1978). Climate response to astronomical forcing, *Earth Planet. Sci. Lett. 41*, 387-394.

Sarnthein, M. (1978). Sand deserts during glacial maximum and climatic optimums, *Nature 171*, 43-46.

Savin, S. M. (1977). The history of the Earth's surface temperature during the past 100 million years, *Ann. Rev. Earth Planet. Sci. 5*, 319-355.

Savin, S. M., R. G. Douglas, and F. G. Stehli (1975). Tertiary marine paleotemperatures, *Geol. Soc. Am. Bull. 86*, 1499-1510.

Shackleton, N. J. (1977). Carbon-13 in Uvigerina: Tropical rainforest history and the equatorial Pacific carbonate dissolution cycles, *The Fate of Fossil Fuel CO_2 in the Oceans*, N. R. Anderson and A. Malahoff, eds., Plenum, New York, pp. 401-427.

Shackleton, N. J., and J. P. Kennett (1975). Paleotemperature history of the Cenozoic and the initiation of Antarctic glaciation: Oxygen and carbon isotope analyses in DSDP Sites 277, 279, and 281, in *Initial Reports of the Deep Sea Drilling Project 29*, U.S. Government Printing Office, Washington, D.C., pp. 743-755.

Shackleton, N. J., and N. D. Opdyke (1973). Oxygen isotope and paleomagnetic stratigraphy of equatorial Pacific core V28-233: Oxygen isotope temperatures and ice volumes on a 10^5 year and 10^6 year scale, *Quat. Res. 3*, 39-55.

Shackleton, N. J., and N. D. Opdyke (1976). Oxygen-isotope and paleomagnetic stratigraphy of Pacific core V28-239, late Pliocene to latest Pleistocene, *Geol. Soc. Am. Mem. 145*, 449-464.

Shackleton, N. J., and N. D. Opdyke (1977). Oxygen isotope and paleomagnetic evidence for early northern hemisphere glaciation, *Nature 270*, 216-219.

Suarez, M. J., and I.M. Held (1979). The sensitivity of an energy balance climatic model to variations in the orbital parameters, *J. Geophys. Res. 84*, 4825-4836.

Tappan, H. (1968). Primary production, isotopes, extinctions, and the atmosphere, *Palaeogeogr. Palaeoclimatol. Palaeoecol. 4*, 187-210.

Thierstein, H. R., and W. H. Berger (1978). Injection events in ocean history, *Nature 176*, 461-466.

van Andel, T. H., and T. C. Moore (1974). Cenozoic calcium carbonate distribution and calcite compensation depth in the central equatorial Pacific, *Geology 2*, 87-92.

van Andel, T. H., J. Thiede, J. G. Sclater, and W. W. Hay (1977).

Depositional history of the South Atlantic ocean during the last 125 million years, *J. Geol. 85*, 651-698.
van Couvering, J. A., W. A. Berggren, R. E. Drake, E. Aquirre, and G. H. Curtis (1976). The terminal Miocene event, *Mar. Micropaleontol. 1*, 263-286.
van Donk, J. (1976). O^{18} record of the Atlantic Ocean for the entire Pleistocene Epoch, *Geol. Soc. Am. Mem. 145*, 147-163.
Vincent, E., and W. H. Berger (1982). Planktonic foraminifera and their use in paleoceanography, in *The Sea*, Vol. 7, C. Emiliani, ed., Wiley-Interscience, New York.
Vincent, E., J. S. Killingley, and W. H. Berger (1980). The Magnetic Epoch-6 Carbon Shift: A change in the ocean's $^{13}C/^{12}C$ ratio 6.2 million years ago, *Mar. Micropaleontol. 5*, 185-203.
Williams, D. F., R. C. Thunell, and J. P. Kennett (1978). Periodic fresh-water flooding and stagnation of the eastern Mediterranean Sea during the late Quaternary, *Science 201*, 147-149.
Wilson, A. T. (1964). Origin of ice ages: An ice shelf theory for Pleistocene glaciation, *Nature 201*, 147-149.
Wilson, A. T. (1969). The climatic effects of large-scale surges of ice sheets, *Can. J. Earth Sci. 6*, 911-918.
Worthington, L. V. (1968). Genesis and evolution of water masses, *Meteorol. Monogr. 8*, 63-67.

The Carbon Cycle—Controls on Atmospheric CO_2 and Climate in the Geologic Past

4

MICHAEL A. ARTHUR
University of South Carolina

INTRODUCTION

The chemical-biogenic sediments deposited in marine settings act as a major buffer for geologically short-term and long-term excursions in atmospheric and ocean chemistry, particularly those of the gases CO_2 and O_2. The chemical and stable isotopic composition of the chemical-biogenic sediments also reflects to some extent the chemistry of the seawater from which the sediments were precipitated. Therefore, the marine sedimentary record can be studied to obtain a record of the chemical history of seawater. This record can be compared to changes in other phenomena such as sea level, positions of land masses, tectonic events, and particularly climate (e.g., Fischer and Arthur, 1977; Berger, 1977, 1979; Berggren and Hollister, 1977).

A common assumption in studies of geochemical cycles is that the ocean reservoir maintains a relatively constant composition (chemical uniformity as opposed to a chemical steady state with respect to equilibria) through time. Although this is a sometimes necessary and simplifying assumption, in the absence of data to the contrary, it is in reality not too satisfactory. There is undoubtedly a complex interplay among atmospheric composition, climate, continental weathering, and the riverine flux of dissolved chemical species to the oceans. In turn, these factors influence or are influenced by changes in rates and mechanisms of ocean circulation and by changes in biological and nonbiological extraction and storage of chemical constituents in marine sediments. The chemical loops just described are to various degrees interdependent. Major perturbations in one flux into, or out of, the system, because of climatic or tectonic forcing (including relative changes in sea level), will spread to the others through a series of feedback mechanisms. The marine sedimentary record—changes in lithology, chemistry, stable isotopic composition, and the biotic constituents of pelagic sediments—is a monitor of changes in ocean chemistry, and through consideration of the feedback mechanism one can deduce possible variations in global climate. The extent to which we can recognize these variations and their causes or effects is dependent on the completeness of the sedimentary and fossil record, on a stratigraphic framework and absolute-time scale adequate for correlations and estimation of the leads and lags in the system, and on the geochemical tools that we have available to us.

The purpose of this chapter is to outline briefly the role of

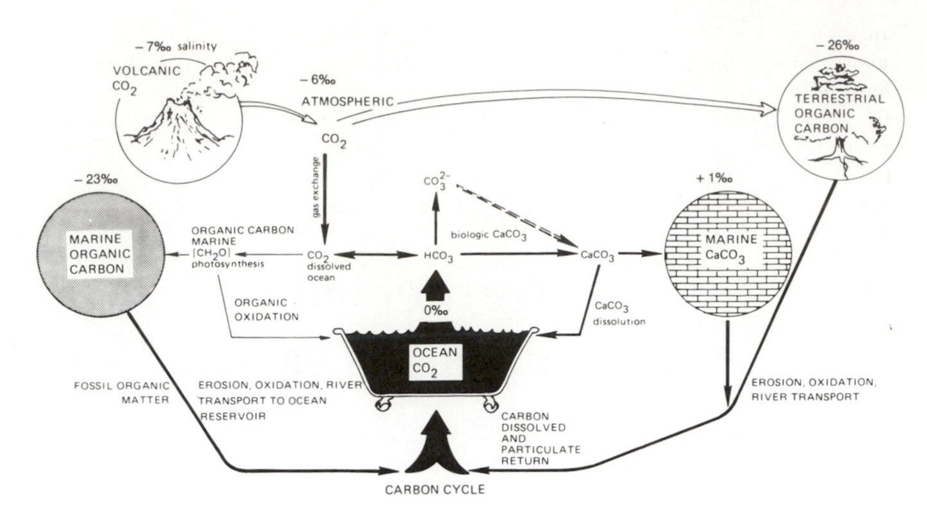

FIGURE 4.1 Main element of the global carbon cycle (from Scholle and Arthur, 1980; see Tables 4.2 and 4.3 for mass of carbon fluxes between reservoirs and carbon isotopic compositions).

ocean chemistry, mainly through its participation in the carbon cycle (Figure 4.1) in buffering changes in pCO_2 and consequent changes in climate. The discussion deals first with the extent of possible excursions in atmospheric pCO_2 and their causes, the operation of the geologic carbon cycle, and the feedback mechanisms that appear to help damp pCO_2 fluctuations (i.e., long- and short-term buffers); then briefly with the record of Cenozoic paleo-oceanography and climate and the possible influence of pCO_2 changes; and finally with geologically sudden events, such as the isolation and evaporation of an ocean basin and their possible influence on global climate.

CARBON DIOXIDE AND CLIMATE CHANGE

It has been difficult to detect any climatic effect of the increased atmospheric CO_2 over the last several decades due to burning of fossil fuels, largely because the predicted effects are within the limits of natural climatic noise (e.g., see Madden and Ramanathan, 1980). However, a CO_2-greenhouse effect is expected, and various climatic models have been constructed to estimate the magnitude of climatic warming associated with excursions in atmospheric pCO_2 (see Marland and Rotty, 1979; NRC Climate Research Board, 1979, for recent reviews). The commonly accepted range for the average global-temperature rise associated with a doubling of atmospheric pCO_2 is about 1.5-3.0°C, although estimates of 0.7-9.6°C have been published. The major problem of determining the potential magnitude of the effect by climate modeling is that feedback mechanisms, such as changes in cloudiness and associated albedo changes, may be improperly modeled. Manabe and Wetherald (1975, 1980), for example, used a sophisticated three-dimensional general-circulation model, but in their early model the degree of cloudiness was fixed, topography was idealized, and there was no seasonal variation imposed. For doubling of atmospheric CO_2 they predicted a 2.9°C increase in global-mean surface temperatures (greater at high latitudes) and an increase in evaporation-precipitation.

These models would not apply well to the Cretaceous or early Cenozoic Earth, when continental configurations were different, global-mean temperatures were much higher, and the oceans were much warmer overall. All in all, the effects of increased pCO_2 in the geologic past are difficult to estimate, but for the sake of discussion we will adopt a change of several degrees for a doubling of pCO_2. This temperature change may, of course, be grossly in error, as there is still difficulty in discerning climatic change because of variations in atmospheric CO_2 versus other factors, such as changes in the latitudinal distribution of continents and their effects on the distribution of albedo, on patterns of surface- and deep-ocean currents, or on both. These problems are briefly dealt with below after an examination of possible natural variations in atmospheric CO_2.

NATURAL SOURCES AND VARIABILITY OF CO_2

We also must consider natural sources of CO_2 to the atmosphere. Berner *et al.* (1981) have demonstrated glacial-interglacial pCO_2 changes on the basis of changes in the CO_2 gas pressure in ice. The pCO_2 apparently was lower during the last glacial. In this regard, Berger (Chapter 3) has discussed the short-term (i.e., 10,000 yr) sudden expulsion of CO_2 that could result from glacial-interglacial changes in the residence time of deep water. Sudden or more rapid overturn of "old" deep water that might occur during the transition from glacial to interglacial periods would inject large amounts of CO_2 into the atmosphere [10^{14} to 10^{18} grams of carbon (g of C)]. Atmospheric CO_2 also can change as the result of changes in overall temperature and salinity (as well as volume) of seawater. Cooling, a decrease in salinity, or both increase the solubility of CO_2 in seawater and thereby reduce the pCO_2 of the atmosphere (Table 4.1), at least on the short term. The glacial to interglacial warming of surface water would have had nearly twice the effect that the salinity decrease would have had, such that the pCO_2 changes from this cause would be minor. However, a change from warm, saline surface- and deep-ocean water in the Eocene to colder, less saline water masses in the Oligocene (Berger, 1977) may have resulted in a more important atmospheric CO_2 decrease. Atmospheric CO_2 must also have varied during glacial-interglacial cycles because of the changes in terrestrial biomass and soil carbon (Shackleton, 1977). The increased oxidation of soil carbon or humus could provide a large source of CO_2 to the atmosphere. Broecker (1981) recently suggested that burial of massive amounts of organic carbon in shelf sediments during sea-level rise following glacial retreat could also be a mechanism for rapid lowering of pCO_2 and removal of phosphate from the ocean. A pCO_2 increase would follow because organic carbon burial rate would then decrease. The buried organic carbon could be oxidized during subsequent glacial lowering of sea level, in such a way that phosphate and CO_2 are returned to the atmosphere. Models by Berger (Chapter 3), Shackleton (1977), and Broecker (1981) each are supported by variations in $\delta^{13}C$ of pelagic microfossils. Currently, there is no available evidence to distinguish between the effects of the three models. These types of short-term CO_2 pulses are not restricted to the Pleistocene, although they may be amplified at that time by the large cyclic variation in climate. We see evidence of similar possible exchanges of carbon dioxide and burial of organic matter reflected in $\delta^{13}C$ values of carbonate across 10^5-yr cycles in the Cretaceous, for example (Figure 4.6). Could the intensity and rapidity of change between glacial and interglacial cycles be, in part, controlled or reinforced by these exchanges of CO_2 (Berger, Chapter 3; Broecker, 1981)?

Oceanic fertility [the availability of phosphorous and nitrogen (Figure 4.2)] also may have changed over time periods of 1-10 million years (m.y.). Decreased fertility could drastically affect the rate of burial of marine organic matter and the ability of the ocean system to absorb "excess" atmospheric CO_2 in this way. Atmospheric CO_2 may have risen during low-fertility episodes (e.g., Tappan, 1968; Berger, 1977). Conversely, times of apparent high fertility [such as those during the deposition of large phosphate deposits (Arthur and Jenkyns, 1981)] would likely result in decreased atmospheric pCO_2 because of increased burial of organic carbon in marine sediments.

Another possible major source of changes in atmospheric CO_2 is volcanism. Vogt (1972, 1979) and Kennett and Thunell (1975) have suggested a major periodicity in volcanism during the Cretaceous-Cenozoic. Could this periodicity result in fluctuations of pCO_2 and changes in global climate? At present, it is estimated, and this is a difficult estimate to make, that degassing of the Earth through volcanism emits 0.09×10^{15} g of C/yr as CO_2 to the atmosphere (Holland, 1978). This amount makes up a predicted deficit caused by operation of the carbon cycle (see Table 4.2 and next section). A doubling of the rate of CO_2 degassing would double atmospheric CO_2 in about 10,000 yr if the CO_2 is not compensated for by other feedback mechanisms. This flux is about 50 times less than the rate of CO_2 addition from the burning of fossil fuels over the past few decades. Thus, over time periods of several million years, increased volcanic discharge could be important to climate. However, the major question here is whether the cooling effect of aerosols ejected into stratosphere during episodes of explosive volcanism would offset the warming effect of increased pCO_2 (e.g., Pollack *et al.*, 1976; Pollack, 1979).

TABLE 4.1 Estimated Flux Rate, Mass, and Isotopic Changes in the Ocean during CO_2 Transfers[a]

Type of Flux	Rate and/or Amount of Carbon Transferred (Duration)[b]	$\Delta\,\delta^{13}C$ Ocean (Total Dissolved Carbon)[b]
Volcanic CO_2 addition (doubling of estimated steady-state degassing rate)	0.8×10^{14} g of C/yr (over 8×10^4 yr)	-1 ‰ (for $\sim 6 \times 10^{18}$ g of C addition)
Oxidation of soil carbon (or burning of fossil fuels)	5.0×10^{14} g of C/yr (over 3×10^3 yr)	-1 ‰ (for $\sim 1.5 \times 10^{18}$ g of C addition)
Net transfer of ocean TDC to marine organic carbon burial	3.0×10^{14} g of C/yr (over 5×10^3 yr) ($<4.3\%$ of ocean TDC; or $\sim 60\%$ of annual TDC input by rivers)	$+1$ ‰ (for 1.5×10^{18} g of C depletion)
Net transfer to and from carbonate reservoirs	~ 0	~ 0

[a]Values determined in order to equal a 1‰ change in ocean TDC over steady-state flux rates (estimated from Tables 4.2 and 4.3).

[b]Changes in total C content of atmosphere as CO_2 induced by changes in temperature and salinity assuming constant surface seawater volume equal to present and 300×10^{-6} atm of CO_2:

$$\Delta pCO_2/\Delta C \text{ (for cooling } 1°C) = -24 \times 10^{15} \text{ g of C } (\sim 4\% \text{ of } pCO_2)/°C;$$

$$\Delta pCO_2/\Delta \text{ salinity (for salinity decrease of 1‰)} = -30 \times 10^{15} \text{ g of C } (\sim 5\% \text{ of } pCO_2)/‰ \text{ salinity.}$$

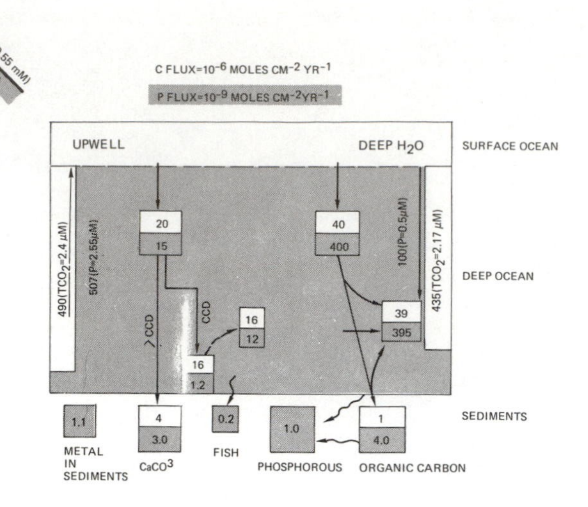

FIGURE 4.2 Linkage between the carbon and phosphorous cycles in the ocean. The availability of phosphorous as a nutrient limits the amount of organic carbon and carbonate that can be produced and buried in the oceans (after Froelich *et al.*, 1981). The importance of different phosphorous (and organic carbon-carbonate) sinks may have varied significantly in the past.

TABLE 4.2 Approximate Present-Day Sizes and Isotopic Compositions of Carbon Reservoirs[a]

Reservoir	Mass of Carbon in 10^{17} g	Average $\delta^{13}C$ (‰)
Ocean-dissolved carbon (primarily HCO_3)	350.0	0
Annual net marine carbonate (biogenic $CaCO_3$)	0.0014	+1.0
Carbonate sediment reservoir	610,000	+1.0
Annual net marine organic carbon (biogenic $C_{H_2}O$)	0.00073	−23
marine biomass	0.018	−23
Organic carbon sediment reservoir	130,000	−23
Atmospheric CO_2	6.0	−7
Land-plant biomass (C_{org})	8.4	−26
soil "humus"	12.0 [10.5-30]	−25
Stream flux (annual)		
Dissolved inorganic carbon	0.0047	−6.5
Dissolved organics	0.0032	−26
Particulate organics	0.0007	−26
Volcanic gases (mantle carbon)	(?) 0.0008 (see Table 4.3)	−7

[a]Modified from Scholle and Arthur, 1980.

LONG-TERM STATUS AND FEEDBACK MECHANISMS IN THE CYCLING OF CARBON

A number of researchers have suggested that the chemistry of the Earth's atmosphere and oceans has remained constant within certain limits to account for the continuity of life and the relatively narrow excursions in the composition of chemical sediments through geologic time (e.g., Holland, 1972, 1974, 1978; Garrels and Perry, 1974). To maintain this long-term geochemical status there must be a variety of efficient feedback mechanisms that buffer expected variations in geochemical cycles due to changes in intensity of tectonism, in continental area and mean altitude, and in climate. These buffering mechanisms may be short term, long term, or both. The operation of the carbon cycle is one of the most important in this regard as it ultimately controls levels of oxygen and carbon dioxide in the atmosphere. The carbon cycle is responsive to changes in climate and oceanography and is coupled to nutrients cycles. If climate is influenced by varying levels of atmospheric CO_2, for example, controls on atmospheric pCO_2 levels are therefore important to understand. This understanding must include adequate knowledge of CO_2 sources and source strengths, the allowable excursions of pCO_2 levels within the limitations imposed on the operation of global carbon cycle by nutrient availability, net productivity and burial of organic carbon, carbonate sedimentation and dissolution, weathering rates and reactions (Figures 4.1 and 4.2), and the rates of CO_2 excursions versus the kinetics of CO_2-consuming processes (see Table 4.3).

We know, for example, that the increase in atmospheric CO_2 levels from burning of fossil fuels and other activities of man has not been immediately taken up [perhaps 48 percent of the total CO_2 released by burning of fossil fuels has been removed from the atmosphere by various mechanisms (Oeschger *et al.*, 1975)]; the CO_2 is produced at a rate of about 5×10^{15} to 6×10^{15} g of C/yr (Rotty, 1977) and increases by about 4 percent per year. Thus, the feedback mechanisms that must act to restore balance are not entirely efficient on the order of a few tens of years. The major control mechanisms on time scales of 10^2 to 10^6 yr are probably weathering reactions, carbonate dissolution in the oceans, and burial of organic carbon in sediments (through changes in net primary productivity or enhanced preservation in sediments). The efficiency of all of these mechanisms is limited by one or more factors. We have only a partial understanding of the carbon-cycle system. Parts of the system can be isolated, but the whole is difficult to integrate. The next section gives an example of how we might approach the CO_2 problem as a subsystem of the global carbon cycle. Short-term buffering mechanisms, such as uptake as dissolved CO_2 in seawater and uptake in increased terrestrial biomass, will not be considered.

Weathering Reactions as a Feedback Mechanism

Weathering reactions, such as decomposition of silicates and carbonates by acid soil, groundwaters, and surface runoff rich in CO_2, may be a major long-term sink for atmospheric CO_2.

TABLE 4.3 Annual Gains and Losses of Atmospheric Carbon (in units of 10^{14} g/yr) Due to Weathering and Sedimentation[a]

	Gains	Losses
Weathering		
Oxidation of elemental carbon	0.09 ± 0.2	
Dissolution of limestones and dolomites		1.6 ± 0.2
Decomposition of Ca and Mg silicates		1.9 ± 0.2
Decomposition of Na and K silicates		0.8 ± 0.2
Sedimentation		
Deposition of elemental carbon		1.2 ± 0.3
Deposition of carbonates	2.2 ± 0.4	
Deposition of Mg silicates	0.8 ± 0.2	
Deposition of Na and K silicates	0.8 ± 0.2	
Net changes		
Due to carbon cycle		0.3 ± 0.1
Due to carbonate cycle		0.5 ± 0.3
Total net loss		0.8 ± 0.4

[a]After Holland, 1978.

An estimated 10 percent (0.43×10^{15} g of C/yr) of the present annual CO_2 flux from the atmosphere is consumed in weathering reactions (Holland, 1978). If atmospheric CO_2 levels increased greatly, rainfall possibly would become more acidic and the amount of CO_2 in soil horizons would increase, thereby leading to higher rates of weathering. However, the increased weathering would lead to a rise in alkalinity and carbonate ion concentration of river water carried to the ocean, which, depending on the concentration of total CO_2 and carbonate ion in seawater, will precipitate and deposit carbonates eventually. Carbonate deposition results in a net flux of CO_2 back to the atmosphere. At present, the net deposition of carbonates results in an estimated gain of about 0.22×10^{15} g of C/yr to the atmosphere and deposition of silicate minerals results in a gain of about 0.16×10^{15} g of C/yr, so that there appears to be a net loss of CO_2 (about 0.05×10^{15} g of C/yr) from the atmosphere owing to the cycle of weathering reactions to silicate and carbonate reconstitution (see Table 4.3).

Weathering may be an important sink for CO_2, but it must operate on longer time scales (i.e., a million years) to control atmospheric CO_2 levels, and it is dependent on such things as continental area, rainfall, temperature, and soil carbon and moisture. However, there is opposition to this idea (E. Sundquist, U.S. Geological Survey, personal communication, 1980), and most weathering reactions may be dependent entirely on the concentration of soil CO_2. If higher pCO_2 results in global warming, possibly in increased rainfall at some latitudes, and in CO_2 fertilization of higher plants, then this combination of effects might ultimately result in larger amounts of soil carbon and greater production of soil CO_2, which, in turn, would promote higher rates of weathering. The controls on soil-CO_2 concentrations and the concept of increasing acidity of rainfall with increasing atmospheric CO_2 require more study.

Organic Carbon Production and Burial

The biosphere also exerts a tremendous influence on atmospheric CO_2 levels. All the atmospheric CO_2 is probably cycled through plants once in every 10 years or less. There is less carbon locked up in the biosphere, including soil humus (a total of more than 2000×10^{15} g of C), than in CO_2 dissolved in oceans (Table 4.2) but more than in the atmosphere. The net burial of organic carbon in sediments accounts for a loss of CO_2 from the atmosphere (and a gain in oxygen). Today, this burial accounts for the net withdrawal of about 0.12×10^{15} g of C/yr of atmospheric CO_2 (Holland, 1978). This transfer of CO_2 more than offsets the estimated gain in CO_2 by oxidative weathering of old organic matter of about 0.09×10^{15} g of C/yr (Holland, 1978). Short-term increases in the size of the marine or terrestrial biota and soil carbon could accommodate short-term increases in atmospheric CO_2 (e.g., Broecker *et al.*, 1979), whereas longer-term control would have to be exerted by increasing the rate of organic carbon burial in sediments. The net carbon-burial increase could occur in peatlands, by an increase in forest-litter accumulation, in coal swamps and lakes, or in marine settings—especially in estuarine environments and on continental slopes. The terrestrial sinks, with the exception of coal (McLean, 1978a), would be relatively short-term reservoirs because of their general susceptibility of subaerial exposure and oxidation. The marine-sediment reservoir is the probable long-term stable sink for organic matter. The major control in productivity of organic matter is nutrient availability.

Changes in nutrient supply may lead, in part, to changes in the burial rate of organic matter. The burial rate of organic matter is also a function of oxygen availability and circulation rates of deep-water masses. Burial of organic matter removes at least part of the nutrients used in organic synthesis, and, without sufficient available nutrients (mainly phosphorous and nitrogen), the organic part of the carbon cycle cannot respond to a CO_2 increase. A large part of the nutrient flux in the oceans today is regenerated from organic-matter oxidation in the water column and within sediments (see Figure 4.2). Burial of terrestrial-organic matter (land plants) is a more efficient CO_2-fixing mechanism in terms of nutrient usage; the average C:N:P ratio is about 510:4.2:1 as opposed to 106:16:1 in unoxidized average marine-organic matter, although much of the P and N may be regenerated before burial. Thus, times of widespread coal deposition (e.g., the Carboniferous) would have efficiently fixed much atmospheric CO_2 into sediments. Burial of terrestrial organic matter in marine sediments is also important. Evidence suggests that, at present, relatively little land-derived organic matter reaches the deep sea beyond the shelf (e.g., Hunt, 1970; Sackett, 1964; Sackett and Thompson, 1963; Sackett *et al.*, 1965; Rogers and Koons, 1969), except in the anoxic Black Sea, which is a large sink for terrigenious organic carbon (e.g., Simoneit, 1977). However, this lack of deposition of terrigenous organic carbon in marine environments may not be the norm because the rapid Holocene rise in sea level has influenced trapping of terrigenous organic matter in nearshore settings. There also may be difficulty in recognizing some terrestrially derived organic matter in marine sediments after early diagnosis on the basis of $\delta^{13}C$ values alone. Evidence from Deep Sea Drilling Project (DSDP) drill sites suggests that during the Early Cretaceous much more terrigenous organic carbon was preserved in deeper marine sediments than at present.

There is some evidence, as yet inconclusive, that greater ambient CO_2 concentrations may fertilize higher plants and may either increase net primary productivity or decrease rates of transpiration (see Strain, 1978, for review). This evidence suggests that land plant productivity might increase with increasing pCO_2, but, as Lemon (1977) pointed out, all data bearing on this problem have come from controlled experiments, and there is no available evidence that natural ecosystems could respond in this way; the process is nutrient-limited. However, this speculation is intriguing, and the mechanism may be of great importance.

In general, the operation of the organic portion of the carbon cycle is critical to controlling pCO_2. Nutrients must be available to allow relatively unrestricted changes in biomass and organic carbon burial in response to increased pCO_2. There is some evidence that nutrient levels in the sea, rates of cycling of nutrient phosphate and nitrogen through ocean waters, or both may have varied significantly in the past (e.g., Piper and Codispoti, 1975; Arthur and Jenkyns, 1981) and that the circulation patterns, supply to surface waters by upwelling, and productivity changed as well (e.g., van Andel *et al.*, 1975; Berger, 1979). Thus, the evidence is not clear that sufficient nutrients are available always to allow the biota to buffer pCO_2 increases. There is possibly a feedback link that provides an increased nutrient flux to the oceans by increased rates of weathering during pCO_2 increase. However, this mechanism involves a time lag, possibly on the order of several million years, and just how weathering rates change with increased atmospheric CO_2 is not yet clear. Times of increased clastic sediment flux to the oceans, possibly from increased tectonic activity and erosion of high-standing land masses, also may result in increased net organic carbon burial because high sedimentation rates enhance organic-matter preservation (Muller and Suess, 1979).

Effectiveness of Carbonate Dissolution in CO_2 Buffering

A final major mechanism of atmospheric CO_2 buffering, and one that could be expected to operate over a time scale of 100,000 years or less (the residence time of carbon in the ocean is about 10,000 years) is dissolution of carbonate minerals in the deep sea. This mechanism has been discussed at length in the literature (e.g., Broecker and Takahashi, 1978; Broecker *et al.*, 1979) and will not be dealt with in detail here. The main area of carbonate dissolution is below the lysocline, a level of increased undersaturation and increased rate of dissolution of calcite in deep-water masses. It is estimated that there are today at least 3600×10^{15} g of C as carbonate readily available for dissolution to about 10-cm depth in deep-sea sediments, with a probable addition of 1-2 $\times 10^{15}$ g of C/yr from annual production. However, the efficiency of increased oceanic dissolved CO_2 levels in dissolution of carbonate depends on the rate-limiting process in dissolution (e.g., burrow-stirring, breakdown of organic coatings, and saturation of pore waters) and on the rate of overturn of oceanic deep waters (currently with a residence time of about 1200 years) that controls the rate of delivery of dissolved CO_2 from the atmosphere to the depths for buffering by carbonate dissolution.

Relatively soluble carbonate minerals such as aragonite and high-Mg calcite deposited in shallow-water environments might provide another small and short-term buffer for increased pCO_2 (e.g., Broecker *et al.*, 1979), but the surface ocean is at present everywhere at least 1.7 times saturated with respect to aragonite. According to Broecker *et al.* (1979) the atmospheric CO_2 would have to increase at least 5.3 times and 8.5 times that of today to cause undersaturation in ocean surface waters for aragonite and calcite, respectively. Therefore, dissolution of shallow-water carbonate sediments is not likely to be too significant as a sink for CO_2 at present and was relatively insignificant in the past unless major pCO_2 increases were allowed. Also, if high-Mg calcite, aragonite, or both were dissolved and reprecipitated as low-Mg calcite, then there is no net CO_2 consumption.

Because dissolution of 1 mole of carbonate consumes 1 mole of CO_2, only about 40,000 yr would be required to dissolve the estimated available deep-water carbonate if the estimated rate of CO_2 degassing by volcanism were to double (e.g., an increase of about 0.09×10^{15} g of C/yr). However, an increase in dissolution rate equal to about 10 percent of the carbonate produced in surface waters each year would also balance the aforementioned CO_2 flux increase. Thus, all other factors being equal, we might expect to see a decrease in the net deep-sea carbonate-accumulation rate with an increase in atmospheric CO_2, as well as an apparent shallowing of the carbonate saturation horizons [the lysocline and calcium carbonate compensation depth (CCD), e.g., Berger, (1977)].

Carbon Isotopes in Pelagic Carbonates as Constraints on Carbon Cycling

Consideration of the possible past excursions of atmospheric CO_2 has been made by examining constraints on changes in ocean chemistry imposed by the composition of marine chemical sediments—mainly evaporites and carbonates—through time and by examining changes in the carbon isotopic composition ($\delta^{13}C$ values) of limestones through time. Holland (1972, 1974, 1978) has shown that during the Phanerozoic, surface seawater has always been saturated or supersaturated with respect to calcite and aragonite. The probability is that the concentration of Ca^{2+}, SO_4^{2-}, and HCO^{3-} have never varied by more than a factor of 2 to 3 in either direction from their present values. Holland has shown that all of these factors constrain pCO_2 variations only by about a factor of 10^3 to 10^4—that is, that pCO_2 has probably always remained between $10^{1.5}$ and $10^{5.5}$ atm (currently $10^{3.5}$ atm).

The $\delta^{13}C$ values of marine limestones are presumed to reflect the carbon isotopic composition of total dissolved carbon in the ocean reservoir (e.g., Broecker, 1974). The average isotopic composition of the reservoir generally represents the balance between deposition of carbonates and organic carbon. The two have very different carbon isotopic compositions (average organic carbon = − 23 ‰ and average carbonate = + 1 ‰; see Figure 4.1). Limestones are isotopically similar to total carbon in the oceanic reservoir. This model assumes an input of dissolved carbon to the oceans of constant mass and isotopic composition. This assumption is necessary because we have no data to substantiate large variations, but in reality the riverine flux may vary as a function of rates of weathering and the proportion of limestone versus organic carbon weathered (or oxidized).

Making the assumption of constant input, we can state the following. Excursions to more positive $\delta^{13}C$ values indicate perhaps that relatively more organic matter is being buried, while burial of carbonate decreases or remains constant. More negative isotopic values imply a shift to larger values in the burial ratio of carbonate to organic carbon. Junge *et al.* (1975), Veizer and Hoefs (1976), Garrels and Perry (1974), and Garrels *et al.* (1976) have suggested that $\delta^{13}C$ values in limestones through time have stayed relatively constant (about 0 ± 2.5‰) implying that there has been little change in the partitioning of carbon between the organic and carbonate reservoirs during the Phanerozoic. This relative constancy suggests an efficient long-term control against major excursions in atmospheric and ocean chemistry. However, these considerations use average values from long time periods (i.e., 30 m.y. to 50 m.y.) and do not have the resolution necessary to detect short-term perturbations. For example, Scholle and Arthur (1980) have detected large and rapid excursions of as much as 4 ‰ in $\delta^{13}C$ in both positive and negative directions of pelagic carbonates of Cretaceous and early Cenozoic age. These excursions may occur in less than a million years and imply rapid changes in the production and burial of organic matter and carbonate (see also Bender and Keigwin, 1979; Vincent *et al.*, 1980). Shackleton (1977) has demonstrated cyclic changes in $\delta^{13}C$ of less than 1 ‰ in benthic foraminifers over a few thousand years during the Late Pleistocene. He attributed these variations to transfer of carbon to and from the terrestrial biosphere to the ocean-atmosphere during glacial-interglacial climatic changes in the amount of about 10^{14} g of C/yr (a total of 10^{18} g of C in 10,000 years or so). Note that this amount is perhaps only a few percent of the rate of addition of CO_2 to the atmosphere by man's activities today. Tree-ring carbon-isotope data suggest that a 1 to 1.5 ‰ decrease has resulted in the $\delta^{13}C$ value of atmospheric CO_2 from burning of isotopically light organic fuels.

DETECTING ATMOSPHERIC CO_2 EXCURSIONS AND CLIMATE CHANGE IN THE CENOZOIC

Is there any hope of detecting variations in atmospheric CO_2 and establishing these as a cause of climate change in the geologic past? The problem amounts to one having several parts: (a) constructing realistic models that demonstrate the significance of climate change related to pCO_2 variations; (b) designing more sophisticated models that incorporate all important factors in the carbon cycle in the land-ocean-atmosphere system (as discussed earlier), and plugging in a variety of possible variations in inputs and outputs in order to provide constraints on interpretation of past geologic data; and (c) collection and integration of geologic and climatic data for comparison with (b) above. Part (a) has already been discussed briefly in previous sections. Part (b) is a difficult enterprise, but various attempts to model the system are in progress (e.g., Bolin *et al.*, 1979; E. Sundquist, U.S. Geological Survey, personal communication, 1980).

Data for part (c) have been slowly accumulating, mainly through paleontologic, lithologic, and geochemical analysis of DSDP cores. Berger (1977) previously inferred atmospheric CO_2 changes during the Cenozoic by examining the known fluctuations in various parameters of the carbon system derived from this data base. However, climate change seems dependent on so many factors that we cannot definitely state that a given climatic event is dependent on a single cause. Many of the factors possibly causing climate changes during the Cenozoic, for example, are coincident (see the other chapters in this volume; Berger, 1977; Berggren and Hollister, 1977; Fischer and Arthur, 1977; Frakes, 1979), and their relative effects must be carefully modeled.

Because of the variety of mechanisms that can induce climate change, a given warming in the geologic past is difficult to attribute to a greenhouse effect resulting from increased atmospheric CO_2. However, there are several signals that, taken together along with evidence of climate warming, might point to increased levels of CO_2 as a cause (e.g., Berger, 1977). These signals are the following: (1) evidence of increased dissolution of carbonate in the deep sea and possible decreases in the rate of accumulation in deep-sea sediments; (2) changes in the $\delta^{13}C$ values of total carbon in the oceanic reservoir as reflected in analyses of pelagic-carbonate bulk samples, or preferably both benthic and planktonic organisms, through the time interval in question; (3) a change in the rate of accumulation of organic matter in marine and/or nonmarine sediments; and (4) negative evidence of a major change in ocean circulation that might result in any of the preceding signals.

The following discussion deals briefly with the Cenozoic climatic and paleo-oceanographic record and some evidence of possible pCO_2 excursions. This is a qualitative treatment only and is intended to illustrate the type of approach necessary and the difficulties in isolating atmospheric pCO_2 variations as causes of climate change. References and discussion of climate changes based on oxygen isotope and paleontologic data can be found in a number of papers in this volume (e.g., Chapters 12, 13, 16, and 18). The paleo-oceanographic data base comes largely from Figures 4.3-4.5, which are compiled from numerous sources (see also Berger, 1979; Arthur, 1979).

The climatic events near and following the Cretaceous-Tertiary boundary may be evidence of an atmospheric CO_2 excursion at that time. The possibility was suggested by McLean (1978b) in an elaborate explanation for the biotic extinctions occurring at the boundary. He suggested that pCO_2 increased because of a failure of marine photosynthetic organisms. This pCO_2 increase resulted in a warming across the boundary during the earliest Paleocene. This warming of a few degrees

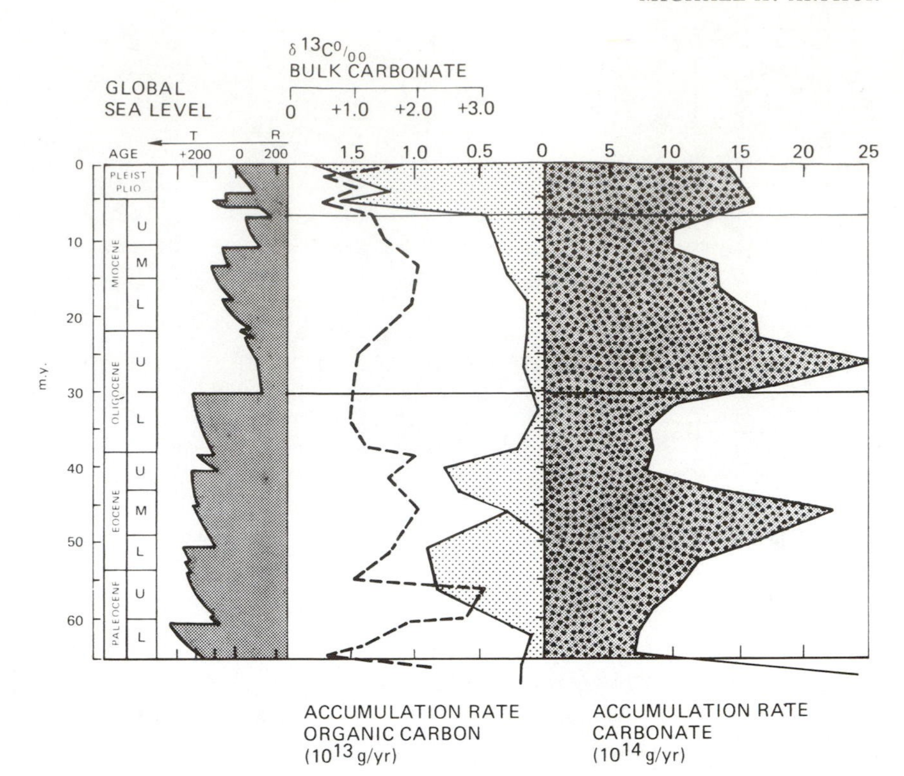

FIGURE 4.3 Accumulation rates of organic carbon (M. A. Arthur, unpublished data) and carbonate (after Worsley and Davies, 1981) in the deep sea during the Cenozoic as compared with average $\delta^{13}C$ values of bulk carbonate and global sea level. The fluctuations reflect changing fertility and possibly changes in atmospheric CO_2.

Celsius has been detected in oxygen isotopic and paleontologic data. Hsü (1980) has also suggested a large increase in pCO_2 at this time associated with an impact of an extraterrestrial object.

The early Paleocene warming lasted 3-5 m.y. The $\delta^{13}C$ values of carbonate were low (average + 1 ‰) at this time, as were $\delta^{13}C$ gradients from surface to deep water inferred from the average difference between planktonic and benthic foraminifers (Figures 4.3 and 4.4). The accumulation rates of carbonate in the deep sea were also at a low point in the early Paleocene, as were those of organic carbon. The CCD was relatively high, especially at the boundary. The early Paleocene (65-62 Ma) is therefore a possible candidate for a possible pCO_2 excursion that leads to climatic warming. A sudden injection of up to 90 × 10^{18} g of CO_2 into the ocean-atmosphere is suggested by the rapid negative $\delta^{13}C$ shift of 1.5 ‰ at the Cretaceous-Tertiary boundary, assuming that the CO_2 had a $\delta^{13}C$ value equivalent to that of volcanic emanations. The effects of this event lasted perhaps 3-5 m.y., suggesting that feedback mechanisms to adjust to the pCO_2 excursion were relatively efficient on a geologic scale. A slight cooling occurred during the mid to early Late Paleocene. It is not clear, however, what caused the amelioration of the possible pCO_2 increase at the Cretaceous-Tertiary boundary. Because deep-sea organic carbon accumulation rates were fairly low and the extent of shallow shelf seas was apparently small, much of the proposed pCO_2 increase might have been accomodated by dissolution of pelagic carbonate (e.g., Worsley, 1974).

A second major warming, not explained by other factors such as continental positions and opening of ocean "gateways," occurred in Late Paleocene-Early Eocene time.

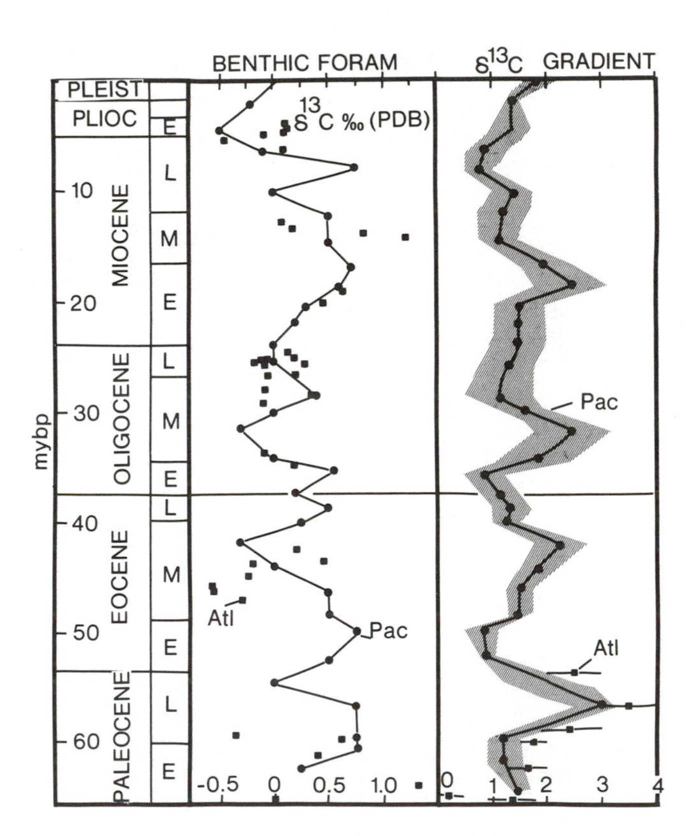

FIGURE 4.4 $\delta^{13}C$ values of calcite of benthic foraminifers and ^{13}C gradients between surface and deep water through the Cenozoic (compiled from Kroopnick *et al.*, 1977; Letolle *et al.*, 1979; Boersma *et al.*, 1979).

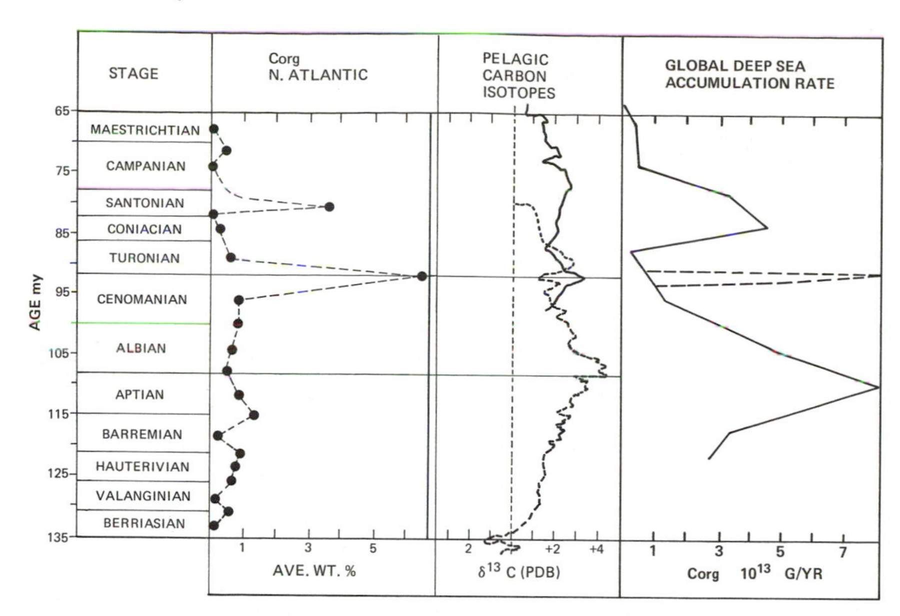

FIGURE 4.5 Accumulation rates of organic matter in the deep sea (M. A. Arthur, unpublished data from DSDP Sites) and $\delta^{13}C$ values of pelagic carbonates (from Scholle and Arthur, 1980). Note major change in both parameters near Aptian-Albian, possibly related to the evaporite episode in the South Atlantic.

There is first an increase then a sudden decrease in bulk pelagic carbonate $\delta^{13}C$ values amounting to over 2 ‰. The $\delta^{13}C$ gradient was highest during the Late Paleocene and also decreased greatly into the Early Eocene. The CCD remained high, and dissolution gradients were also relatively high (van Andel, 1975; van Andel *et al.*, 1975). The accumulation rate of carbonate in the deep sea was only slightly higher than that in the early Paleocene. Organic carbon accumulation rates, however, increased greatly in the Early Eocene. There was certainly a Late Paleocene-Early Oligocene increase in oceanic fertility as evidenced by large phosphorite deposits and high accumulation rates of organic carbon both on the shelves and in the deep sea (Arthur and Jenkyns, 1981). This combination of evidence again seems to be a possible candidate for warming because of an atmospheric CO_2 increase. It is intriguing that Vogt (1979) has suggested a substantial peak in volcanicity that occurred in the Late Paleocene (centered at about 56 Ma). Could this volcanicity have been the cause of the climatic warming and changes in paleo-oceanographic parameters? The Early Eocene was also one of the warmest periods, as interpreted from terrestrial floras (see, e.g., Chapter 16).

The aforementioned pCO_2 excursion and climatic optimum possibly induced increased weathering rates on land, and this feedback loop brought increased dissolved Ca^{+2}, bicarbonate, silicates, and phosphate to the oceans. The Middle Eocene is marked by cooling, possibly because of decreased pCO_2, by a gradual increase in pelagic carbonate $\delta^{13}C$ values; by widespread biogenic silica-rich sediments, by low organic-carbon accumulation rates; and by the highest accumulation rates of carbonate in the deep sea (even with a relatively high CCD) of any time in the Early Cenozoic. Again, about 5 m.y. seems to be the response time of the carbon cycle to dampen the effects of a possible pCO_2 rise.

Climatic cooling in the Oligocene generally has been explained by the increased isolation of the Antarctic and by the development of the circum-polar current (see Chapter 13). The cooling may also be due to low atmospheric CO_2, although there is no good evidence for this. A drop in the deep-sea accumulation rate of organic carbon and carbonate accompanied a deepening of the CCD, a decrease in carbonate-dissolution rates, and low $\delta^{13}C$ values, which suggest low fertility and a high ratio of preservation of carbonate to organic matter. These relations suggest, but do not prove, a period of relatively low-atmospheric CO_2 in latest Eocene through probably Late Oligocene. In fact, deep-sea carbonate accumulation rates reached a maximum in the Late Oligocene. However, part of this increase may have been due to greater supply of carbonate to the oceans during major sea-level regression (e.g., Worsley and Davies, 1981).

However, an early to middle Miocene warming has not been satisfactorily explained by other mechanisms. This warming episode peaked at about 17-15 Ma and coincides with a second major peak in volcanism shown by Vogt (1979). This peak also coincided with a sharp rise in the CCD in nearly all ocean basins, an apparent increase in dissolution rates (e.g., van Andel *et al.*, 1975), and a decrease in deep-sea carbonate accumulation rates, as well as an elevated $\delta^{13}C$ gradient and more positive $\delta^{13}C$ values. Deep-sea accumulation rates of organic carbon remained low, as in the Oligocene, but were high around the continental margins and began to increase in the deep sea in the Late Miocene. Pelagic carbonate $\delta^{13}C$ values also dropped sharply in the Late Middle Miocene, and $\delta^{13}C$ gradients were lower. These factors again suggest that atmospheric CO_2 increase could have been responsible for climatic warming. By Late Miocene time, again on the order of about 5-7 m.y. later, the system recovered and cooling began. Both deep-sea organic carbon and carbonate accumulation rates picked up at about 7 Ma.

These relationships are highly speculative but suggest something about the role of $p\ CO_2$ in climate change in the geologic past. Contrary to this, however, the Plio-Pleistocene peak in volcanism noted by Vogt (1979) and Kennett and Thunell (1975) appears to have had the opposite effect—that is, inducing cooling—or no effect at all.

EVAPORITE DEPOSITION EVENTS AND GLOBAL CLIMATE CHANGE

Changes in ocean chemistry can modulate or change climate in other ways as well. Rapid and large-scale deposition of evaporites in isolated small ocean basins may have a substantial effect on ocean chemistry, and, in addition, may directly or indirectly influence global climate through its effects on the sulfur and carbon cycles. Garrels and Perry (1974) have pointed out that precipitation of major evaporite bodies may require large transfers in the sulfur and carbon reservoirs. Precipitation of calcium sulfate at higher than steady-state rates in an evaporite basin requires transfer of Ca^{2+} from the carbonate to the evaporite reservoir and results in a net gain of CO_2 to the atmosphere. The CO_2 gain, in steady state, should be compensated for by net gain in burial of organic carbon, by increased dissolution of carbonates, or in both. However, in the event that evaporite deposition is extremely rapid, these balancing processes may not be able to work effectively in the short term to remove the CO_2 excess. The CO_2 spike to the atmosphere may then lead to climate warming, depending on the extent of the CO_2 anomaly.

Early Cretaceous Evaporites and a Global Warm Episode

An example of this type of "internal" control on global climate may have occurred during the early Cretaceous when the northern South Atlantic (Angola-Brazil Basin) became the site of massive evaporite deposition as it was effectively isolated from the rest of the world ocean at intermediate latitudes under high evaporation rates. Arthur and Kelts (1979) have suggested that the Angola-Brazil Basin was isolated for 2 m.y., and within that time period a basin 500 km wide by 2000 km long was filled with between 2 and 3 km of evaporites. Assuming at least 30 percent $CaSO_4$ and 70 percent NaCl within the evaporites, they suggest that nearly 1.4×10^{21} g of $CaSO_4$ and 3.0×10^{21} g of NaCl were deposited in the geologically brief 2 m.y. period. Assuming modern rates of river input of Na^+, Cl^-, Ca^{2+}, and SO_4^{2-}, and an initial oceanic reservoir of those elements equal to that of today, this chemical extraction means that oceanic salinity could have been decreased by 4 to 5 ‰ (see also Hay, 1979) and that the oceanic sulfate reservoir of sulfate and calcium would have been drawn down significantly. A trend to much lighter $\delta^{34}S$ values during Aptian time (e.g., Claypool *et al.*, 1980) may be evidence for this drawdown. The cycling of carbon would have been affected as well. An abrupt CCD rise in Aptian time (Thierstein, 1979) and a trend toward more positive $\delta^{13}C$ in marine carbonates (Figure 4.6) may have occurred because of an increase in the rate of dissolution of carbonate resulting from a decrease in the oceanic Ca^{2+} concentrations and an increase in CO_2, and because of an increase in the rate of burial of organic carbon. The increased rate of burial of organic carbon may have resulted from enhanced preservation under anoxic or near-anoxic conditions. Stable stratification related to salinity contrasts between surface and deep waters may have been one mechanism for the development of poorly oxygenated deep waters (Ryan and Cita, 1977; Roth, 1978; Thiersten and Berger, 1978; Arthur and Natland, 1979); some of the most saline deep water may have been derived from periodic spillage from the evaporatic northern South Atlantic and from epicontinental and shelf seas in low latitudes. This mechanism might essentially provide the feedback to rid the system of the supposed CO_2 excess by enhancing burial of marine organic carbon. However, if the Aptian-Albian oceans were relatively nutrient depleted as proposed by Roth (1978) and Arthur and Kelts (1979), plankton productivity would be low, and this would not be an effective way to draw down atmospheric CO_2. The burden of fixing this CO_2 might have fallen on the terrestrial plants; in support of this theory it has been argued that a significant proportion of organic carbon buried in Aptian-Albian deep-sea sediments is of terrestrial derivation (Tissot *et al.*, 1980). This is probably a more slowly operating feedback mechanism, thus pCO_2 concentrations may have risen fairly rapidly and been only slowly lowered following the Aptian evaporite episode. The Albian climatic optimum (Savin, 1977) may have resulted from this CO_2 excess. A similar scenario could be envisioned for the Permian evaporite episode.

The "Messinian Event" and Global Cooling

The preceding ideas are preliminary and are based on intuitive rather than rigorously systematic interpretations of the

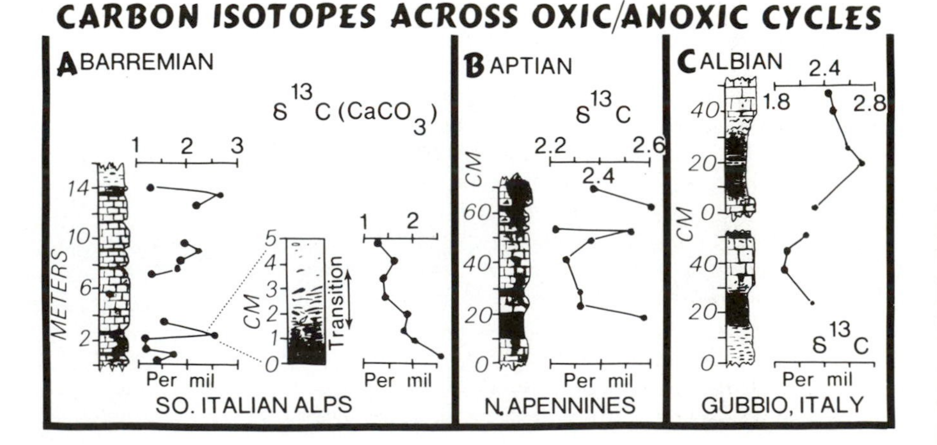

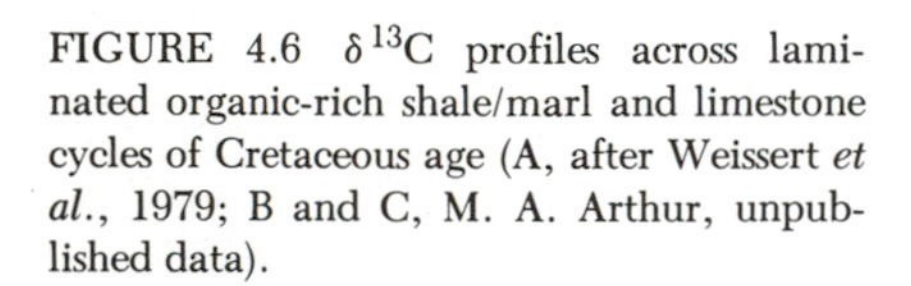

FIGURE 4.6 $\delta^{13}C$ profiles across laminated organic-rich shale/marl and limestone cycles of Cretaceous age (A, after Weissert *et al.*, 1979; B and C, M. A. Arthur, unpublished data).

linkages between major geochemical cycles. However, the discussion provides some insight as to how the carbon cycle and global climate may have responded to sudden perturbations of internal forcing mechanisms such as tectonic isolation of major ocean basins. But climatic warming may not be a necessary consequence of rapid and massive evaporite precipitation. The Messinian (latest Miocene) event may provide an illustration of the opposite effect, that is, climatic deterioration, because of the salinity crisis brought on by the isolation of the Mediterranean Tethys and the deposition of as much as 1.5×10^6 km^3 of evaporites within about 1 m.y. This amount is about three quarters of that deposited in the Aptian northern South Atlantic. Ryan (1973) has suggested that the overall change in ocean salinity may have led to the increased production of sea ice in high southern latitudes that precipitated a sudden latest Miocene cooling rather than a warming by a CO_2 excursion. In fact, the high rate of burial of organic matter in deep-sea sediments just prior to and associated with the Messinian event may have led to a net atmospheric-CO_2 drawdown. This drawdown may also have contributed to the latest Miocene cooling. However, there is still some doubt as to whether the evaporite deposition preceded the cooling or followed a pronounced regression related to increased glaciation on Antarctica (see Wright and Cita, 1979). Clearly, similar events may have different climatic consequences, depending perhaps on the global climate configuration at the time. The latitudinal distribution of continents during the Early Cretaceous contrasts greatly with that of the Late Miocene, a time when a greater proportion of the land mass was concentrated in higher latitudes. The position of Antarctica over the South Polar region is probably critical in the difference in global climate between the Early Cretaceous and the Late Miocene. Also, the Miocene oceans seem to have been much more fertile and possibly circulated more rapidly than the Early Cretaceous oceans. This apparent increase in nutrient availability may have allowed more rapid depletion of any pCO_2 increase by burial of marine organic matter in the Miocene oceans.

CONCLUSIONS

At present, separation of cause from effect is difficult when examining possible changes in ocean chemistry and their relations to climate during the last 120 m.y., as is attributing any given climate change in the past to changes in atmospheric pCO_2. The chemistry of the ocean plays a dominantly passive role in modulating climate largely through its thermal inertia and its part in the carbon cycle. Conversely, ocean chemistry and circulation can certainly change in response to climatic events. Abrupt changes in the depth distribution and accumulation rate of organic carbon and carbonate in the ocean basins, and in the carbon isotopic composition of pelagic-calcareous microfossils indicate major changes in ocean-water mass structure, rates of oceanic overturn, atmospheric CO_2 flux, and fertility. These changes within the ocean system in turn appear related to those in global climate.

Excess atmospheric CO_2 from volcanic or other sources may be titrated by dissolution of carbonate in deeper water. Increased rates of burial of organic carbon of either terrestrial or marine origin in marine sediment also facilitate removal of excess CO_2. The ability of the ocean-chemical system to maintain steady-state atmospheric CO_2 levels in this way, however, is largely dependent on overall nutrient availability and on rates of replenishment and levels of oxygenation of deep water, among other things. These factors bear a complicated and varying relation to climate as well.

Changes in ocean chemistry also may trigger climatic variation. For example, rapid and massive precipitation of evaporites in isolated basins involves major chemical transfers affecting the carbon cycle and perhaps leading to increased levels of atmospheric CO_2. However, the relative influence of such events and the direction of resultant climatic changes appear to depend on the prevailing climatic regime at the time.

More definite reconstructions of the relationship between changes in atmospheric CO_2, ocean chemistry, and climate will depend on collecting and analyzing large amounts of data on accumulation rates of organic matter and carbonate in marine sediments through time, on the bulk chemistry of sediment, and on the carbon isotopic composition of organic matter and carbonate. Time series of such data, in conjunction with those of isotopic paleotemperatures and faunal studies, will aid in examining the interrelationships of climate and ocean chemistry and will allow better estimation of leads and lags in the system. These data must be integrated into sophisticated computerized models relating changes in climate to those in oceanic and atmospheric chemistry. But in order to evaluate the role of atmospheric CO_2 changes in changing global climate, we must also be able to evaluate critically the climatic (and ocean chemical) effects of changing sea levels, continental distribution, tectonic and volcanic activity, and varying ocean gateways. At present, the number of degrees of freedom allow only a speculative and qualitative approach to the relationship between climate and atmospheric CO_2 in the geologic past.

ACKNOWLEDGMENTS

I am grateful for the thorough and enlightening reviews of an earlier manuscript by George E. Claypool and Eric Sundquist and for discussions of aspects of the CO_2 climate problem with them and with Eric Barron, Kemy Kelts, and Pete Scholle.

REFERENCES

Arthur, M. A. (1979). Paleo-oceanographic events—Recognition, resolution, and reconsideration, *Rev. Geophys. Space Phys. 17*, 1474-1494.

Arthur, M. A., and H. C. Jenkyns (1981). Phosphorites and paleoceanography, in *Ocean Geochemical Cycles*, W. H. Berger, ed., Oceanol. Acta, Supplement.

Arthur, M. A., and K. R. Kelts (1979). Evaporites, black shales and perturbations of ocean chemistry and fertility, *Geol. Soc. Am. Abstr. Programs 11*, p. 381.

Arthur, M. A., and J. H. Natland (1979). Carbonaceous sediments in the North and South Atlantic: The role of salinity in stable stratification of early Cretaceous basins, in *Results of Deep Drilling in the Atlantic Ocean, Proceedings of the Second Maurice Ewing Symposium 3*, M. Talwani, W. W. Hay, and W. B. F. Ryan, eds., American Geophysical Union, Washington, D.C., pp. 375-401.

Bender, M. L., and L. D. Keigwin Jr. (1979). Speculations about the upper Miocene change in abyssal Pacific dissolved bicarbonate $\delta^{13}C$, *Earth Planet. Sci. Lett. 45*, 383-393.

Berger, W. H. (1977). Carbon dioxide excurions and the deep sea record: Aspects of the problem, in *The Fate of Fossil Fuel CO_2 in the Ocean*, N. R. Andersen and A. Malahoff, eds., Plenum, New York, pp. 505-542.

Berger, W. H. (1979). The impact of deep sea drilling on paleoceanography, in *Deep Drilling Results in the Atlantic Ocean, Continental Margins and Paleoenvironment, Maurice Ewing Series 3*, M. Talwani, W. W. Hay, and W. B. F. Ryan, eds., American Geophysical Union, Washington, D.C., pp. 297-314.

Berggren, W. A., and C. D. Hollister (1977). Plate tectonics and paleocirculation: Commotion in the ocean, *Tectonophysics 38*, 11-48.

Berner, W., H. Oeschger, and B. Stauffer (1981). Information on the CO_2 cycle from ice core studies, *Radiocarbon 22*, 227.

Boersma, A., N. J. Shackleton, M. Hall, and Q. Given (1979). Carbon and oxygen isotope records at DSDP Site 384 (North Atlantic) and some Paleocene paleotemperatures and carbon isotope variations in the Atlantic Ocean, in *Initial Reports of the Deep Sea Drilling Project 43*, U.S. Government Printing Office, Washington, D.C., pp. 695-718.

Bolin, B., E. T. Degens, S. Kempe, and P. Ketner, eds. (1979). *The Global Carbon Cycle*, SCOPE Rep. 13, Wiley, New York, 491 pp.

Broecker, W. S. (1974). *Chemical Oceanography*, Harcourt Brace Jovanovich, New York, 214 pp.

Broecker, W. S. (1981). Glacial to interglacial changes in ocean chemistry, in *CIMAS Symposium*, E. Kraus, ed., U. of Miami, Miami, Florida.

Broecker, W. S., and T. Takahashi (1978). The relationship between lysocline depth and *in situ* carbonate ion concentration, *Deep Sea Res. 25*, 65-95.

Broecker, W. S., T. Takahashi, H. J. Simpson, and T. H. Peng (1979). Fate of fossil fuel carbon dioxide and the global carbon budget, *Science 206*, 409-418.

Claypool, G. E., W. T. Holser, I. R. Kaplan, H. Sakai, and I. Zak (1980). The age curves of sulfur and oxygen isotopes in marine sulfate and their mutual interpretation, *Chem. Geol.*

Fischer, A. G., and M. A. Arthur (1977). Secular variations in the pelagic realm, in *Deep Water Carbonate Environments*, H. E. Cook and P. Enos, eds., Soc. Econ. Paleontol. Mineral. Spec. Publ. 25, pp. 19-50.

Frakes, L. (1979). *Climates Throughout Geologic Time*, Elsevier, Amsterdam, 310 pp.

Froelich, P. N., M. L. Bender, N. A. Luedtke, G. R. Heath, and P. Devries (1982). The marine phosphorous cycle, *Am. J. Sci. 282*, 474-511.

Garrels, R. M., and E. H. Perry (1974). Cycling of carbon, sulfur, and oxygen through geologic time, in *The Sea*, Vol. 5, E. D. Goldberg, ed., Wiley-Interscience, New York, pp. 303-336.

Garrels, R. M., A. Lerman, and F. MacKenzie (1976). Controls of atmospheric O_2 and CO_2: Past, present, and future, *Am. Sci. 64*, 306-315.

Hay, W. W. (1979). Impact of Deep Sea Drilling Project on paleo-oceanography (abs.) *Am. Assoc. Petrol. Geol. Bull. 63*, p. 464.

Holland, H. D. (1972). The geologic history of sea water: An attempt to solve the problem, *Geochim. Cosmochim. Acta 36*, 637-651.

Holland, H. D. (1974). Marine evaporites and the composition of sea water during the Phanerozoic, in *Studies in Paleo-Oceanography*, W. W. Hay, ed., Soc. Econ. Paleontol. Mineral. Spec. Publ. 20, pp. 187-192.

Holland, H. D. (1978). *The Chemistry of the Atmosphere and Oceans*, Wiley, New York, 351 pp.

Hsü, K. J. (1980). Terrestrial catastrophe caused by cometary impact at the end of Cretaceous, *Nature 285*, 201-203.

Hunt, J. M. (1970). The significance of carbon isotope variations in marine sediments, in *Advances in Organic Geochemistry 1966*, G. B. Hobson and G. C. Speers, eds., Pergamon, Oxford, pp. 27-35.

Junge, C. E., M. Schidlowski, R. Eichmann, and H. Pietrek (1975). Model calculations for the terrestrial carbon cycle: Carbon isotope geochemstry and evolution of photosynthetic oxygen, *J. Geophys. Res. 80*, 4542-4552.

Kennett, J. P., and R. C. Thunell (1975). Global increase in explosive volcanism, *Science 197*, 497-503.

Kroopnick, P. M., S. V. Margolis, and C. S. Wong (1977). $\delta^{13}C$ variations in marine carbonate sediments as indicators of the CO_2 balance between the atmosphere and oceans, in *The Fate of Fossil Fuel CO_2 in the Ocean*, N. R. Anderson and A. Malahoff, eds., Plenum, New York, pp. 295-322.

Lemon, E. (1977). The land's response to more carbon dioxide, in *The Fate of Fossil Fuel CO_2 in the Oceans*, N. R. Anderson and A. Malahoff, eds., Plenum, New York, pp. 97-130.

Letolle, R., C. Vergnaud-Grazzini, and C. Pierre (1979). Oxygen and carbon isotopes from bulk carbonates and foraminiferal shells at DSDP Sites 400, 401, 402, 403 and 406, in *Initial Report of the Deep Sea Drilling Project 48*, U.S. Government Printing Office, Washington, D.C., pp. 741-755.

Madden, R. A., and V. Ramanathan (1980). Detecting climate change due to increasing carbon dioxide, *Science 209*, 763-768.

Manabe, S., and R. T. Wetherald (1975). The effects of doubling the CO_2 concentration on the climate of a general circulation model, *J. Atmos. Sci. 32*, 3-15.

Manabe, S., and R. T. Wetherald (1980). On the distribution of climate change resulting from an increase in CO_2 content of the atmosphere, *J. Atmos. Sci. 37*, 99-118.

Marland, G., and R. M. Rotty (1979). Carbon dixoide and climate, *Rev. Geophys. Space Phys. 17*, 1813-1824.

McLean, D. M. (1978a). Land floras: The major late Phanerozoic atmospheric carbon dioxide/oxygen control, *Science 200*, 1060-1062.

McLean, D. M. (1978b). A terminal Mesozoic "greenhouse," lessons from the past, *Science 201*, 401-406.

Muller, P. J. and E. Suess (1979). Productivity, sedimentation rate, and sedimentary organic matter in the oceans—organic preservation, *Deep Sea Res. 26A*, 1347-1362.

NRC Climate Research Board (1979). *Carbon Dioxide and Climate: a Scientific Assessment*, National Academy of Sciences, Washington, D.C., 22 pp.

Oeschger, H., V. Siegenthaler, V. Schotterer, and A. Gugelmann (1975). A box diffusion model to study the carbon dioxide exchange in nature, *Tellus 27*, 168-192.

Piper, D. Z., and L. A. Codispoti (1975). Marine phosphate deposits and the nitrogen cycle, *Science 188*, 15-18.

Pollack, J. B. (1979). Climate change on the terrestrial planets, *Icarus 37*, 479-553.

Pollack, J. B., O. B. Toon, C. Sagan, A. Summers, B. Baldwin, and W. Van Camp (1976). Volcanic explosions and climatic change: A theoretical assessment, *J. Geophys. Res. 81*, 1071–1083.

Rogers, M. A., and C. B. Koons (1969). Organic carbon $\delta^{13}C$ values from Quaternary marine sequences in the Gulf of Mexico: A reflection of paleotemperature changes, *Trans. Gulf Coast Assoc. Geol. Soc. 19*, 529-534.

Roth, P. H. (1978). Cretaceous nannoplankton biostratigraphy and oceanography of the northwestern Atlantic Ocean, in *Initial Reports of the Deep Sea Drilling Project 44*, U.S. Government Printing Office, Washington, D.C., pp. 731-759.

Rotty, R. M. (1977). Global carbon dioxide production from fossil fuels and cement, A.D. 1950-A.D. 2000, in *Fate of Fossil Fuel CO_2 in the Oceans*, N. R. Anderson and A. Malahoff, eds., Plenum, New York, pp. 167-181.

Ryan, W. B. F. (1973). Geodynamic implications of the Messinian crisis of salinity, in *Messinian Events in the Mediterranean*, C. W. Drooger, ed., K. Ned. Akad. Wet., Amsterdam, pp. 26-38.

Ryan, W. B. F., and M. B. Cita (1977). Ignorance concerning episodes of oceanwide stagnation, *Mar. Geol. 23*, 197-215.

Sackett, W. A. (1964). The depositional history and isotopic organic carbon composition of marine sediments, *Mar. Geol. 2*, 173-185.

Sackett, W. M., and R. R. Thompson (1963). Isotopic organic carbon composition of recent continental derived clastic sediments of Eastern Gulf Coast, Gulf of Mexico, *Bull. Am. Assoc. Petrol. Geol. 47*, 525.

Sackett, W. M., W. R. Eckelmann, M. L. Bender, and A. W. H. Be (1965). Temperature dependence of carbon isotope composition in marine plankton and sediments, *Science 148*, 235-237.

Savin, S. M. (1977). The history of the Earth's surface temperature during the last 100 million years, *Ann. Rev. Earth Planet. Sci. 5*, 319-355.

Scholle, P. A., and M. A. Arthur (1980). Carbon isotopic fluctuations in pelagic limestones: Potential stratigraphic and petroleum exploration tool, *Bull. Am. Assoc. Petrol. Geol. 64*, 67-89.

Shackleton, N. J. (1977). Carbon-13 in Uvigerina: Tropical rainforest history and the equatorial Pacific carbonate dissolution cycles, in *The Fate of Fossil Fuel CO_2 in the Ocean*, N. R. Anderson and A. Malahoff, eds., Plenum, New York, pp. 401-427.

Simoneit, B. R. T. (1977). The Black Sea, a sink for terrigenous lipids, *Deep Sea Res. 24*, 813-830.

Strain, B. R. (1978). *Report of the Workshop on Anticipated Plant Responses to Global Carbon Dioxide Enrichment*, Dept. of Botany, Duke U., Durham, N.C.

Tappan, H. (1968). Primary production, isotopes, extinctions a the atmosphere, *Paleogeogr. Paleoclimatol. Paleoecol. 4*, 187-210.

Thierstein, H. R. (1979). Paleoceanographic implications of organic carbon and carbonate distribution in Mesozoic deep-sea sediments, in *Deep Drilling Results in the Atlantic Ocean: Continental Margins and Paleoenvironment, Maurice Ewing Series 3*, M. Talwani, W. W. Hay, and W. B. F. Ryan, eds., American Geophysical Union, Washington, D.C., pp. 249-279.

Thierstein, H. R., and W. H. Berger (1978). Injection events in Earth history, *Nature 276*, 461-464.

Tissot, B., G. Demaison, P. Masson, J. R. Delteil, and A. Combaz (1980). Paleoenvironment and petroleum-potential of middle Cretaceous black shales in Atlantic basins, *Bull. Am. Assoc. Petrol. Geol. 64*, 2051-2063.

van Andel, T. H. (1975). Mesozoic/Cenozoic calcite compensation depth and the global distribution of calcareous sediments, *Earth Planet. Sci. Lett. 26*, 187-195.

van Andel, T. H., G. R. Heath, and T. C. Moore, Jr. (1975). Cenozoic history and paleoceanography of the central equatorial Pacific Ocean, *Geol. Soc. Am. Mem. 143*, 1-134.

Veizer, J., and J. Hoefs (1976). The nature of $^{18}O/^{16}O$ and $^{13}C/^{12}C$ secular trends in sedimentary carbonate rocks, *Geochim. Cosmochim. Acta 40*, 1387-1395.

Vincent, E., J. S. Killingley, and W. H. Berger (1980). The magnetic Epoch-6 carbon shift: A change in the oceans $^{13}C/^{12}C$ ratio 6.2 million years ago, *Mar. Micropaleontol. 5* 185-203.

Vogt, P. R. (1972). Evidence for global synchronism in mantle plume convection and possible significance for geology, *Nature 240*, 338-342.

Vogt, P. R. (1979). Global magmatic episodes: new evidence and implications for the steady-state mid-oceanic ridge, *Geology 7*, 93-98.

Weissert, H., J. McKenzie, and Hochuli (1979). Cyclic anoxic events in the Early Cretaceous Tethys Ocean, *Geology 7*, 147-151.

Worsley, T. R. (1974). The Cretaceous-Tertiary boundary event in the ocean, in *Studies in Paleo-Oceanography*, W. W. Hay, ed., Soc. Econ. Paleontol. Min. Spec. Publ. *20*, pp. 94-125.

Worsley, T., and T. Davies (1981). Paleoenvironmental implications of oceanic carbonate sedimentation rates, in *Symposium and Results of Deep-Sea Drilling*, R. Douglas, E. L. Winterer, and J. Warme, eds., Soc. Econ. Paleontol. Min. Spec. Spec. Publ. *30*.

Wright, R., and M. B. Cita (1979). Geo and biodynamic effects of the Messimean salinity crisis in the Mediterranean, *Palaeogeogr. Palaeoclimatol. Palaeoecol. 29*, 215-222.

5 Solar, Astronomical, and Atmospheric Effects on Climate

JAMES B. POLLACK
NASA Ames Research Center

INTRODUCTION

Significant climatic changes have taken place on the Earth over a very broad range of time scales ranging from decades to billions of years [Figures 5.1(a) and 5.1(b)]. There is probably not a single cause for these changes but rather a number, which, however, are individually effective over limited time domains (Pollack, 1979). Here, we will be concerned with solar, atmospheric, and astronomical factors that may have played a role in some of the climatic variability that characterizes pre-Pleistocene times.

The sections of this paper are organized by the climatic factors of interest. In each case, we define the manner in which the factor *may* have influenced past climates, provide an estimate for the characteristic time scales over which the factor is thought to vary, and give a summary of some of the research that has been done in relating this factor to climatic change, with an assessment of its likely importance. Spacecraft missions and ground-based observations have provided evidence that climatic changes have occurred on other objects in the solar system. For example, liquid water may have once flowed across the now desertlike surface of Mars. Therefore, where relevant, we also consider the possible influence of the above factors for the climate and its variability on other solar-system objects.

SOLAR VARIABILITY

From almost the beginning of the solar system, 4.6 billion years (b.y.) ago, the Sun's luminosity has been steadily increasing with time, according to almost all models of the Sun's evolution. Recent calculations suggest that the fractional increase in the Sun's output over the entire period is about 25 to 30 percent (Newman and Rood, 1977; R. Strothers as quoted in Canuto and Hsieh, 1978). A representative calculation of the long-term change in the Sun's output is given in Figure 5.2.

If no other factor varied, the lower solar output in the past would imply that the temperature of the Earth was progressively lower as we proceed further back in time. In fact, quantitative assessments of the amount of cooling give rise to a serious paradox, as first pointed out by Sagan and Mullen (1972). Global radiation models, in which the CO_2 content and

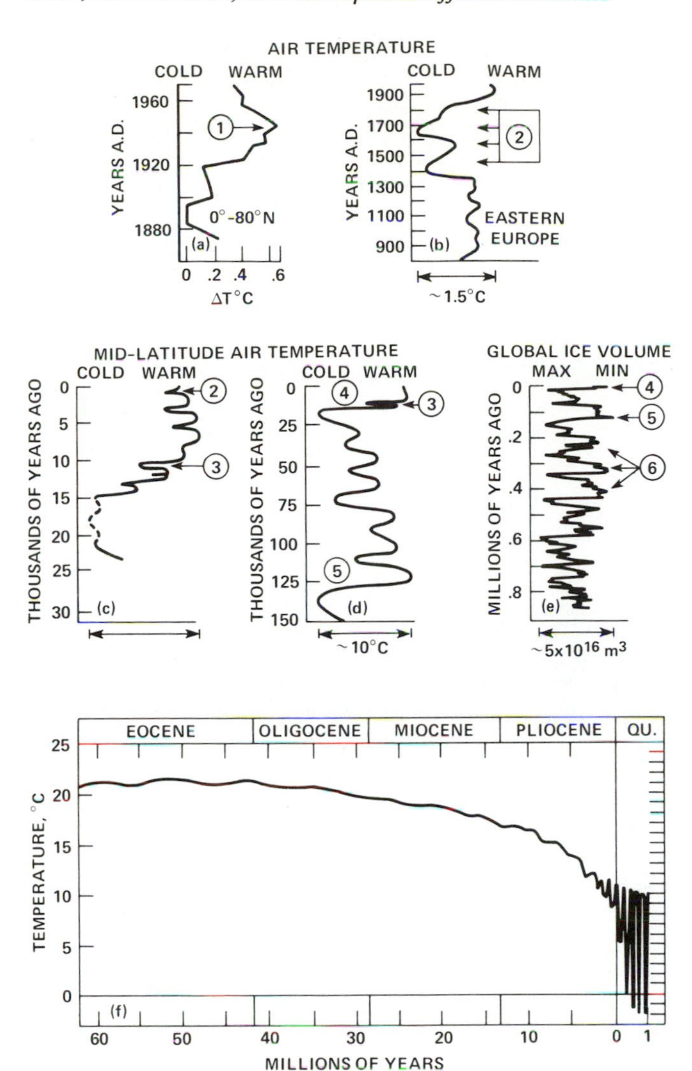

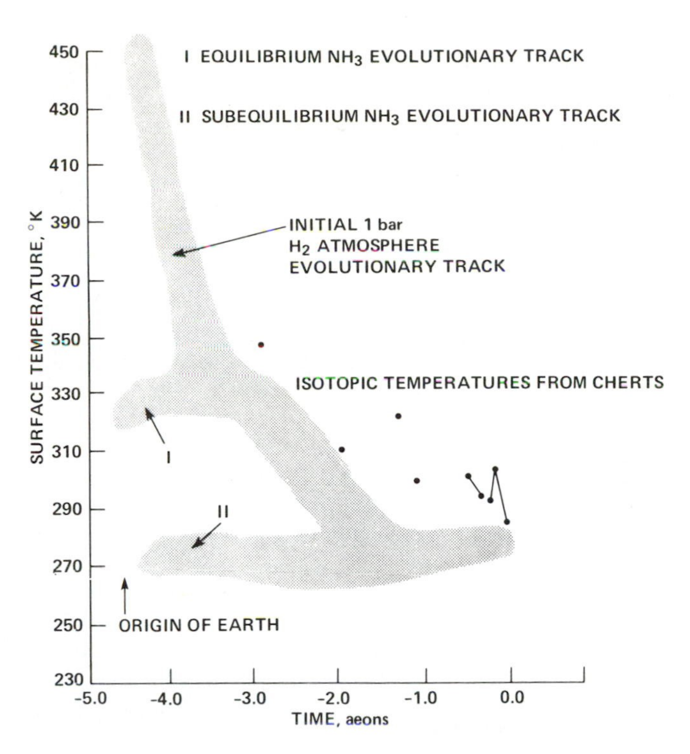

FIGURE 5.1 (a) *Left* Estimates of the change in the globally averaged surface temperature of the Earth over time scales ranging from the last century to the last 60 million years. These estimates are based on a variety of proxy records except for those of the last century, which are based on direct measurements [from Pollack (1979)]. (b) *Above* Surface temperature of the Earth over the last several billion years. The filled circles show temperatures inferred from isotopic studies of cherts by Knauth and Epstein (1976). The shaded tracks represent three possible evolutionary tracks for the temperature, as given by Sagan and Mullen (1972).

relative-humidity profile of the atmosphere are held constant with time, indicate that the Earth should have been totally covered with ice from 4.5 to 2.3 b.y. ago as a result of the lower solar luminosity then. But microfossils and stromatolites provide evidence that life has existed on the Earth for the last 3.5 b.y., and the occurrence of sedimentary rocks in the oldest geologic provinces implies that there have been water oceans on the Earth for the past 3.8 b.y. (W. Schopf, University of California, Los Angeles, and H. D. Holland, Harvard University, private communication). Allowance for some climatic feedback processes, such as the ice-albedo mechanism of Budyko (1969), only accentuates this paradox (e.g., Ghil, 1976). Unfortunately, even the sign of other key feedback mechanisms, such as one involving the radiative properties of water clouds, is unknown. Hence, the net response of the complete climate system to a change in solar luminosity cannot be calculated with precision at present.

Before discussing possible solutions to this problem in the next section, it is important to assess how well the Sun's evolution can be predicted. It is almost universely agreed among astronomers that fusion of hydrogen into helium in the deep interior of the Sun is the ultimate source of the energy it radiates to space. Detailed descriptions of these nuclear transformations have been worked out, with the so-called p-p chain being the dominant sequence for the Sun. An observational test of this theory is provided by attempts to detect neutrinos that are emitted at several steps of the transformation chain. Unfortunately, while solar neutrinos have apparently been detected (Rowley *et al.*, 1980), the flux of them is about a factor of 3 or 4 smaller than that predicted by conventional models of the Sun (Newman and Rood, 1977). This problem is somewhat alleviated by the fact that the neutrinos studied come from a relatively minor branch of the fusion chain. It has been suggested recently that neutrinos may have a small nonzero mass, in which case they may partially transmute from one type of neutrino to other types on the path from the Sun to the Earth. Consequently, fewer "standard" neutrinos may be detected. Finally and most to the point, virtually all solar models, including some very exotic ones that have been devised to resolve the neutrino problem, are characterized by long-term changes in solar output that are very similar to that of the standard models. This degree of agreement is due to the luminosity variation's being fundamentally related to the change of the Sun's mean molecular weight that accompanies the conversion of hydrogen into helium. Hence, once we accept that the fusion of hydrogen into helium has been the source of the Sun's output over almost its entire history, then it is difficult to avoid the temporal variations of output shown in Figure 5.2.

There is, however, one additional factor that needs to be considered before we accept the above long-term changes in solar output. These changes occur over such long time scales that the universe itself has varied significantly over them. There is the possibility that such cosmological factors as the universal gravitational constant, G, has varied as the universe has expanded. Canuto and Hsieh (1978) have developed cosmologies of this type and have found that the temporal history of the solar flux at the orbit of the Earth is quite different for these models than for the more conventional ones. In one model, in which G varies and matter is created, the Sun's luminosity increases even more rapidly with time than for the "standard" case. In a second model in which G varies but no matter is created, the Sun's luminosity *decreases* with time. In this latter case, we may have just the opposite problem to the one considered above: too high a temperature (greater than 373 K) for life to exist on the early Earth.

In summary, according to almost all models, the Sun's luminosity has varied by several tens of percent over the history of the Earth. For cosmologies in which G does not vary, the lower solar output of the Sun in the past needs some other variation to prevent the Earth from being entirely covered with ice over much of its lifetime, contrary to the geologic record. Even cosmologies with varying G are characterized by significant temporal changes in the Sun's output.

The evolution of the Sun's luminosity also presents a paradox for Mars, at least for constant-G cosmologies. Certain types of channels that are present on the surface of that planet appear to have been carved by running water in the past, although they are now dry. While it is not necessary to invoke an altered climate to explain the occurrence of certain classes of fluvial channels, especially the large "outflow" channels, a much warmer climate may be required by other classes, especially the smaller gullies, which are found ubiquitously on the older terrains (Pollack, 1979). But since these terrains were formed several billion years ago, a colder and not a warmer climate is expected, owing to the long-term changes in solar output.

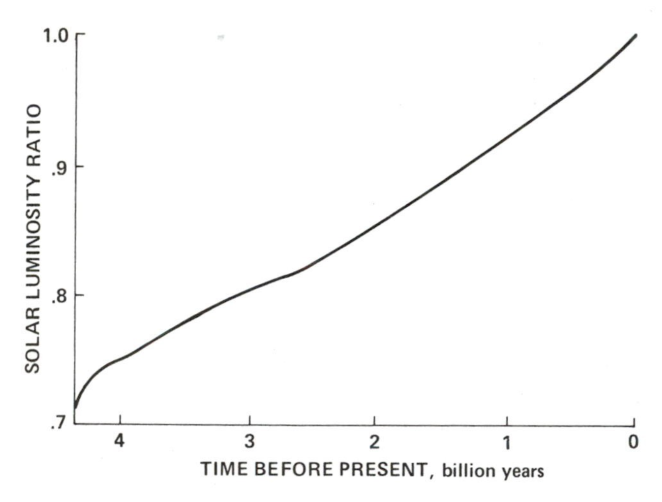

FIGURE 5.2 Solar luminosity as a function of time. The luminosity values have been normalized by the current value. From Stothers as quoted in Canuto and Hsieh (1978).

CHANGES IN ATMOSPHERIC COMPOSITION

The composition of the Earth's atmosphere is controlled by a series of thermodynamic, biological, and geologic factors. For example, the amount of water vapor in the atmosphere is buffered by the much larger amount of water in the oceans, with the partitioning between these two reservoirs being controlled by the saturation vapor-pressure curve of water. Hence, the amount in the atmosphere depends exponentially on the surface temperature. Almost all the oxygen in the Earth's atmosphere is the result of photosynthesis by living organisms, chiefly phytoplankton, with its abundance being controlled over long time scales ($>10^8$ yr) by a balance between losses suffered during chemical weathering of reduced surface material and gains resulting from the burial of reduced carbon compounds [see Figure 5.3(a)]. Much more carbon dioxide is locked up in carbonate rocks than exists in the atmosphere, with the latter being determined over long time scales ($>10^8$ yr) by a balance between losses due to the chemical weathering of silicate rocks and gains due to volcanic outgassing of juvenile CO_2 and CO_2 derived from the thermalization of buried carbonate rocks [cf., Figure 5.3(b)].

There are good reasons for believing that the atmospheric composition has varied with time. First, very little oxygen was presumably present in the early Earth's atmosphere at epochs prior to the evolution of photosynthetic organisms: the reduced compounds contained in volcanic effluents probably overwhelmed the oxygen produced from the photodissociation of water vapor followed by the escape of hydrogen from the top of the atmosphere (Kasting *et al.*, 1979). Interpretations of the geologic record suggest that little oxygen was present prior to 2.1×10^9 yr ago, with the oxygen partial pressure subsequently rising more or less monotonically from that point up until a time close to the end of the Pre-Cambrian ($\sim 6 \times 10^8$ yr ago), when a partial pressure close to the present value was achieved (Windley, 1977). The rise in oxygen was also accompanied by a progressive buildup in the ozone content of the atmosphere (Kasting *et al.*, 1979). Thus, during much of the Pre-Cambrian, a lot more biologically damaging solar UV light reached the surface than today, although organisms vary considerably in the doses they can tolerate. The rise of oxygen may have set the stage for the invasion of the continents by plants about 4×10^8 yr ago both by leading to increased shielding of the surface from UV radiation and by providing the basis for an energy efficient metabolism that permitted the development of complex organisms (Kasting *et al.*, 1979; Pollack and Yung, 1980).

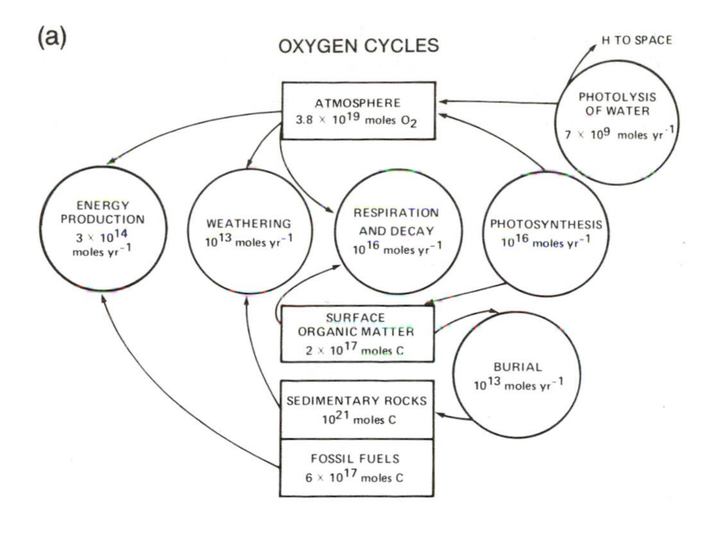

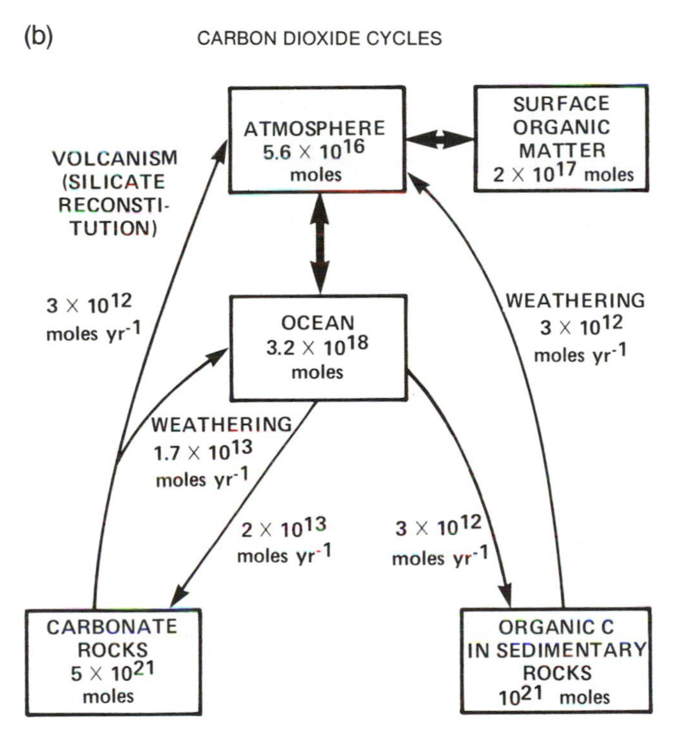

FIGURE 5.3 Geochemical cycles of oxygen (a) and carbon (b). The rectangular boxes show the amount of material stored in major reservoirs: the numbers within circles in (a) and along narrows in (b) indicate the fluxes between reservoirs. From Walker (1977).

The very early atmosphere of the Earth (~ 3.8-4.6×10^9 yr ago) may have been reducing in character in order for the chemical steps that led to the origin of life to have occurred (Miller and Orgel, 1974). While a number of years ago this biological requirement was thought to imply a fully reducing atmosphere consisting only of H_2, NH_3, CH_4, and other hydrogen-containing gases, it is now known that complex organic molecules can be biologically synthesized in mildly reducing atmospheres. There is even a minority opinion that such syntheses are possible in neutral atmospheres that interact with catalytic clays (Baur, 1978). Crude estimates of the early composition of the Earth's atmosphere can be obtained from considerations of constraints placed by thermodynamic equilibrium on the composition of volcanic gases and by the loss of certain gases through the top and bottom boundaries of the atmosphere. These studies suggest that N_2 and CO_2 were the dominant N- and C-containing gases in the Earth's early atmosphere, with minor amounts of H_2 ($\sim 10^{-3}$) and CO also being present (Pollack and Yung, 1980).

Variations in the CO_2 content of the atmosphere can occur owing to changes in the factors that control its geochemical cycle. For example, the lithosphere may have been thinner and the volcanic outgassing rate higher during the early history of the Earth (Holland, 1978). As a result, the CO_2 content of the atmosphere may have been higher then.

We now consider the way in which altered atmospheres may have helped to counter the lower solar luminosity in the past and hence have prevented the Earth from being totally covered with ice. In order for this to occur, there needs to be an augmentation in the concentration of gases that are optically active in the thermal infrared region. Such an augmentation results in an enhanced greenhouse effect. Water vapor is not able by itself to fill this role because its abundance is determined by its saturation vapor curve, however it will roughly double the warming caused by the increase in another gas: the initial warming leads to more water vapor in the atmosphere and thus an even stronger greenhouse effect.

Three types of altered atmospheres have been suggested to counteract the lower solar luminosity in the past: ones containing large amounts of hydrogen (1 bar) (Sagan and Mullen, 1972), trace amounts of ammonia (few to tens of parts per million), and enhanced amounts of CO_2 (up to a 1000-fold increase) (Owen *et al.*, 1979). Because hydrogen can readily escape from the top of the atmosphere, with the loss rate being proportional to its mixing ratio, it seems unlikely that hydrogen was a major component of the Earth's atmosphere over any extended period, and hence the first of these altered atmospheres has a serious problem (Pollack and Yung, 1980). The trace amounts of ammonia cited above for the second model are consistent with its large solubility in water. However, at the above concentration, photodissociation by solar-UV radiation quickly will irreversibly convert NH_3 to N_2. For example, the total N_2 content of the present atmosphere can be generated from NH_3 photolysis in only 10^7 yr (Kuhn and Atreya, 1979). Furthermore, much of the initial N-containing gases may have been N_2 rather than NH_3, as indicated above. Conceivably, ways of shielding NH_3 from solar-UV radiation can be found (see e.g., Pollack and Yung, 1980). But clearly, this second model faces serious difficulties.

The third model, the one involving enhanced CO_2, appears to be the most attractive one at present. The factors that control the geochemical cycle of CO_2 are such that one can imagine ways in which much more CO_2 was present in past atmospheres of the Earth, e.g., an enhanced amount of volcanic outgassing, as mentioned above. Furthermore, the geologic record places only very loose constraints on the level of CO_2 in the atmosphere over the last 3.8×10^9 yr (H. D. Holland,

Harvard University, personal communication). Specifically, the absence of alkali-containing carbonates implies that the CO_2 partial pressure was less that about 0.1 bar during this interval. Naturally, it still remains to be demonstrated that there were orders of magnitude more CO_2 in the Earth's atmosphere several billion years ago than today (see Figure 5.4, which summarizes the results of Owen *et al.*, 1979).

An increased abundance of CO_2 in the past atmospheres of the Earth may have been due to the combination of the way the Earth's interior has evolved with time and to feedback relationships between surface temperature and the atmospheric mixing ratio of CO_2. With regard to the former, we have already cited the possibility of the Earth's lithosphere being thinner in the past. Such a possibility is indicated by modern models of the temperature history of the Earth's interior (Schubert *et al.*, 1979). An extreme example of a feedback relationship between surface temperature and the amount of atmospheric CO_2 is provided by considering a totally ice-covered Earth (Moroz and Mukhin, 1980). In this event, little chemical weathering of silicate rocks would occur and essentially no carbonate rocks would be formed. Hence, the CO_2 content of the atmosphere would steadily rise as volcanoes injected new CO_2 into it. The surface temperature would increase, because of the enhanced greenhouse effect, to the melting point of water. Walker *et al.* (in press) have suggested a similar type of feedback process for an Earth dominated by liquid water. They suggest that weathering rates depend monotonically on both the partial pressure of CO_2 and temperature. Hence, a lowering of surface temperature leads to a higher partial pressure of CO_2, in order to keep the geochemical cycle in balance; in turn, the enhanced CO_2 results in a stronger greenhouse effect, which partially counteracts the reduced temperature.

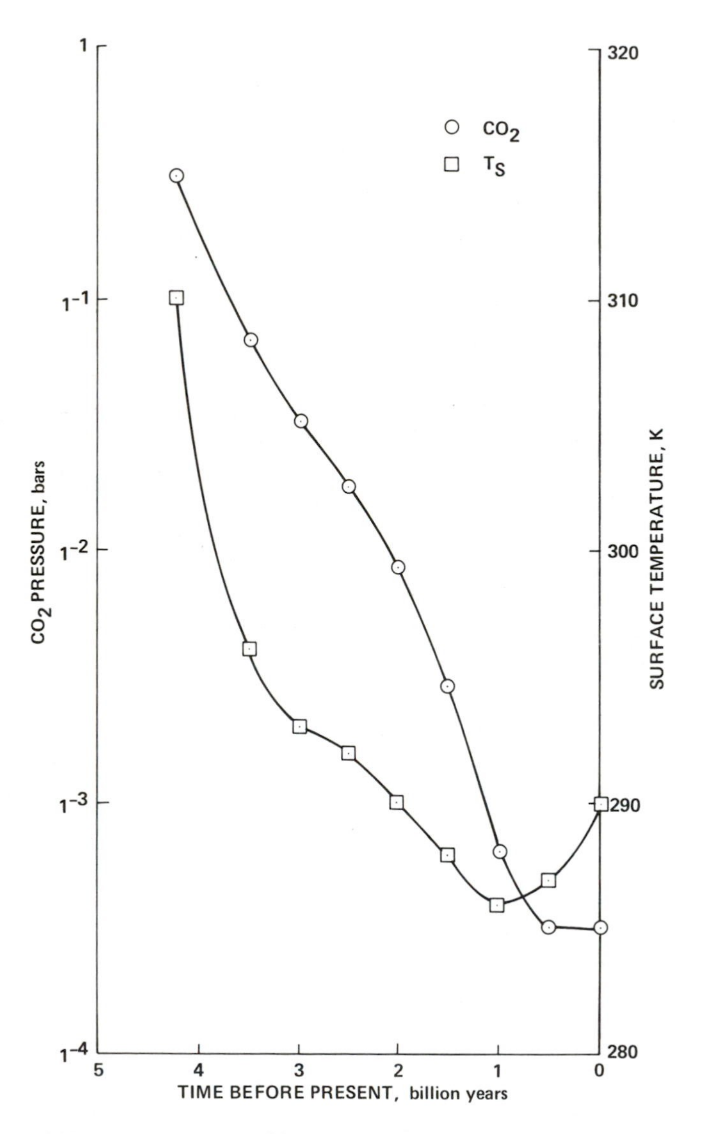

FIGURE 5.4 One possible time history of the partial pressure of CO_2 and the resulting variation in the mean surface temperature of the Earth. This figure is based on the model proposed by Owen *et al.* (1979) to resolve the solar-luminosity paradox.

It is worth noting that changes in the CO_2 content of the Earth's atmosphere may not only have occurred secularly over the entire history of the planet but may have also occurred episodically in amounts that could have caused shorter-term climatic fluctuations. If, hypothetically, the outgassing rate of CO_2 was doubled and the weathering rate stayed the same, the CO_2 content of the atmosphere and ocean would double in only about 4×10^5 yr (Walker *et al.*, in press). The comparable response time for the CO_2 rock reservoir is about 3×10^8 yr. Thus, changes in the level of volcanism and/or the characteristics of continents (total area, location, and topography) may also have been accompanied by changes in the amount of CO_2 in the atmosphere and thus the globally averaged surface temperature. Changes in the CO_2 content of the atmosphere of a factor of 2 or more are required to produce sizable changes in surface temperature. For example, doubling the CO_2 content of the present atmosphere, as may occur over the next several decades or century because of man's activities, would cause the surface temperature to increase on the average by several degrees Celsius, with larger changes occurring in the polar regions.

An enhanced amount of CO_2 in past atmospheres of Mars may be the means by which the surface temperature of that planet was increased to the melting point of water, thus setting the stage for the formation of certain types of fluvial channels on the older terrain (Pollack, 1979). Greenhouse calculations suggest that about 1 bar of CO_2 is required, i.e., an amount that is a factor of 150 larger than the current atmospheric content of CO_2 and an amount that is comparable with the total quantity of CO_2 outgassed over the history of the planet (Pollack, 1979; Cess *et al.*, 1980; Pollack and Yung, 1980). Such a large partitioning of CO_2 into the early atmosphere of Mars is not implausible, since, in the absence of liquid water, the weathering rates there may have been much smaller than for the Earth. Indeed, the ultimate consequence of the large buildup of CO_2—the occurrence of liquid water—may have brought about the demise of the large atmosphere through increased weathering. Because Mars is less massive than the Earth by a factor of about 10, its lithosphere may have thickened much more rapidly than the Earth's lithosphere, and, as a result, little recycling of CO_2 may have occurred for Mars (Pollack and Yung, 1980).

OSCILLATIONS IN ORBITAL AND AXIAL CHARACTERISTICS

As a result of gravitational perturbations by the other planets, most notably Jupiter and Saturn, the obliquity of the Earth's axis of rotation, i, and the eccentricity, e, of its orbit undergo quasi-periodic variations. Also, the orientation of its axis steadily precesses around the normal to its orbital plane because of solar and lunar torques. The characteristic period over which i varies is 41,000 yr, with i varying between about 22.0 and 24.5° (its current value is 23.4°). The dominant period for e is 413,000 yr, with there also being a number of secondary periods clustered around 100,000 yr. e varies between 0 and 0.06, with its current value being 0.017. The precession period is 25,000 yr (Imbrie and Imbrie, 1980).

The above astronomical variations result in seasonal and latitudinal redistributions of solar energy, although only a minor change in the annually and globally averaged solar energy results (Pollack, 1979; Imbrie and Imbrie, 1980). In particular, the obliquity variations lead to a 10 percent peak-to-peak change in the annually averaged solar insolation near the poles, and eccentricity variations result in a 25 percent peak-to-peak modulation in the summertime insolation in a given hemisphere, with the two hemispheres being out of phase. These modulations may be particularly important at times when the continents in one hemisphere are marginally able to have ice sheets as a result of their location and the mean surface temperature.

The last several million years have been characterized by a succession of alternating glacial and interglacial epochs. Indeed, the last major ice age, the Wisconsin, ended only about 12,000 yr ago. Strong evidence that the above astronomical variations are in part responsible for these ice ages is given by comparisons of the characteristic frequencies and phases found in well-dated sea cores with those expected from the astronomical theory (Hays *et al.*, 1976). In particular, periods of 41,000, 23,000, and 19,000 yr characterize certain climate indicators in the cores, which correspond quite closely to those expected from the obliquity variations and the combined eccentricity-precession variations. However, there is still a good deal of uncertainty as to whether the dominant period of the cores, a 100,000-yr period, should be associated with the corresponding eccentricity period. According to *linear* theories of the relationship between the astronomical variations and climate, only the combined eccentricity-precession variable is relevant. Conceivably, the nonlinear response characteristics of ice sheet growth and decay permit the eccentricity alone to affect climate, although details of the relationship need to be worked out (Imbrie and Imbrie, 1980).

Regardless of the outcome of the above problem, the astronomical variations are at most a necessary, but not a sufficient, cause of ice ages. Prior to several million years ago for a time span of several hundred million years, no large continental glaciations in subpolar and mid-latitude regions occurred, despite the occurrence of astronomical variations with about the same amplitudes as those for the last several million years. In all probability, the drift of the continents toward the poles set the stage for the Pleistocene ice ages and presumably for earlier ones as well (Pollack, 1979).

The amplitude of the astronomical variations in eccentricity are set by the initial orbits of the planets in the early history of the solar system and is not expected to change in any significant way subsequently (W. R. Ward, Jet Propulsion Laboratory, personal communication). However, the amplitudes of the obliquity variations depend on the relationship between the precession frequency, ω_p, and the characteristic frequencies of the planetary perturbations, ω_i. When $\omega_p \gg \omega_i$ for all i, the amplitudes of the obliquity variations are relatively small and can be significantly augmented only if ω_p decreases to a value comparable with ω_i. At present, the torque exerted by the Moon on the Earth's equatorial bulge accounts for about two thirds of its precession rate, with the rest due to the Sun's torque. During past epochs when the Moon was closer to the Earth, ω_p was presumably larger and hence the obliquity variations were somewhat smaller. A dramatic event may occur in the future when the moon moves somewhat further from the Earth. ω_p will increase to a 50,000-yr period, becoming comparable with one of the ω_i. When that occurs ($\sim$ few $\times$ 10^8 yr from now), the amplitude of the obliquity oscillations will increase by more than an order of magnitude and enormous climatic changes will occur (W. R. Ward, Jet Propulsion Laboratory, personal communication).

It is highly unlikely that the mean obliquity has varied significantly over the age of the Earth. In principle, such a change could occur because of core-mantle coupling: the mantle has a bigger bulge and is more strongly affected by the lunar and solar torques. Only very gradually through viscous interactions does the mantle drag the core along. Such a coupling would gradually *decrease* the mean obliquity with time. However, the time constant for the coupling may well exceed the age of the Earth.

Because its ω_p is comparable with some of the ω_i, the obliquity variations for Mars have a peak-to-peak amplitude of about 20°, i.e., almost an order of magnitude larger than that for the Earth (Ward, 1974). Furthermore, its eccentricity variations are about a factor of 2.5 larger than those for the Earth. These much larger astronomical variations for Mars may have played an important role in generating a quasi-periodic sequence of sedimentary layers in its polar regions (Cutts, 1973; Pollack, 1979). These layers are believed to consist of a mixture of water ice and suspendable dust particles, with the astronomical variations modulating both the deposition rate and the ratio of the two constituents. For example, at times of low obliquity, the atmospheric pressure may drop because of absorption of CO_2 on surface dust grains (CO_2 is the dominant constituent of the Martian atmosphere). As a result, the frequency of dust storms may decline sharply (Pollack, 1979).

COLLISIONS WITH APOLLO ASTEROIDS

Apollo asteroids are small, stray bodies whose orbits cross the orbital plane of the Earth. At present, some 30 such objects are known, with diameters ranging from 0.2 to 8 km. When allowance is made for the large incompleteness of the telescopic searches for them, it is estimated that there are approximately 750 Apollo asteroids whose diameters exceed 1 km (Wetherill, 1979).

It is inevitable that some of the Apollo asteroids will collide with the Earth; because of orbital perturbations by the planets, the semimajor axes of their orbits and the normals to their orbital planes precess. As a result, every few thousand years their orbits intersect that of the Earth. Almost all of the time, the Apollo asteroid intersects the Earth's orbit when the Earth is not at the intersection point. But there is a well-defined, nonzero probability for the Earth being at the intersection point, in which case the asteroid collides with the Earth. The average lifetime of a given Apollo asteroid before such a collision occurs is about 2×10^8 yr (Wetherill, 1979). Hence, collisions with Apollo asteroids with diameters of 1 km or greater happen about once every 2.5×10^5 yr (Wetherill, 1979) and with ones having a diameter of at least 10 km about once every 10^8 yr. Because the current population of Apollo asteroids represents one in which an equilibrium has been established between losses through collisions with planets and ejection from the solar system and gains from objects added from the asteroid belt and old cometary nuclei, the above collision frequencies hold throughout most of the Earth's history. However, during the first 7×10^8 yr of the Earth's history, the bombardment rate was probably orders of magnitude larger, based on our knowledge of lunar chronology (Hartmann, 1972).

A collision on land with an Apollo object results in the creation of a crater, whose diameter is about 20 times that of the impacting object and whose initial depth at the center of the crater is about one quarter the crater's diameter. The Apollo object is almost totally volatized, and a mass equal to about 100 times that of the object is ejected from the crater. Some of the ejecta is highly comminuted rock, and this material may reach high altitudes and stay in the atmosphere for an extended time. Much water as well as pulverized rock may be injected into the atmosphere if the impact occurs in an ocean.

While collision with an Apollo asteroid is clearly a catastrophic event in the immediate vicinity of the impact, the more important issue is whether such collisions can perturb the climate in a significant way on a global scale. Some evidence that collisions with the largest Apollo objects ($\sim$10 km) had a profound impact on the global climate has been given by Alvarez *et al.* (1980). In an effort to detect the presence of extraterrestrial material in the geologic record, they determined the abundance of the element iridium in sedimentary sections that spanned the time period from the Upper Cretaceous through the Lower Tertiary. This element was chosen because its abundance in the Earth's crust is orders of magnitude smaller than its "solar abundance" value (presumably much of the Earth's iridium has been segregated to its deep interior) and because very small amounts of it can be accurately measured with neutron activation techniques. Alvarez *et al.* (1980) found very large enhancements of the concentration of iridium in the acid insoluble clay fraction at precisely the Cretaceous-Tertiary boundary (see also Chapter 8). An enhancement factor of 30 and 160 characterized sections from Italy and Denmark, respectively.

The above result is of considerable interest because the iridium anomaly occurs precisely at the point in the geologic record where mass extinctions, including those of the dinosaurs, occurred 65 million years ago. Thus, the extraterrestrial event responsible for the iridium enrichment may have been responsible for the extinction. A supernova origin for the enrichment was ruled out because no observable amount of ^{244}Pu was detected and because the isotopic ratio, ^{191}Ir/^{193}Ir, was essentially identical to the terrestrial one. The enrichment was attributed by Alvarez *et al.* (1980) to an impact of a 10-km Apollo asteroid, whose iridium abundance was presumably comparable with the solar-abundance value. The above size is consistent with both the total amount of iridium found and the expected frequency of such collisions.

Alvarez *et al.* (1980) offer the following scenario for the manner in which an impact of a large Apollo object could have resulted in the extinction at the Cretaceous-Tertiary boundary: Some fraction of the ejecta was very fine dust, which reached the stratosphere and remained there for 3 to 5 yr, by analogy with the large enhancement of the stratospheric aerosol population following the Krakatoa volcanic explosion. So much dust was placed in the stratosphere by the Apollo impact event that essentially no sunlight reached the surface for the next 3 to 5 yr. As a result, photosynthesis ceased for this time period, causing whole food chains to collapse and leading to the extinction of organisms dependent directly or indirectly on *living* plants and plants, such as phytoplankton, that lacked seeds or their equivalent. Unfortunately, the above scenario rests on a misunderstanding of the enhancement of the stratospheric aerosol population following the Krakatoa event: the fine dust remained in the stratosphere for only a month or two, with almost all of the subsequent perturbation being due to the slow conversion of injected sulfur gases to sulfuric acid aerosols (Lazrus *et al.*, 1971; Pollack *et al.*, 1976). Nevertheless, it is quite clear that study of the various effects that accompany the impact of a large Apollo object—changes in atmospheric temperature and ozone abundance as well as opacity—is a fruitful topic of research and that it may have relevance to the extinction of the dinosaurs as well as possibly other similar occurrences.

PASSAGE THROUGH DENSE INTERSTELLAR CLOUDS

The density of the interstellar medium of gas and dust varies considerably, with gas densities ranging from about 1 atom or molecule/cm^3 to about 10^6 atoms or molecules/cm^3. As a very crude approximation, about 1 percent of the mass is partitioned into solid grains, whose characteristic size is about 0.1 μm. During the course of its motion around the galactic center, the solar system may pass through a spiral arm, in which dense interstellar clouds are preferentially located and, in the process, encounter one or several dense clouds. Hydrogen is the most abundant element in the interstellar medium, with it being principally in the form of molecular hydrogen in the dense clouds. Below, we briefly consider the possible climatic consequences of accretion of molecular hydrogen by, alternatively, the Sun and the Earth. It is to be noted that, even during the passage through a region having a density of 10^6 H_2/cm^3, the opacity of dust grains between the Earth and the Sun is sufficiently small so as not to cause a noticeable reduction in the solar flux at the orbit of the Earth.

Hoyle and Lyttleton (1939) and McCrea (1975) have considered the effect on the Sun's luminosity of its accreting molecular hydrogen during the solar system's passage through a dense interstellar cloud. The Sun's luminosity increases at such times because of the release of gravitational energy by the accreted matter, with the amount of enhancement being proportional to the rate of accretion and thus the density of the cloud. For typical velocities of the Sun relative to the interstellar medium (about 5-25 km/sec), McCrea (1975) finds that cloud densities of about 10^5 to 10^7 H_2/cm^3 are required in order to increase the Sun's luminosity by 1 to 100 percent. Oddly enough, the increased luminosity is invoked as a means of triggering an ice age, in accord with a theory of Sir George Simpson. Whereas Simpson's theory is not currently widely held, there is no doubt that an increase in the Sun's luminosity of 1 percent or more would be climatically significant. McCrea (1975) claims that the solar system passes through an interstellar cloud whose density exceeds 10^5 H_2/cm^3 at intervals of about 10^8 yr. However, more recent considerations of the properties of the interstellar medium by Talbot and Newman (1977) indicate that the above interval may be significantly longer, in which case accretion of interstellar matter by the Sun may not have climatic relevance.

The statistical analysis of Talbot and Newman (1977) indicates that the Earth passes through an interstellar cloud with a density of 10^3 H_2/cm^3 about once every 3×10^8 yr. McKay and Thomas (1978) have examined the possible chemical and climatic consequences of the Earth accreting H_2 from such a cloud. During such time, the duration of which is about 10^5 yr, the solar wind is confined to within the Earth's orbit and the Earth accretes H_2 at a globally average rate of about 7×10^9 $H_2/cm^2/sec$. This value may be compared with the present rate of 10^8 $H_2/cm^2/sec$ at which H is escaping from the top of the Earth's atmosphere. According to McKay and Thomas (1978), the interstellar H_2 molecules are thermalized and diffuse down to an altitude of 140 km, below which they are destroyed by reactions with O, leading to the formation of odd-hydrogen compounds, including H, H_2O, OH, and HO_2.

There are several possible major effects of the above flux of H_2 into the Earth's atmosphere and its subsequent chemical conversions (McKay and Thomas, 1978). First, the ionospheric F-region may be largely destroyed owing to the great efficiency of H_2 in affecting electron recombination. Second, there may be a substantial depletion of ozone at altitudes above 50 km owing to the great increase in the abundance of odd-hydrogen species, which dominate the catalytic destruction of ozone there. Third, and most important, there may be a large increase in the water-vapor mixing ratio near the mesopause (about 100 ppm), which may engender the appearance of a dense, global cloud of water and ice in the upper mesosphere. At present, only a very thin ice cloud occurs near the mesopause, and it is located only in the polar regions. According to McKay and Thomas's calculations, the optical depth of the enhanced mesospheric ice clouds reaches a value on the order of 0.1, resulting directly in a decrease in the globally averaged surface temperature of about 1°C. They claim that such a cooling may trigger an ice age. However, as noted earlier, the position of the continents plays an important role in setting the stage for ice ages. The probability of the Earth passing through a dense interstellar cloud when the continents are properly aligned is extremely small. Nevertheless, the above cooling is still climatically significant and merits further study. Finally, the above chemistry and its consequences apply to times when oxygen was present in substantial quantities in the Earth's atmosphere and, thus, is not relevant prior to about 2×10^9 yr ago.

SUPERNOVA EXPLOSIONS

Supernova explosions are cataclysmic events in the lives of some stars whereby their outer envelopes are ejected at high velocities into space. During the first hundred days or so following the explosion, the light output from the supernova exceeds by many orders of magnitude that of the pre-explosion star, and it spans a broad range of the electromagnetic spectrum. In addition, a large flux of cosmic rays accompanies the expanding envelope or supernova remnant.

The climate of the Earth may be affected by a nearby supernova explosion, principally through its effects on the ozone layer. Ruderman (1974) suggested that the terrestrial ozone layer would be severely depleted by both the gamma rays generated during the early phases of the supernova event (approximately first 100 days) and by the cosmic rays generated in the envelop of the remnant, when they swept past the Earth. However, Whitten *et al.* (1976) have pointed out some significant revisions to Ruderman's proposal brought about by newer astrophysical data and have carried out a more detailed calculation of the possible ozone reduction. They suggest that the amount of energy contained in the gamma-ray portion of the supernova's spectrum is too small to affect the ozone layer appreciably and that the cosmic rays contained in the supernova envelope are chiefly due to the trapping of cosmic rays from the interstellar medium. As a consequence of the latter, the Earth is exposed to an enhanced level of cosmic rays over a longer period of time (about 10^3-10^4 yr) than suggested by Ruderman.

The amount of ozone in the atmosphere is reduced by the cosmic rays because of their producing large quantities of nitrogen oxides in the stratosphere, which act to destroy the ozone catalytically. According to the calculations of Whitten *et al.* (1976), ozone reductions of 20 and 50 prcent result from enhancements of the cosmic-ray flux at the Earth of factors of 10^2 to 10^3, respectively, which can be expected from supernova explosions situated 10 and 5 parsecs from the Earth. Such reductions will increase the amount of potentially biologically damaging solar-UV radiation reaching the Earth's surface and will affect the atmospheric temperatures, as indicated below. However, Whitten *et al.* (1976) estimate that the mean interval between such nearby supernova explosions is about 2×10^9 and 2×10^{10} yr for the 20 and 50 percent depletions, respectively.

Hunt (1978) has calculated the effects of the reduction in ozone on the atmospheric temperature structure. He finds that temperatures in the stratosphere are reduced by 5 and 11°C and surface temperatures are reduced by 0.1 and 0.3°C, by ozone reductions of 20 and 50 percent, respectively. He also points out that the large temperature changes in the stratosphere may affect the troposphere through dynamical coupling of these two regions.

CONCLUSIONS

Of the various factors discussed above, the ones that would appear to have most profoundly affected the Earth's climate during the pre-Pleistocene period are changes in the solar output due to long-term (about 10^9 yr), evolutionary effects, long-term changes in the composition of the terrestrial atmosphere, and impacts by large Apollo asteroids. The first of these factors tended to cause a progressive cooling with increasing time into the past, which may have been counteracted by the second factor. Large increases in the carbon dioxide content of the early Earth appears to be the most promising means of engendering the desired enhanced greenhouse effect. In addition, episodic variations in the carbon dioxide content of the atmosphere appear to be possible due to variations in tectonic processes, with shorter-term climatic oscillations accompanying these changes. Collisions with 10-km-sized Apollo objects about every 10^8 yr may have been the determining factor for the mass extinctions that characterize the geologic record, although details of the mechanism for this relationship need to be worked out and additional cores need to be studied.

The finding that climatic changes have occurred on other solar system objects, particularly Mars, adds a new and important dimension to studies of long-term climatic changes on the Earth. Furthermore, some of the same climatic factors that are important for the Earth also appear to have been involved in these extraterrestrial climate changes.

REFERENCES

Alvarez, L. W., W. Alvarez, F. Asaro, and H. V. Michel (1980). Extraterrestrial cause for the Cretaceous-Tertiary extinction: Experiment and theory, *Science 208*, 1095-1108.

Baur, M. E. (1978). Thermodynamics of heterogeneous iron-carbon systems: Implications for the terrestrial reducing atmosphere, *Chem. Geol. 22*, 189-206.

Budyko, M. I. (1969). Climatic change, *Sov. Geogr. 10*, 429-457.

Canuto, V., and S. H. Hsieh (1978). Scale covariant cosmology and the temperature of the Earth, *Astron. Astrophys. 65*, 389-391.

Cess, R. D., V. Ramanathan, and T. Owen (1980). The Martian paleoclimate and enhanced carbon dioxide, *Icarus 41*, 159-165.

Cutts, J. A. (1973). Nature and origin of layered deposits of the Martian polar regions, *J. Geophys. Res. 78*, 4231-4249.

Ghil, M. (1976). Climatic stability for a Sellers-type model, *J. Atmos. Sci. 33*, 3-20.

Hartmann, W. K. (1972). Paleocratering of the moon: Review of post-Apollo data, *Astrophys. Space Sci. 17*, 48-64.

Hays, J. D., J. Imbrie, and N. J. Shackelton (1976). Variation in the Earth's orbit: Pacemaker of the ice ages, *Science 194*, 1121-1132.

Holland, H. D. (1978). The evolution of seawater, in *The Early History of the Earth*, B. F. Windley, ed., Wiley, New York, pp. 559-568.

Hoyle, F., and R. A. Lyttleton (1939). *Proc. Cambridge Philos. Soc. 35*, 405-415.

Hunt, G. E. (1978). Possible climatic and biological impact of nearby supernovae, *Nature 271*, 430-431.

Imbrie, J., and J. Z. Imbrie (1980). Modelling the climatic response to orbital variations, *Science 207*, 943-953.

Kasting, J. F., S. C. Liu, and T. M. Doanhue (1979). Oxygen levels in the prebiological atmosphere, *J. Geophys. Res. 84*, 3097-3107.

Knauth, L. P., and S. Epstein (1976). Hydrogen and oxygen isotope ratios in nodular and bedded charts, *Geochim. Cosmochim. Acta 40*, 1095-1108.

Kuhn, W. R., and S. K. Atreya (1979). Ammonia photolyses and the greenhouse effect in the primordial atmosphere of the Earth, *Icarus 37*, 207-213.

Lazrus, A., B. Gandrud, and R. D. Cadle (1971). Chemical composition of air infiltration samples of the stratospheric sulfate layer, *J. Geophys. Res. 76*, 8083-8088.

McCrea, W. H. (1975). Ice ages and the galaxy, *Nature 255*, 607-609.

McKay, C. P., and G. E. Thomas (1978). Consequences of a past encounter of the Earth with an interstellar cloud, *Geophys. Res. Lett. 5*, 215-218.

Miller, S. L., and L. E. Orgel (1974). *The Origins of Life on Earth*, Prentice-Hall, Englewood Cliffs, N.J.

Moroz, V. I., and L. M. Mukhin (1980). About the initial evolution of atmosphere and climate of the Earth type planets, report of U.S.S.R. Space Research Institute.

Newman, M. J., and R. T. Rood (1977). Implications of solar evolution for the Earth's early atmosphere, *Science 198*, 1035-1037.

Owen, T., R. D. Cess, and V. Ramanathan (1979). Enhanced CO_2 greenhouse to compensate for reduced solar luminosity on early Earth, *Nature 277*, 640-642.

Pollack, J. B. (1979). Climatic change on the terrestrial planets, *Icarus 37*, 479-553.

Pollack, J. B., and Y. L. Yung (1980). Origin and evolution of planetary atmospheres, *Ann. Rev. Earth Planet. Sci. 8*, 424-487.

Pollack, J. B., O. B. Toon, C. Sagan, A. Summers, B. Baldwin, and W. Van Camp (1976). Volcanic explosions and climatic change: A theoretical assessment, *J. Geophys. Res. 81*, 1071-1083.

Rowley, J. K., B. T. Cleveland, and R. Davis, Jr. (1980). Brookhaven National Laboratory preprint 27190.

Ruderman, M. A. (1974). Possible consequences of nearby supernova explosions for atmospheric ozone and terrestrial life, *Science 184*, 1079-1081.

Sagan, C., and G. Mullen (1972). Earth and Mars: Evolution of atmospheres and surface temperature, *Science 177*, 52-56.

Schubert, G., P. Cassen, and R. E. Young (1979). Subsolidus convective cooling histories of terrestrial planets, *Icarus 38*, 192-211.

Talbot, R. J., and M. J. Newman (1977). Encounters between stars and dense interstellar clouds, *Astrophys. J. Suppl. 34*, 295-308.

Ward, W. R. (1974). Climatic variations on Mars. 1. Astronomical theory of insolation, *J. Geophys. Res. 79*, 3375-3386.

Walker, J. C. G. (1977). *Evolution of the Atmosphere*, Macmillan, New York, 318 pp.

Walker, J. C. G., P. B. Hays, and J. F. Kasting (in press). Long-term stabilization of Earth's surface temperature by the greenhouse effect of carbon dioxide.

Wetherill, G. (1979). Apollo objects, *Sci. Am. 240*, 54-65.

Whitten, R. C., J. Cuzzi, W. J. Borucki, and J. H. Wolfe (1976). Effect of nearby supernova explosions on atmospheric ozone, *Nature 263*, 398-400.

Windley, B. F. (1977). *The Evolving Continents*, Wiley, New York.

Continental Glaciation through Geologic Times

6

JOHN C. CROWELL
University of California, Santa Barbara

INTRODUCTION

For about two and a half billion years the Earth's surface temperatures have been near the transition between water in its three phases: ice, liquid, and vapor. During this long span of time, ice has accumulated to form huge glaciers on the continents from time to time, separated by intervals when they dwindled and disappeared (Figure 6.1). Here we examine briefly the geologic documentation of the record of waxing and waning of continental glaciers and discuss hypotheses to explain the ice ages.

NATURE OF THE GLACIAL RECORD

Glaciers scour into hard rock and cut characteristic landforms such as U-shaped valleys, stoss and lee shapes, and roches moutonnées. In addition, the surfaces of these landforms are scratched by stones frozen fast within the ice as the glacier inexorably scrapes across bedrock. Such landforms along with faceted and striated stones, after the ice has long melted, have been confidently identified beneath sedimentary rock as old as 2200 million years (m.y.). They provide convincing evidence of glaciation and have been identified in association with strata of all the major ice ages on Earth (Figure 6.1).

Such surfaces are at places overlain by unsorted mixtures of blocks, stones, sand, and mud, which, when hardened into rock, are known descriptively as diamictite (or mixtite). If diamictite is closely associated with ice-sculptured landforms and pavements, and especially if it also contains glacially faceted and striated stones, it is interpreted genetically as consolidated glacial moraine or till (tillite). In contrast, beds and lenses of diamictite are formed where unsorted debris slides downslope into a sedimentary basin from nearby sources and is in no way associated with glaciers (Crowell, 1957; Dott, 1963). The interpretation of diamictite as tillite and the recognition of its glacial origin involve the reconstruction of paleogeography based primarily on the study of associate stratal facies. Where did the ice lie with respect to outwash plains, river systems, shorelines, and the sea? Observations from widespread outcrops are interpreted to reveal the geography over a large region. They must come from beds that are close time equivalents. Distinction is also needed between

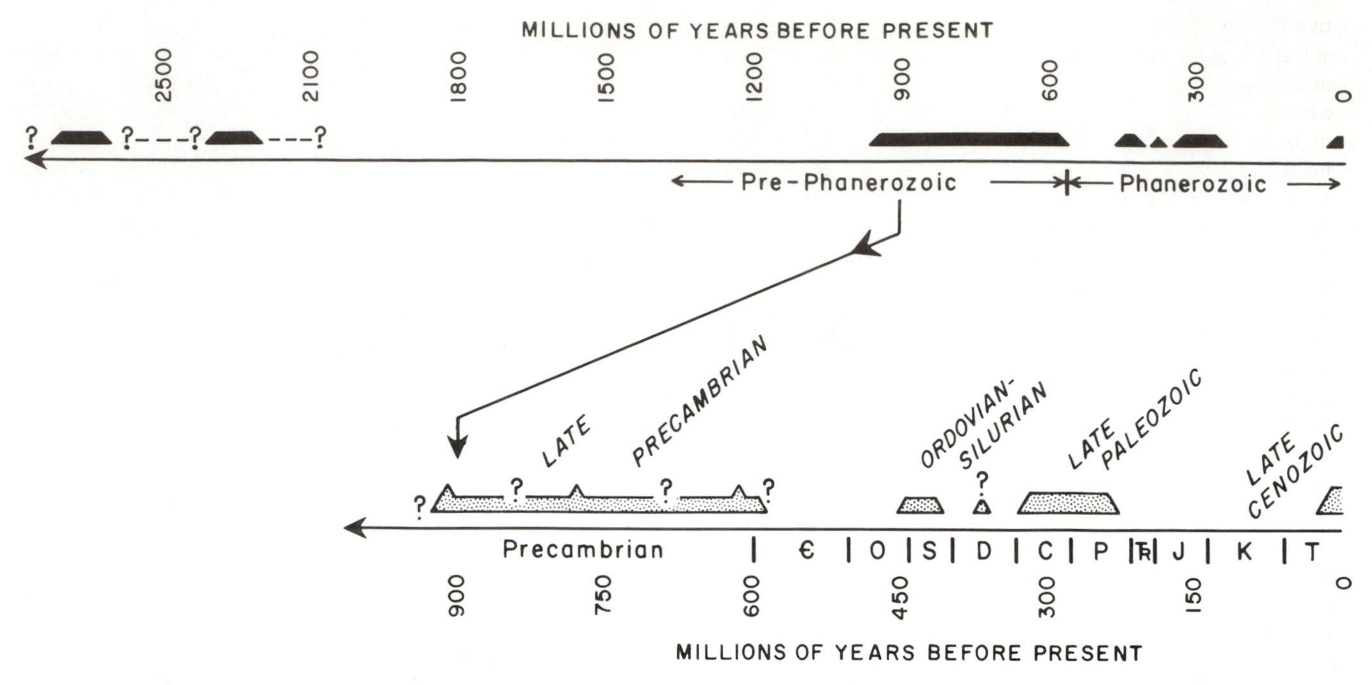

FIGURE 6.1 Occurrence in time of ancient ice ages plotted linearly, to two scales. O, Ordovician; S, Silurian; D, Devonian; C, Carboniferous; P, Permian; Tr, Triassic; J, Jurassic; K, Cretaceous; T, Tertiary.

huge continental ice sheets, smaller ice caps, piedmont ice sheets or aprons, and alpine glaciers confined to mountain ranges.

Where glaciers reach the sea, icebergs calve and float away, carrying within them clots of debris including large boulders. This debris drops to the seafloor when the ice melts and may accumulate in an environment where currents are weak and where thin-bedded or laminated deposits are being laid down. Ancient deposits of laminated shales with sporadic large stones, or clots of unsorted debris, may therefore be properly interpreted as the result of rafting. If these dropstones in turn are large and angular and are composed of mixed rock types carried from great distances, and especially if some of them are glacially faceted and straited, we can infer that the rafts were thick icebergs from calving glaciers rather than thin sheets of shore ice or sea ice. Root mats and kelp holdfasts may also raft oversized stones and debris to environments of laminated deposition but are not so efficient as ice. In the Precambrian, however, plants did not yet exist, so ice rafting is confidently inferred.

Another sedimentary facies indicative of glaciation consists of thick deposits of massive diamictite containing widely dispersed stones. The stones may have been emplaced either by rafting or by downslope mixing. If some of these stones are glacially faceted and straited, it is especially reassuring for the glacial interpretation. Such deposits accumulated around the margins of Antarctica in the Ross and Weddell Seas during the late Cenozoic. Here glacial detritus is deposited in water deepening irregularly toward the deep sea and in part beneath ice shelves extending seaward from the eroding land (Kurtz and Anderson, 1979). Similar facies are identified in Late Paleozoic strata in the Falkland Islands, in the southern Cape Ranges of South Africa, in the Sentinel Range of Antarctica, and in Precambrian strata of the western United States (Crowell and Frakes, 1975; Blick, 1976).

Features preserved within the sedimentary strata range from those features clearly indicative of glaciation where erosion and deposition occur directly beneath glaciers in contrast to features at great distance from the glaciers themselves. Near-glacier features such as eskers, kettle holes, englacial and subglacial stream deposits, and patterned ground formed by seasonal freezing and thawing in terrains marginal to glaciers, for example, are identified in some ancient rocks. Some features, however, are difficult to interpret. Twisted and folded sand lenses within massive diamictite may have formed either by the bulldozing of an advancing glacier or by sliding down submarine slopes where nearby glaciers played no part. Continuing research on the modern and Pleistocene glacial and periglacial environments will help in characterizing ancient environments.

Studies of the Late Cenozoic Ice Age, both of the land record of the advance and retreat of ice sheets and of the distribution of ice-rafted debris in deep-sea cores, provide helpful clues in interpreting the more ancient rock record. Ice-rafted dropstones and debris have been recovered from the middle latitudes of all oceans. The conclusion follows that currents can carry icebergs great distances. Debris now dropped from melting icebergs on the Grand Banks south of Newfoundland, for example, have drifted in currents mainly from west Greenland glaciers. The nearest land (Newfoundland and

Nova Scotia) is not the source because at present these regions are free from ice caps. Ancient strata containing glacial dropstones demonstrate the existence of glaciers somewhere. If the beds can be dated adequately, the timing of the glaciation can be determined, but the ice source can be located with confidence only when much is known about the paleogeography and the patterns of ancient ocean currents. This confidence will depend in turn on confidence in our environmental interpretations from the strata, on our regional tectonic reconstructions, and on our ability to correlate strata over great distances so that we are dealing with relatively synchronous shreds of data. As we go back farther and farther into the past, strata containing the evidence have been eroded away or covered up, or deformed and metamorphosed, so that the older the beds the less able we are to deal with time synchroneity and the less confidence we have in our reconstructed geography.

In examining the record of ancient ice ages, it may be difficult to distinguish between continental ice sheets, ice caps, and mountain glaciers. Tectonic reconstructions may indicate that plate convergence took place near the preserved strata, and so we may infer with confidence that alpine glaciers prevailed. Such a situation exists today along the coast of Alaska, where glaciers in high mountains next to the sea debouch debris onto a narrow coastal shelf and then seaward into deep sedimentation troughs. If, on the other hand, the record lies at the margin of a shield area, and deposition took place within a sedimentary prism marginal to it, an ice sheet is more likely. This situation existed around the periphery of the Scandinavian ice caps during the Pleistocene Epoch. Clast types within the diamictites or dropstone facies may also be helpful in such interpretations. Shields yield mature sedimentary rocks from a variety of granitic, gneissic, migmatitic, and similar sialic types; arcs rising inboard on continents from a convergent plate boundary are likely to yield calc-alkaline volcanic and granitic debris to make immature sediments. The regional tectonic environment must therefore be reconstructed carefully as the glacial record is evaluated.

In working out the timing of such paleogeographic reconstructions, stratigraphic methods are first employed to find the age of the beds that contain the glacial record. For Phanerozoic beds, fossils are useful, but fossils within glacial beds are rare at best and may have been reworked from older deposits. The strata that are indisputably glacial are usually terrestrial and only locally contain marine fossils, which are most helpful for worldwide correlation. The strata lying above pavements scoured by glaciers are probably those laid down as lodgment till as the glaciers dwindled and retreated and as the sediments encroached landward across previously glaciated terrain. Their age will therefore document the ending of the glaciation rather than its maximum or its beginning. As an ice cap grows into a widespread ice sheet, the sediments brought to its margin are quickly worked over by outwash streams before reaching the sea. Sediments laid down at the same time, but at a distance within a basin where the sedimentary record is more complete, will contain little if any imprint of the glaciation. Stones and sand grains have been worn by river, beach, and marine processes so that no vestige of the glacial history may remain. Only when an ice sheet has grown so that tongues of ice reach the sea will icebergs calve to drift away in currents, to leave on melting a record within a distant sedimentary section. Sedimentary sections peripheral to glacial centers therefore need special scrutiny. In fact, since the direct evidence of continental glaciation is sparse at best, we need to employ indirect observations such as results from changes in sea level and attempt to interweave arguments from many sources, including paleomagnetism, as we reconstruct the geography during ancient ice ages (Crowell, 1978).

ICE AGES THROUGH TIME

Ice ages on Earth are first documented about 2300 m.y. ago (Ma), followed by an immensely long interval of about 1300 m.y., when, so far as we now know, none took place (Figure 6.1). Several glaciations occurred in the Late Precambrian and several during Paleozoic time until the Late Permian about 240 Ma. Continental ice sheets then disappeared and only began to wax again beginning about 25 Ma. We are still living during an ice age, but during one of the mild interglacial fluctuations within it. Continental glaciers have therefore come and gone on the Earth (Hambrey and Harland, 1981).

An early widespread glaciation between 2500 and 2100 Ma is documented in what is now southern Ontario, and the ice sheet may have reached as far as southern Wyoming (Young, 1970). In Canada, the Huronian glacial sequence includes several tillite horizons and dropstone facies and locally overlies a striated pavement. In South Africa a similar glaciation may also be recorded (Visser, 1971), but investigations have not gone far enough to demarcate paleogeography, and the dating is uncertain—somewhere between 2720 and 2200 Ma. Possible tillites are also reported from western Australia (Trendall, 1976). At present we can be reasonably sure that glaciation affected North America and perhaps other distant regions between 2700 and 2100 Ma but probably not continuously for this entire time, a span of 600 m.y., which is as long as all of Phanerozoic time. There may have been several glaciations during this long interval, and outstanding questions remain, concerning not only the nature, extent, and timing of these most ancient of recorded ice ages but also why there is no record of continental glaciation during the ensuing billion years.

In the late Precambrian, continental glaciation flourished intermittently from about 950 Ma to about 560 Ma (the beginning of the Paleozoic) (Harland and Herod, 1975; Williams, 1975). During this interval of about 400 m.y., ice sheets lay upon all the continents (with the possible exception of Antarctica) and at places waxed and waned more than once. Unfortunately, dating through either fossils or isotopic methods is uncertain so that we do not know whether distant strata are of the same age. Data now in hand, however, suggest three glaciation peaks: one about 940 Ma, another near 770 Ma, and a third near 615 Ma (Williams, 1975). Two of the three glacial culminations are recorded within the same stratal sequence in regions around the present North Atlantic; in central and southwestern Africa; and in Brazil, western North America, and Australia.

Near the end of the Ordovician Period, and lasting into the Silurian, continental glaciation left an incontrovertible imprint on Gondwana, especially in northern Africa (Beuf *et al.*, 1971). The record extends in scattered localities from northern Europe to South Africa and from the Saharan region to Bolivia and Peru (Harland, 1972; Frakes, 1979; Crowell *et al.*, 1980). This refrigeration was introduced by cold-water faunas of Early Ordovician age in northern Europe, culminated during the latest Ordovician (Ashgillian) in Saharan regions, and then reached into Andean South America during the Early Silurian (Wenlockian and part of Ludlovian). This ice age is therefore recorded over a time span of about 50 m.y. between 400 and 450 Ma. In addition, probable glacial sediments of mid-Devonian age (about 365 Ma) are reported from the Amazon Basin (Caputo *et al.*, 1972).

The Late Paleozoic Ice Age, recorded on all Gondwana continents, lasted for about 90 m.y., from 330 to 240 Ma (Crowell and Frakes, 1970, 1975; Crowell, 1978). It began in South America, culminated over much of Gondwana, and ended in Australia. On each continent the glacial record is preserved around the margins of deeply eroded sialic shields and within local basins. In Australia, six or more ice centers are suggested, but paleontological control does not allow determining whether they were all flourishing at the same time or at somewhat different times. In Antarctica, glaciers primarily lay adjacent to the now Transantarctic Mountains and also just south of the Sentinel Range. In Africa, several ice lobes reached into South Africa and Namibia from the interior, and farther north ice extended into the Congo Basin from the east and also occupied the vicinity of Gabon. On the east, lobes extended into southern Madagascar. In South America, ice tongues from Africa reached into Brazil and Uruguay and left a convincing and detailed record in the Parana Basin. Centers also accumulated over Paraguay and adjacent Bolivia, the Pampean Ranges, at several places in Argentina, and in the Falkland Islands. In Asia, ice caps bordered several of the basins in northeastern and central India and along a belt marginal on the south to the present Himalayan Ranges and were sited as far northwest as central Pakistan. In the Verkoyansk region of northeastern Siberia, mid-Permian beds may also record refrigeration. Although the record of the Late Paleozoic glaciation is widespread it is likely that ice caps came and went during the long span of 90 m.y. inasmuch as sea level was not drastically lowered, as would have been the case with one huge ice cap. Moreover, cyclic sedimentation (cyclothems) on distant continents, as in North America and Europe, suggest that slight fluctuations in sea level were caused both by the waxing and waning of ice caps and by the replacement by new ice centers as old ones melted away.

The Late Cenozoic Ice Age began slowly and locally in the Paleogene with the growth of the Antarctic Ice Cap, and by early-Miocene time, about 22 Ma, worldwide refrigeration had set in (Frakes, 1979, p. 223). Cooling continued, so that by 3 Ma sea-level glaciation is recorded in Iceland and within cores recovered from the floor of the Labrador Sea. The pattern of ice sheets around the Arctic and reaching southward, so characteristic of the Pleistocene Period, had already developed.

CAUSES OF GLACIATION

Explanations for ice ages fall into two groups: those based on extraterrestrial influences and those resulting from changes on Earth itself (Crowell and Frakes, 1970). Variations in solar radiation conceivably could cause glaciations, but as yet there is no evidence that fluctuations in solar heat output have been greater than a few percent (Jerzykiewicz and Serkowski, 1968). Heat arriving at the Earth might also be reduced if the Earth were to pass through a dust cloud, but there is no evidence for this either. Although we do not yet understand the extent of normal variations in the Sun's heat flux, we must at present look to changes on Earth for the basic causes of ancient ice ages.

Variations in the geometry and mechanics of the Earth-Sun system are now well established (Broecker, 1965; Imbrie and Imbrie, 1980). The precession of the equinoxes with a period of about 21,000 yr, changes in the obliquity of the ecliptic over 40,000 yr, and variations in the eccentricity of the orbit over about 96,000 yr introduce climatic variations that are detected in the Late Cenozoic record and largely account for the glacial and interglacial stages. Such orbital influences on the amount and pattern of distribution of heat received by the Earth have certainly operated back into the readable geologic past. They are considered here as part of the noise superimposed upon changes taking place on the Earth itself. Whether there are significant longer cycles seems doubtful at present. The record of ancient ice ages as plotted in Figure 6.1 apparently is too irregular to contain effects of some cosmic cycle.

The changing arrangement of continents has long been advocated as influential in bringing about climatic change including glaciation and with the evolution of the concepts of plate tectonics has received new attention (Crowell and Frakes, 1970). With continents in polar or subpolar sites, and otherwise appropriately located so that uplands acrete snow from year to year to build up an ice cap, glaciation ensues. Studies in paleomagnetism show that during the Paleozoic, for example, continents with a record of glaciation also occupied a high-latitude position (McElhinny, 1973). As the united supercontinent of Gondwana drifted across the south poles, glacial centers roughly followed the migrating poles. Despite incompleteness of paleomagnetic data and controversy concerning interpretation, recently published apparent wander paths (Figure 6.2) cross glacial sites in chronological order throughout the Paleozoic Era (Morel and Irving, 1978). For glaciations during the Paleozoic and Late Cenozoic, the polar hypothesis is supportable.

Between Late Permian and Late Paleogene times, however, there is no record of continental glaciation, even though Antarctica occupied a position at or near the south pole. Dur-

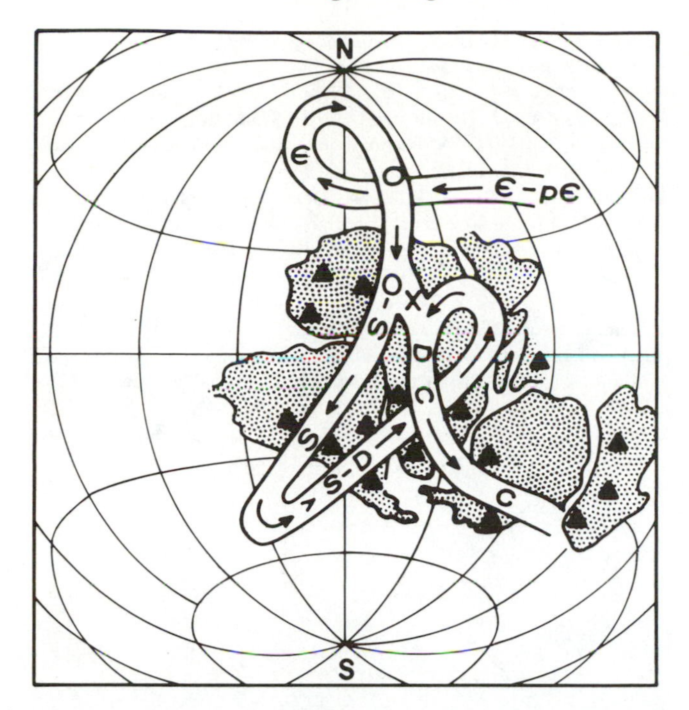

FIGURE 6.2 Apparent polar wander path for Gondwana for the Paleozoic Era. Redrawn and simplified from Morel and Irving (1978, Figure 1). Glacial strata, with ages appropriate to the apparent polar wander path, are shown with black triangles. Data on occurrence of glacial strata from many sources. X, alternative path. The close association of the glacial beds to the south polar position, for the different geologic periods (as indicated by the letters as in Figure 6.1), supports the concept that continental ice sheets developed when their sites were in near-polar positions.

ing most of this long interval the north pole lay within an Arctic Ocean or nearby and in a position not too different from that of today (Smith and Briden, 1977). Other factors therefore must have played a role in accounting for the lack of glaciation. Among these are sea-level changes. With marked raising of sea level so that continents are widely flooded, the albedo of the Earth is reduced, with the result that the Earth is warmed. Perhaps this is related to a speeding up of seafloor spreading (Pitman, 1978). In addition, during the Mesozoic Era, continents were arranged differently from that before and after so that the oceanic circulation followed different paths in both low latitudes and high latitudes. These changing continental arrangements in turn affected the heat exchange between the equatorial regions and the polar regions. With warm water in high latitudes, evaporation is facilitated so that snow on nearby continental sites is enhanced (Crowell and Frakes, 1970). When the continents are arranged so that the latitudinal flow is reduced and longitudinal flow enhanced, a different pattern of heat exchange between equatorial and polar regions follows.

The air-ocean system that controls climate on land is complex and beset with feedbacks. Many interrelated factors interweave to account for climate, and as these factors change, so does the climate. For example, the Earth's albedo will rise when clouds are plentiful, but the average temperature at the surface may increase because of trapped infrared radiation. Albedo will increase when inland seas are frozen and when the ground is snow covered in the middle latitudes.

Back in time for the last half billion years the air-ocean-continent system apparently operated dynamically as it does today, although we still do not understand the complex interweaving of the many forcing factors. Back that far the record of climate and life on Earth, including glaciation, can be explained in terms of processes operating today, in nearly the same balance, but with still ill-defined variability. In the late Precambrian, however, a body of evidence from paleomagnetism suggests that continental glaciation occurred at low latitudes (Tarling, 1974; McWilliams and McElhinny, 1980), for which as yet no satisfactory explanation has come forth. In fact, an important reason for studying the climatic history during the long interval of recorded Earth history is to document the extreme ranges in flucuations of the air-ocean-continent system. Documentation of this record will then provide mankind with better understanding of the causes of variations, including those to be expected in the future. Such investigations will aid significantly in arriving at a satisfactory theory to explain climate and especially climate change.

REFERENCES

Beuf, S., B. Biju-Duval, O. DeCharpal, P. Rognon, O. Gariel, and A. Bennacef (1971). *Les Grès du Paléozoïque Inférieur au Sahara—Sédimentation et Discontinuitiés, Évolution Structurale d'un Craton*, Inst. Fr. Pétroles-Sci. Tech. Pétroles 18, 454 pp.

Blick, N. H. (1976). Late Proterozoic glaciation: Evidence in Sheeprock Mountains, Utah, *Geol. Soc. Am. Abstr. Programs 8*, 783-784.

Broecker, W. S. (1965). Isotope geochemistry and the Pleistocene climate record, in *The Quaternary of the United States*, H. E. Wright, Jr., and D. G. Frey, eds., Princeton U. Press, Princeton, N.J., pp. 737-753.

Caputo, M. W., R. Rodrigues, and D. N. N. de Vasconcellos (1972). Nomenclatura estratigrafica da bacia do Amazonas, *An. XXVI Congr. Bras. Geol., Soc. Bras. Geol. 3*, 35-46.

Crowell, J. C. (1957). Origin of pebbly mudstones, *Geol. Soc. Am. Bull. 68*, 993-1010.

Crowell, J. C. (1978). Gondwanan glaciation, cyclothems, continental positioning, and climate change, *Am. J. Sci. 278*, 1345-1372.

Crowell, J. C., and L. A. Frakes (1970). Phanerozoic ice ages and the causes of ice ages, *Am. J. Sci. 268*, 193-224.

Crowell, J. C., and L. A. Frakes (1975). The late Paleozoic glaciation, in *Gondwana Geology*, K. S. W. Campbell, ed., Australian National U. Press, Canberra, pp. 313-331.

Crowell, J. C., A. C. Rocha-Campos, and R. Suarez-Soruco (1980). Silurian glaciation in central South America, in *Gondwana Five*, M. S. Cresswell and P. Vella, eds., Proc. Fifth International Gondwana Symposium, Wellington, New Zealand, February 1980, pp. 105-110.

Dott, R. H., Jr. (1963). Dynamics of subaqueous gravity depositional processes, *Am. Assoc. Petroleum Geol. Bull. 47*, 104-128.

Frakes, L. A. (1979). *Climates Throughout Geologic Time*, Elsevier, Amsterdam, 310 pp.

Hambrey, M. J., and W. B. Harland (1981). *Earth's pre-Pleistocene Glacial Record*, Cambridge U. Press, Cambridge, 1004 pp.

Harland, W. B. (1972). The Ordovician ice age, *Geol. Mag. 109*, 451-456.

Harland, W. B., and K. N. Herod (1975). Ice ages: Ancient and modern, *Geol. J. Spec. Issue No. 6*, A. E. Wright and F. Moseley, eds., Sell House Press, Liverpool, pp. 189-216.

Imbrie, J., and J. Z. Imbrie (1980). Modeling the climate response to orbital variations, *Science 207*, 943-953.

Jerzykiewicz, M., and K. Serkowski (1968). A search for solar variability, in *Causes of Climatic Change*, J. M. Mitchell, Jr., ed., Am. Meterol. Soc. Meterol. Monogr. 8, no. 30, pp. 142-143.

Kurtz, D. D., and J. B. Anderson (1979). Recognition and sedimentologic description of Recent debris flow deposits from the Ross and Weddell Seas, Antarctica, *J. Sedimentary Petrol. 49*, 1159-1170.

McElhinny, M. W. (1973). *Palaeomagnetism and Plate Tectonics*, Cambridge U. Press, Cambridge, 358 pp.

McWilliams, M. O., and M. W. McElhinny (1980). Late Precambrian paleomagnetism of Australia: The Adelaide geosyncline, *J. Geol. 88*, 1-26.

Morel, P., and E. Irving (1978). Tentative paleocontinental maps for the early Phanerozoic and Proterozic, *J. Geol. 86*, 535-561.

Pitman, W. C., III (1978). Relationship between eustacy and stratigraphic sequences, *Geol. Soc. Am. Bull. 89*, 1389-1403.

Smith, A. G., and J. C. Briden (1977). *Mesozoic and Cenozoic Paleocontinental Maps*, Cambridge U. Press, Cambridge, 63 pp.

Tarling, D. H. (1974). A paleomagnetic study of Eocambrian tillites in Scotland, *J. Geol. Soc. London 130*, 163-177.

Trendall, A. F. (1976). Striated and faceted boulders from the Turee Creek Formation—evidence for a possible Huronian glaciation on the Australia continent, *Geol. Surv. W. Australia, Ann. Rept.*, pp. 88-92.

Visser, J. N. J. (1971). The deposition of the Griquatown Glacial Member in the Transvaal Supergroup, *Trans. Geol. Soc. South Africa 74*, 186-199.

Williams, G. E. (1975). Late Precambrian glacial climate and the Earth's obliquity, *Geol. Mag. 112*, 441-544.

Young, G. M. (1970). An extensive early Proterozoic glaciation in North America? *Palaeogeogr. Palaeoclimatol. Palaeocol. 7*, 85-101.

Ocean Circulation, Plate Tectonics, and Climate

7

GARRETT W. BRASS, E. SALTZMAN, J. L. SLOAN II, *and* J. R. SOUTHAM
Rosenstiel School of Marine and Atmospheric Science

WILLIAM W. HAY
Joint Oceanographic Institutions, Inc.

W. T. HOLSER
University of Oregon

W. H. PETERSON
Cooperative Institute for Marine and Atmospheric Studies

INTRODUCTION

Studies of climate frequently involve identifying a plausible forcing mechanism (e.g., solar fluctuations), hypothesizing the response, and attempting to verify the hypothesis with data. This is difficult because the mechanisms are many and frequently small in amplitude, the response complex, and the data meager. For changes in oceanic mixing on the geologic time scale, a significant forcing function has been the change in area, location, and configuration of marginal seas due to plate-tectonic motions and eustatic sea level fluctuations. The hypothesized response is that these changes influence the formation of oceanic deep water and cause the thermohaline circulation of the ocean to differ substantially from one age to another. We suggest that during much of the geologic past (especially in the late Cretaceous) bottom water was produced by the sinking of warm, salty water formed by evaporation in low-latitude marginal seas rather than by the sinking of cold water formed in polar and subpolar marginal seas. The possibility of the formation of warm saline bottom water (WSBW) and its role in climate was first suggested, to our knowledge, by Chamberlin in 1906.

We begin by discussing thermohaline circulation and the formation of oceanic deep water and suggest that marginal seas play a dominant role in deep water formation as well as in determining the characteristics of deep water. In the next two sections we present a simple model that provides a conceptual framework for evaluating the potency of marginal seas as deep-water sources and for understanding the formation of WSBW. The substantial geologic evidence for the occurrence of WSBW is the topic of the fourth section. We then show that paleogeography indicates that the configuration of marginal seas was favorable for the formation of WSBW at those times in the past when there is evidence in the geologic record (presented in fourth section) for the occurrence of WSBW. Finally, we present discussions of the consequences of WSBW on oceanic and atmospheric chemistry and on the climatic consequences of WSBW.

THERMOHALINE CIRCULATION AND DEEP-WATER FORMATION

Thermohaline circulation is generally taken to be that circulation driven by density differences imposed at the ocean surface by interaction with the atmosphere. The bulk of the present ocean is filled with water from only a few source regions of restricted surface areas (e.g., the Norwegian Sea, the Weddell Sea, the Mediterranean Sea, and the Red Sea-Persian Gulf). This is verified by the fact that these deep waters essentially retain characteristics of their source regions. Simple box models using both radiocarbon and stable tracers (see Broecker *et al.*, 1960; and Bolin and Stommel, 1961) yield estimates of the residence time of the deep water of about 1000 yr. If the ocean is assumed to be in a quasi-steady state (i.e., changes, if any, occur on time scales much greater than 1000 yr), the sinking of deep water in a few small regions must be compensated for by rather slow upwelling in the bulk of the ocean. If this upwelling were uniformly distributed (which it almost certainly is not) the upwelling velocity, based on the 1000-yr residence time, would be on the order of a few meters per year. In the present-day ocean the upwelling of dense, cold, deep water must be compensated for by a downward diffusive flux of negative buoyancy (heat in the present-day ocean) if the ocean is in a steady state.

Because of the small spatial scales and temporal variability of convective processes such as deep-water formation, there are few direct observations of deep-water formation. Furthermore, the small magnitude of the upwelling velocity precludes its direct observation. Finally, the details of oceanic diffusive processes are poorly known; therefore, most knowledge of thermohaline circulation and deep-water formation is based on indirect observation (e.g., the radiocarbon models mentioned above) and theoretical considerations. Because of these problems, knowledge of present-day thermohaline circulation is rudimentary compared with knowledge of the wind-driven circulation.

Rossby (1965) suggested that because the convective buoyancy transfer processes are so much more efficient than the diffusive processes, convective deep-water formation must be confined to small regions in order to maintain steady-state conditions. Observations indicate (see Gordon, 1975) that most deep water is formed in marginal seas or over continental shelves. Apparently, the trapping of water allows it to undergo intensive interaction with the atmosphere and thereby create a substantial buoyancy deficit. This suggests that accidents of geography may determine the sites of deep-water formation.

A SIMPLE PLUME MODEL

Models of thermohaline circulation have tended to treat the formation of deep water as a tacit assumption or as a response to global-scale forcing. In light of the apparent importance of marginal seas to deep-water formation, one of us (Peterson, 1979) has formulated a simple steady-state thermohaline convection model driven by imposed buoyancy sources. Because so little is known about thermohaline circulation, a simple horizontally averaged model exploring how buoyancy sources determine the stratification and vertical circulation seemed appropriate. The essential assumption of the model is that dense water sources have a sufficient buoyancy deficit to drive turbulent plumes that entrain water from the ocean's interior, thus increasing the plume's volume transport and decreasing their density as they penetrate. The present-day Mediterranean outflows lend credence to the entrainment assumption. The density of the outflow at the Straits of Gibraltar is greater than any density in the Atlantic Ocean (Wüst, 1961), but the Mediterranean outflow terminates at about 1200 m and spreads throughout virtually the entire North Atlantic as an intermediate salinity maximum. If the Mediterranean water were not diluted by entrainment it would sink to the bottom of the Atlantic Ocean.

The model is an extension of the pioneering work of Baines and Turner (1969) and can be envisioned as the filling of a large box from one or more steady buoyancy sources. The buoyancy sources drive turbulent plumes that interact with a laterally well-mixed interior by turbulent entrainment of non-turbulent interior fluid into the plume. The interior equation is a simple vertical balance of advection and diffusion of buoyancy. This interior balance is the same one much used by some geochemists (e.g., Craig, 1969), with the important exception that the vertical velocity is determined by the source strength and the plume dynamics.

The buoyancy flux is the product of the volume flux and the density difference between the plume and the interior. A buoyancy source is characterized by its initial buoyancy flux (F_0), defined as follows:

$$F_0 = g\frac{(\rho_e - \rho)}{\rho_0}Q,$$

where Q is the volume flux, g is the acceleration due to gravity, ρ is the plume density, ρ_e is the environmental density, and ρ_0 is a reference density (a constant).

In the case of multiple plumes it is the one with the greater initial buoyancy flux not the one with the greater initial density that penetrates to the bottom and controls the stratification. The buoyancy fluxes of plumes with lesser F_0 vanish at intermediate levels, causing them to cease sinking and spread out horizontally. The model was extended to include sloping side walls and wall drag; these additions modify the details of plume termination but do not affect the qualitative results of the model.

Experiments with a two-plume model revealed some interesting behavior when the same total buoyancy flux was distributed differently between the two plumes. When the initial source strengths were between 10 and 90 percent apart, the pycnocline depth and depth of termination of the weaker plume were nearly the same. If the initial source strengths were nearly the same, a small perturbation could result in a dramatic change in the stratification and characteristics of the bottom water. We suggest below that this type of behavior may be reflected in the isotope record. Furthermore, it may well be that this type of flip-flop behavior has ramifications for climate stability (see last section).

FORMATION OF WARM SALINE BOTTOM WATER

Evidence from the measurement of oxygen isotopes in the tests of benthic foraminifera indicates that in the past deep water has been as warm as 15°C (see next section). Peterson's (1979; preceding section) model provides a conceptual framework for evaluating the relative potential of various marginal seas for the production of deep water. We suggest that at some time in the past, the strongest initial buoyancy flux, and therefore the bottom water, originated from a marginal basin subject to high evaporation rates and where the density deficit was caused by high salinity rather than low temperature. It is as if the present-day Mediterranean Sea became much larger or the evaporation rate increased such that its buoyancy flux became greater than that from any of the polar sources. According to our preliminary estimates, the present-day polar deep-water sources have about four times the initial buoyancy flux of the present-day Mediterranean. It is interesting to note that within the Mediterranean itself, the saltiest summer water is not sufficiently dense to sink; but winter cooling increases the density still more, and the resulting deep water has a temperature near 13°C, 10° cooler than typical summer surface temperatures (Lacombe and Tchernia, 1972).

Weyl (1968) noted that cold deep-water formation would be inhibited by a slight lowering of surface salinity. Freshening of the surface layer in the regions of cold deep-water formation would cause bottom-water temperatures to increase by drastically reducing the buoyancy flux from polar sources. Weyl (1968) suggested perturbations in the water-vapor transport between the Atlantic and Pacific Oceans as a mechanism for lowering the surface salinity. It has been shown by Lazier (1973) that in the middle 1960's a decrease in the salinity of the outflow from the Arctic Ocean caused a freshening of the surface layer of the Labrador Sea, which in turn caused about a five year hiatus in the usual wintertime deep convection.

Another mechanism for freshening the surface layers may be removal of salt in basins where evaporites are being precipitated.

ISOTOPIC EVIDENCE FOR THE OCCURRENCE OF WARM SALINE BOTTOM WATER

Measurement of oxygen isotopes on the tests of benthic foraminifera from deep sea cores (see Chapter 18) indicates that the temperature of bottom water has decreased from approximately 15°C at the end of the Cretaceous to 3°C at present (see Figure 7.1). Superimposed on the generally decreasing isotopic temperature trend is a series of step changes. The abruptness of these changes may be due to hiatuses in sedimentation, or they may reflect a series of shorter time-scale transitions in bottom-water temperature. If these stepwise changes are eventually confirmed, Peterson's competing plume model suggests an explanation. Deep water is produced by multiple sources, both warm and cold. The temperature and other characteristics of the bottom water are determined by the deep-water source area that produces the largest buoyancy flux. These source strengths have varied through time as a consequence of changes in areas, location, and configurations of marginal seas owing to plate motions and eustatic sea-level changes. In Cretaceous times, warm water sources were dominant over cold-water sources, whereas today cold bottom water is produced at high latitudes. This change required at least one transition from a regime dominated by WSBW to one dominated by cold polar bottom water. Because the plume termination depths are sensitive to small differences in source strengths, several transitions from one regime to the other might be expected as a result of variations in buoyancy fluxes from competing deep water sources during the time when the source strengths were nearly the same.

PALEOGEOGRAPHY OF EPICONTINENTAL SEAS

A simple box model of an evaporative basin shows that the buoyancy flux depends only on the evaporative flux (i.e., area of basin times evaporation rate) (Peterson *et al.*, 1981; Brass *et al.*, 1982) until the onset of halite precipitation, when the buoyancy flux begins to decrease because of salt removal. An evaporative basin insufficiently concentrated to deposit salt is the most effective source of WSBW.

Epicontinental seas as sources of WSBW provide the link by which the solid Earth forces oceanic and atmospheric circulation. Eustatic sea-level fluctuations can be caused by variations in the global seafloor spreading rate (Hays and Pitman, 1973). The area of continent flooded by a given increase in sea level is a function of the global hypsography at that time. The size, and hence drainage efficiency, of the continents directly controls the shape of the hypsographic curve (Harrison *et al.*, 1981; Hay *et al.*, 1981). Thus, times of increased seafloor spreading and continental breakup generate large areas of epicontinental seas both by raising sea level and by lowering the elevation of the continents.

Epicontinental seas producing WSBW must be located within the zone of net evaporation (10-40° N and S) associated with the descending branches of the atmospheric Hadley cell circulation. The distribution of Mesozoic and Cenozoic evaporite deposits strongly suggests that this zone has remained stationary during the last 120 million years (m.y.). Figure 7.2 shows the areas of flooded continents and marginal seas in 10° latitude intervals over the last 100 m.y., as measured from the paleogeographic maps of Barron *et al.* (1981). The most dramatic change during this time interval has been the decrease in area of shallow seas in the high evaporation belt (10-40°).

Because area is one of the important factors determining the buoyancy flux from evaporative basins, the decrease in the area of evaporative marginal seas over the last 100 m.y. strongly suggests that the rate of production of WSBW has declined over the same time. The shape of the curve of marginal

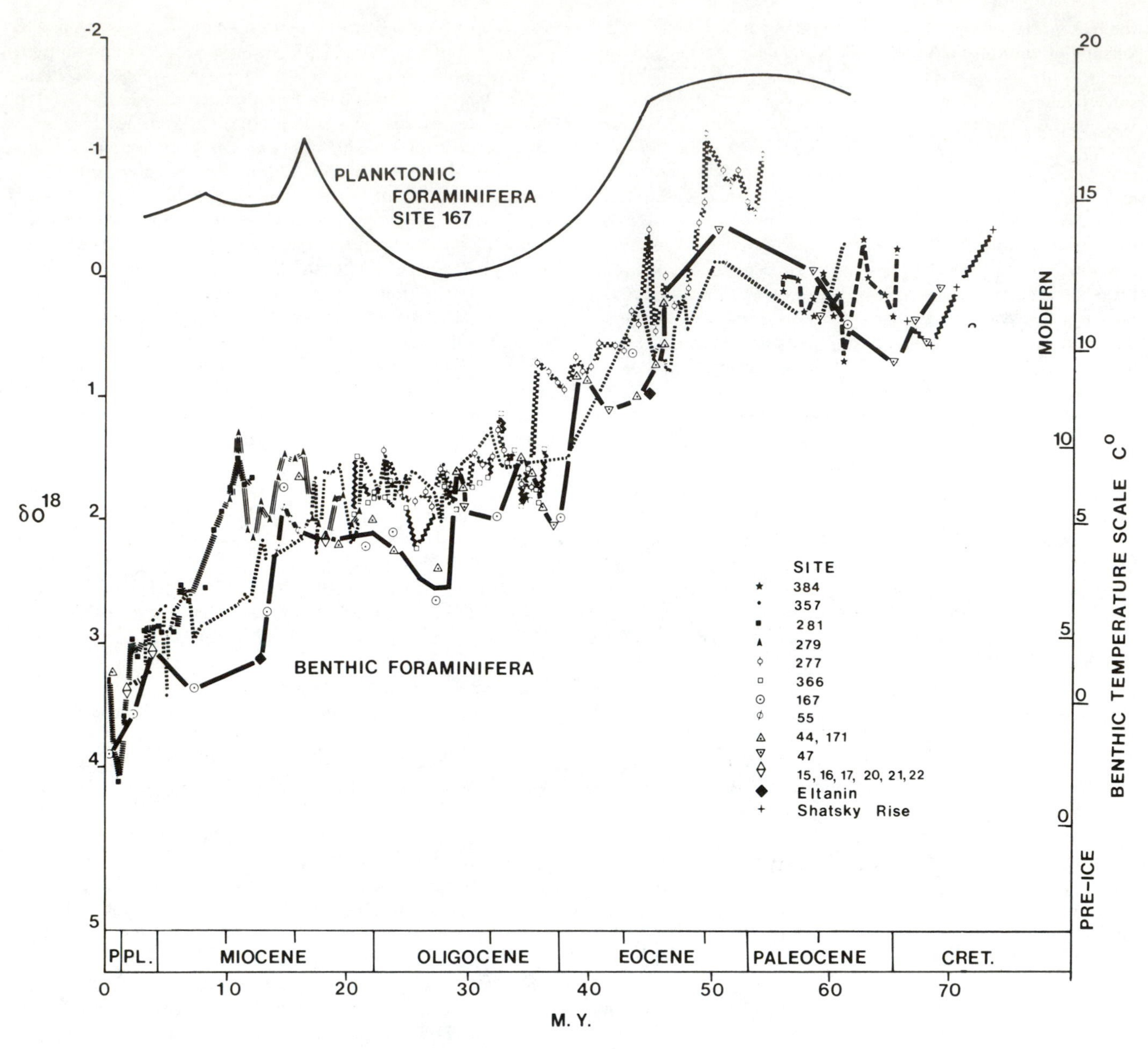

FIGURE 7.1 Oxygen isotope paleotemperatures from Savin (1977), copyright Annual Reviews, Inc.

sea area versus time is similar in shape to the oxygen isotopic temperature record curve from benthic foraminifera. The decrease in area of evaporative marginal seas and the decrease in the temperature of bottom water during the Cenozoic suggests a transition in the mode of deep-water formation in which cooling at high latitudes has played an increasingly important role.

Changes in the latitudinal distribution of epicontinental seas due to the motions of the lithospheric plates also occur. This movement may transport flooded regions into or out of the net evaporation belt. These movements appear to have had little effect on the areas producing WSBW in the Mesozoic and Cenozoic but may have been more important in earlier times.

CONSEQUENCES OF WARM SALINE BOTTOM WATER FOR OCEANIC AND ATMOSPHERIC CHEMISTRY

Our hypothesis, that plate motions and sea-level changes provide the mechanisms responsible for variations in the production of deep water, has many consequences for the chemistry of the ocean and atmosphere. The chemical state of the ocean at times when WSBW was dominant would have been very different from that which exists at present, and evidence of these differences should be present in the chemistry and isotopic composition of marine sediments. The solubilities of many gases are strongly dependent on temperature and salinity, and the increased temperature of WSBW at its source (about 15°C versus −2°C for present-day cold bottom water), where it is in contact with the atmosphere, reduces the concentration of dissolved gases such as molecular oxygen and carbon dioxide in water subsequently transported to the deep sea. As a consequence, WSBW variations should leave a signature in the accumulation rates of organic carbon and carbonate in the deep sea.

The distribution of both oxygen and carbon dioxide within the ocean is the result of the formation and destruction of organic matter, exchange with the atmosphere, and physical transport processes. As a result of respiration by animals and bacteria, oxygen is depleted in most subsurface ocean waters. The longer a water mass is isolated from the atmosphere the lower its oxygen content becomes. This fact has enabled oceanographers to use O_2 as a circulation tracer.

Anoxic conditions develop in deep water when the consumption of oxygen by organisms exceeds the rate of oxygen supply. When anoxic conditions exist, organic carbon accumulates at a higher rate. Deep-sea anoxic events have been observed in the geologic record by many authors (e.g., Degens and Stoffers, 1974; Thierstein and Berger, 1978). Many mechanisms have been proposed to explain these events, including stagnation resulting from a layer of fresh or brackish surface water or expansion of the oxygen-minimum layer by inhibited oceanic circulation. These events may also be explained within the context of our model as resulting from a decrease in ventilation of the deep ocean because of reduced oxygen solubility in the source regions without requiring any change in circulation rate. The oxygen content of WSBW may have been similar to the oxygen content of the Mediterranean deep water, which south of France and in the Adriatic contains about 220 μmoles per kilogram of O_2 (Miller *et al.*, 1970). In contrast, North Atlantic deep water near its source contains 310 μmoles of O_2 per kilogram (Bainbridge, 1980). The Mediterranean escapes anoxic conditions owing to the short residence time of its deep waters (100 yr, Lacombe and Tchernia, 1972) and its low nutrient content (Miller *et al.*, 1970), which restricts biological productivity.

The CO_2 system is of interest because changes in atmospheric CO_2 may affect climate by altering the Earth's thermal balance. Dissolved inorganic carbon is present in seawater in the form of bicarbonate and carbonate ions in addition to the dissolved gas. The ocean is a much larger reservoir of CO_2 than is the atmosphere and ultimately determines the atmospheric CO_2 content.

Significant variations in rates of accumulation of carbonates and organic carbon in the ocean basins suggest corresponding variations in atmospheric CO_2 in the past. The decrease in solubility of CO_2 with increasing temperature suggests that WSBW production leads to an increase in atmospheric CO_2. Thus, at times in the geologic past when WSBW production was large, atmospheric CO_2 levels may have been larger than at present and may have played a major role in forcing climate.

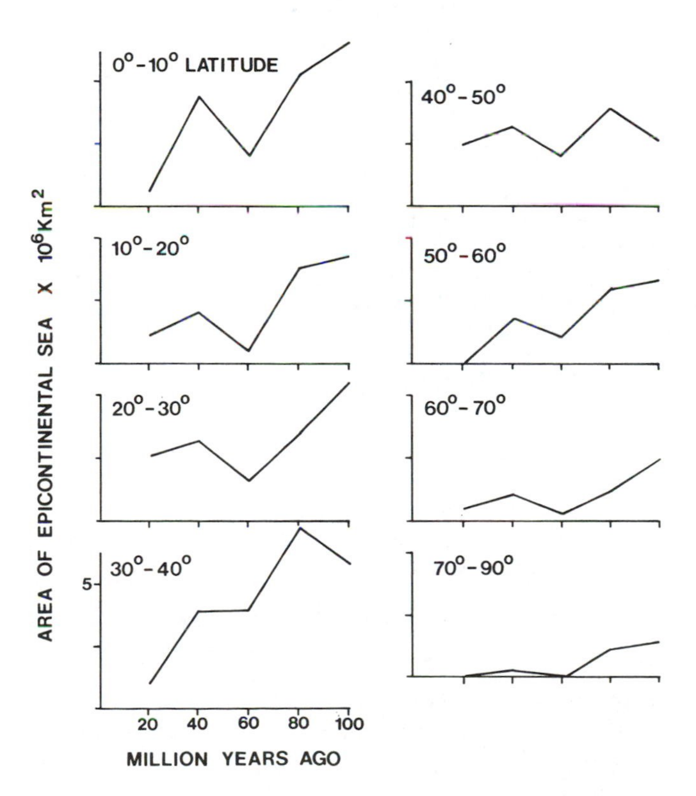

FIGURE 7.2 Area of epicontinental seas versus age, by latitude band (from the maps of Barron, 1980).

CLIMATIC CONSEQUENCES OF WARM SALINE BOTTOM WATER

The terrestrial record shows that the late Cretaceous climate was more equable (i.e., had a lesser meridional gradient) with much milder conditions at high latitudes than at present, as evidenced by the occurrence of tropical and temperate faunas and floras in higher latitudes than at present (Barron, 1980). We believe that the change to the present climate may be a consequence of the decrease in WSBW production, which is, in turn, a consequence of tectonic and eustatic activity. Two mechanisms may explain the effects of WSBW production on climate: (1) modification of heat transport via both the atmosphere and ocean and (2) changes in the transparency of the atmosphere to incoming solar and outgoing infrared radiation due to increases in the atmospheric water vapor and CO_2.

Incoming solar radiation is more intense in low than in high latitudes. The outgoing reradiation is also more intense in low latitudes; however, the difference is much less, and there is a net heating in low latitudes. This heating imbalance requires a poleward transport of heat across latitude to maintain the entire Earth's mean temperature distribution in steady state. This transport is accomplished by sensible heat transport in the atmosphere and ocean and by latent heat transport in the form of water vapor in the atmosphere. The poleward transport of sensible heat in the atmosphere for the same circulation intensity would be curtailed by an equable climate because of the reduced meridional temperature gradient. Lorenz (1967) noted that in the present-day atmosphere most of the mid-latitude poleward transport of both heat and moisture is accomplished by eddy transports. These eddies are the mid-latitude storm systems that are driven primarily by baroclinic instability, which represents the conversion of available potential energy into kinetic energy. There would be less available potential energy in an equable climate, and, therefore, this mechanism should be less effective. Manabe and Wetherald (1980) suggested that large-scale monsoonal circulations might accomplish the mid-latitude transport of moisture. Our model requires an increase in net evaporation from low latitude marginal seas to produce WSBW and implies a freshening of surface water at high latitudes to inhibit the sinking of cold water. The link between these two phenomena is higher poleward transport of water vapor in the atmosphere with its equivalent latent heat. Manabe and Wetherald studied the climatic structure in a model atmosphere forced by an increase in CO_2 content and found a striking increase in latent heat transport. One of the causes for the reduced thermal gradient was the poleward retreat of highly reflective snow cover. "Another important reason for the general reduction of the meridional temperature gradient is the large increase in the poleward transport of latent heat due to the general warming of the model atmosphere" (Manabe and Wetherald, 1980, p. 102). They concluded that the increased poleward latent heat transport may account in part for the equable Mesozoic climate (Manabe and Wetherald, 1980, p. 117). We suggest that the forcing mechanism for an increase in CO_2 abundance may be changes in the area of marginal seas in the high net evaporation regions and the production of WSBW.

It is difficult to assess the importance of oceanic sensible heat transport when the ocean is filled with WSBW. The present-day oceanic heat transport is not well known and is usually measured indirectly as a residual required to complete a heat balance (e.g., Vonder Haar and Oort, 1973). Bryan (1962) suggested that the vertical thermohaline circulation is more important in the oceanic heat budget than is the wind-driven horizontal circulation. Thus, WSBW formed in low latitudes might well accomplish some of the required poleward heat transport.

We have emphasized steady-state situations based on the notion that climate tends to reside in one of many possible quasi-steady states until it is perturbed into another quasi-steady state by some event. It is important to understand the steady-state WSBW system before studying the transitions between WSBW and cold bottom water. However, it is clear that the transitions would be accompanied by substantial perturbations in the heat budgets and would have dramatic consequences for oceanic and atmospheric chemistry. An understanding of how these transitions occur will be a valuable contribution to understanding how climate evolves.

From these discussions it is evident that the ocean of the geologic past may have had substantially different thermohaline circulation and bottom water characteristics. It has been suggested (e.g., Weyl, 1968) that the modern ocean provides a stabilizing influence on climate because of the difference in characteristic time scales of the ocean and the atmosphere (about 1000 yr versus a few months). It is also reasonable to suggest that a substantial change in the characteristics of the ocean should have substantial effect on the climate.

ACKNOWLEDGMENTS

This chapter is a contribution of the Rosenstiel School of Marine and Atmospheric Sciences. This research was supported by grants from the National Science Foundation, the Office of Naval Research, the National Oceanic and Atmospheric Administration, and the Petroleum Research Fund of the American Chemical Society.

REFERENCES

Bainbridge, A. E. (1980). *GEOSECS Atlantic Expedition, Vol. 2, Sections and Profiles*, U.S. Government Printing Office, Washington, D.C., 198 pp.

Baines, W. D., and J. S. Turner (1969). Turbulent buoyant convection from a source in a confined region, *J. Fluid Mech. 37*, 51-80.

Barron, E. J. (1980). Paleogeography and climate, 180 million years to the present, Ph.D. Dissertation, U. of Miami, Miami, Fla., 270 pp.

Barron, E. J., C. G. A. Harrison, J. L. Sloan, and W. W. Hay (1981). Paleogeography, 180 million years ago to the present, *Eclogae Geol. Helv. 74*, 443.

Bolin, B., and H. Stommel (1961). On the abyssal circulation of the world ocean-IV, *Deep Sea Res. 8*, 95-110.

Brass, G. W., J. R. Southam, and W. H. Peterson (1982). Warm saline bottom water in the ancient ocean, *Nature 296*, 620-623.

Broecker, W. S., R. Gerard, M. Ewing, and B. C. Heezen (1960). Natural radiocarbon in the Atlantic Ocean, *J. Geophys. Res. 65*, 2903-2931.

Bryan, K. (1962). Measurements of meridional heat transport by ocean circulation, *J. Geophys. Res. 67*, 3404-3414.

Chamberlin, T. C. (1906). On a possible reversal of deep-sea circulation and its influence on geologic climates, *J. Geol. 14*, 363-373.

Craig, H. (1969). Abyssal carbon and radiocarbon in the Pacific, *J. Geophys. Res. 74*, 5491-5506.

Degens, E. T., and P. Stoffers (1974). Stratified waters as a key to the past, *Nature 263*, 22-26.

Gordon, A. R. (1975). General ocean circulation, in *Numerical Models of Ocean Circulation*, proceedings of a symposium held at Durham, N.H., October 17-20, 1972. National Academy of Sciences, Washington, D.C.

Harrison, C. G. A., G. W. Brass, E. Saltzman, J. Sloan III, J. Southam, and J. M. Whiteman (1981). Sea level variations, global sedimentation rates and the hypsographic curve, *Earth Planet. Sci. Lett. 54*, 1-16.

Hay, W. W., E. J. Barron, J. L. Sloan II, and J. R. Southam (1981). Continental drift and the global pattern of sedimentation, *Geol. Rundsch. 70*, 302-315.

Hays, J. D., and W. C. Pitman (1973). Lithospheric plate motions, sea level changes and climatic and ecological consequences, *Nature 246*, 18-22.

Lacombe, H., and P. Tchernia (1972). Caractères hydrologiques et circulation des eaux en Méditerranée, in *The Mediterranean Sea*, D. J. Stanley, ed., Dowden, Hutchinson and Ross, Stroudsburg, Pa., pp. 25-36.

Lazier, J. R. N. (1973). The renewal of Labrador Sea water, *Deep Sea Res. 20*, 341-353.

Lorenz, E. N. (1967). *The Nature and Theory of the General Circulation of the Atmosphere*, WMO No. 218, T.P. 115, World Meteorological Organization, Geneva.

Manabe, S., and R. T. Wetherald (1980). On the distribution of climate changes resulting from an increase in CO_2 content of the atmosphere, *J. Atmos. Sci. 37*, 99-118.

Miller, A. R., P. Tchernia, and H. Charnock (1970). *Mediterranean Sea Atlas*, Woods Hole Oceanographic Inst., Woods Hole, Mass., 190 pp.

Peterson, W. H. (1979). A Steady Thermohaline Convection Model, Ph.D. Dissertation, U. of Miami, Fla., 160 pp.

Peterson, W. H., G. W. Brass, and J. R. Southam (1981). The formation of warm saline bottom water in ancient oceans, *Ocean Modeling 38*, 1-7.

Rossby, H. T. (1965). On thermal convection driven by nonuniform heating from below: An experimental study, *Deep Sea Res. 22*, 853-873.

Savin, S. M. (1977). The history of the Earth's surface temperature during the past 100 million years, *Ann. Rev. Earth Planet. Sci. 5*, 319-355.

Thierstein, H. R., and W. H. Berger (1978). Injection events in ocean history, *Nature 276*, 461-466.

Vonder Haar, T. H., and A. H. Oort (1973). New estimate of annual poleward energy transport by northern hemisphere oceans, *J. Phys. Oceanogr. 3*, 169-172.

Weyl, P. K. (1968). The role of the oceans in climate change: A theory of the ice ages, *Meteorol. Monogr. 8*, 37-62.

Wüst, G. (1961). On the vertical circulation of the Mediterranean Sea, *J. Geophys. Res. 66*, 3261-3271.

The Terminal Cretaceous Extinction Event and Climatic Stability

8

HANS R. THIERSTEIN
Scripps Institution of Oceanography

WHY THE LATEST CRETACEOUS?

The most outstanding feature of late Phanerozoic climatic evolution is the gradual cooling of the world's high-latitude areas. This cooling trend started about 60 million years ago (Ma) and has been particularly well documented in changes of the oxygen isotopic composition of fossil carbonate shells (Savin, 1977). The time just prior to the onset of Cenozoic cooling may therefore be considered as the climatic state farthest removed from the present or Pleistocene one and thus may represent the youngest non-Pleistocene climatic end-member. A second major source of interest in the late Cretaceous derives from the fact that it preceded one of the most severe evolutionary breaks recorded in the Phanerozoic history of life (Russell, 1979). Because climate today is the most important single factor determining the abundance and distribution of living organisms, paleontologists have long been including climatic feedback mechanisms in their proposals on possible causes for the terminal Cretaceous mass-extinctions.

THE CRETACEOUS-TERTIARY BOUNDARY EVENT

Until quite recently, the terminal Cretaceous extinctions were a subject that was discussed almost exclusively by paleontologists. The main topic of interest was the disappearance of several major groups of organisms that had dominated Mesozoic fossil assemblages, such as the dinosaurs, most of the cephalopods, the rudists, and the inoceramids (e.g., Schindewolf, 1954), as well as the marked reduction in the diversity observed in virtually all fossil communities between the Cretaceous and the Paleocene (e.g., Newell, 1962). The discovery that a rapid evolutionary turnover was also recorded in oceanic plankton communities (Luterbacher and Premoli Silva, 1964; Bramlette, 1965) appeared to confirm the postulated global extent of the extinction event and led to a rapid diversification of ideas on possible causes, which included geochemical as well as climatic feedback mechanisms (Bramlette, 1965; Tappan, 1968; Worsley, 1971; Gartner and Keany,

1978; Thierstein and Berger, 1978; Gartner and McGuirk, 1979; Alvarez *et al.*, 1980).

Which processes are involved in these models? How compatible are the proposals with the available evidence? What are some of the crucial gaps in our knowledge?

EVOLUTION AND CHRONOLOGY

There are several questions of major concern: Are the mass-extinctions real? Are they synchronous? How rapidly did they occur? Were they selective?

The reality and suddenness of the mass-extinction has been established beyond doubt for the planktonic foraminifera and for the calcareous phytoplankton by high-resolution stratigraphy (Figure 8.1) and by the verification of continuous sedimentation provided by paleomagnetic dating (e.g., Kent, 1977). A number of apparently complete sedimentary sections have been identified by these methods and have led to an estimation of the time of evolutionary turnover of less than a few thousand years (Thierstein, in press).

Deep-water benthic foraminifera show no evolutionary change across the Cretaceous-Tertiary boundary (Beckmann, 1960). However, it appears to be real wherever it could be well documented (e.g., Birkelund, 1979; Hakansson and Thomsen, 1979). Gradual evolutionary changes, mainly expressed as declining diversities through the latest Cretaceous, have been related to a gradual deterioration of unspecified nature in neritic environments (Kauffman, 1979), which could be related to local or global sea-level changes. Such changes may be real, but their relationship to the extinction event, if any, is obscure.

Claims of a gradual late Cretaceous evolutionary decline of the terrestrial large vertebrate communities in North America have been shown to be unsubstantiated when sample sizes are taken into account (Russell, 1979). Recent attempts to correlate the youngest fossilized dinosaur remains with the paleomagnetic stratigraphy have either been marred with controversies over local correlation (Butler *et al.*, 1977; Alvarez and Vann, 1979) or were based on rocks with only marginally preserved remanent magnetization (Lerbekmo *et al.*, 1979; Danis *et al.*, in press). There is evidence for time transgressiveness of the dinosaur extinctions relative to the rapidly diversifying mammalian faunas, with some late reptilian survivors reported from the Paleocene of southern China (R. A. Sloan, U. of Miami, personal communication, 1979). Whether the evolution of mammals or the extinction of the dinosaurs are time transgressive and predate or postdate the plankton extinctions is obviously one of the more important issues, which will eventually have to be resolved by magnetic stratigraphy.

The severity of the change in the terrestrial vegetation is also controversial (Krassilov, 1978). Some of the best-documented data, however, appear to indicate at least a moderate change, representing climatic cooling and an increase in seasonality (Russell, 1979).

There is growing evidence that the extinctions were selective. A geologically instantaneous extinction of virtually the entire late Cretaceous oceanic calcareous phytoplankton has been well documented (Figure 8.1); this was followed by blooms of neritic taxa such as *Braarudosphaera* and the dino-

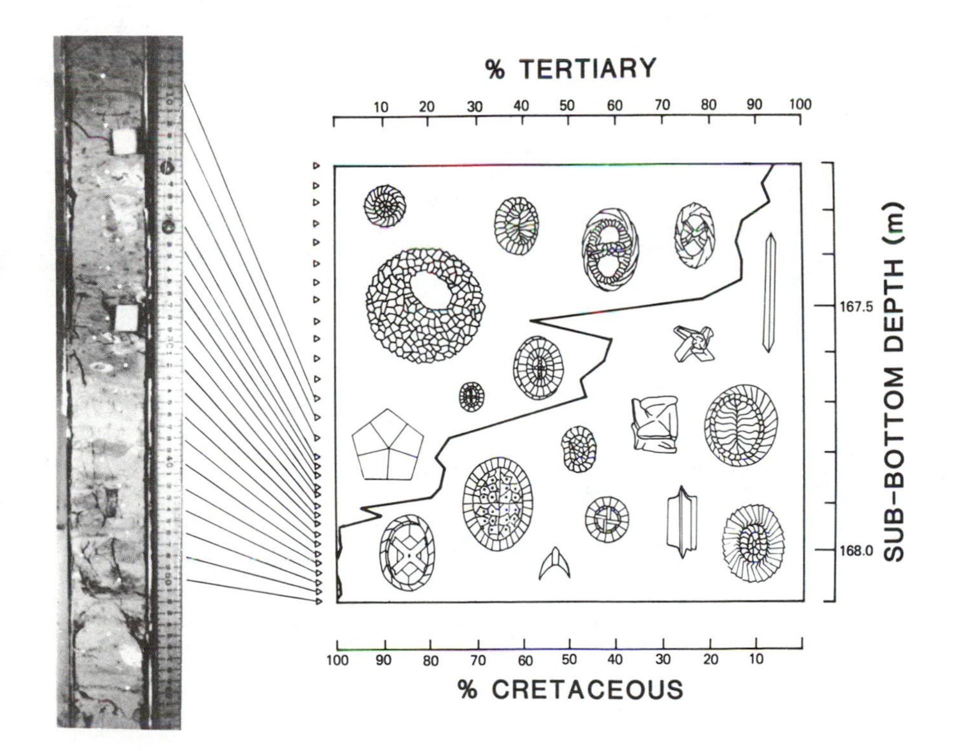

FIGURE 8.1 Replacement of the late Cretaceous by early Tertiary calcareous nannofossils within 70 cm of sediment thickness at DSDP Site 384, western North Atlantic (after Thierstein and Okada, 1979).

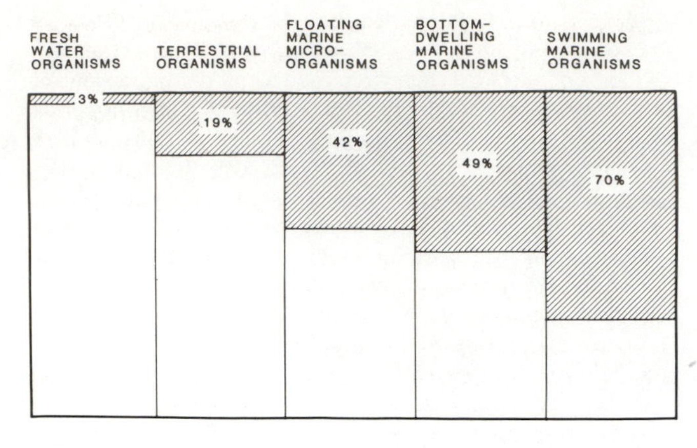

FIGURE 8.2 Terminal Cretaceous reduction in generic diversity of major fossil groups in different environments (after Russell, 1979).

flagellate cyst *Thoracosphaera*. The Maastrichtian-Paleocene reduction in the number of genera in different environments was clearly most severe in the oceanic realm and was less significant in terrestrial and particularly lacustrine environments (Figure 8.2). A reduction in generic diversity, however, must be considered a rather poor indicator of evolutionary change, as it may include taxonomic and sampling bias and possibly conceal a much larger evolutionary change at the specific level. For example, the ratio of Maastrichtian to Paleocene calcareous nannoplankton genera in the well-preserved sequence at Deep Sea Drilling Project (DSDP) Site 384 (Okada and Thierstein, 1979) is 30 : 24, a reduction of only 20 percent, although the assemblages are composed of entirely different species. More detailed data on the changes at the specific level for most groups of organisms are clearly required.

TECTONICS AND SEA LEVEL

There is widespread evidence for sea-level changes close to the Cretaceous-Tertiary boundary. However, little agreement appears to exist on their timing and direction in various parts of the world (Russell, 1979). Changing water depths certainly would exert a strong control on the local stratigraphic succession of fossil assemblages in shallow-water environments (Kauffman, 1979). The rates of pre-Pleistocene eustatic sea-level changes, estimated by Pitman (1978) to be less than 1 cm per 1000 yr (excluding buildup and melting of continental ice) are considered too slow to be solely responsible for the quite sudden extinction events; such sea-level changes may have been instrumental, however, in triggering positive climatic feedback mechanisms or the isolation of ocean basins and their reconnection with the world oceans.

HISTORY OF CARBONATE DEPOSITION

Bramlette's (1965) proposal of nutrient starvation in oceanic surface waters due to low erosion rates caused by tectonic quiescence was the first mechanism proposed that involved changes in the geochemical cycling of elements. Subsequently, Tappan (1968) pointed out that such a lowering of oceanic productivity may have resulted in additional environmental stress by significantly altering the oxygen-carbon dioxide balance in oceans and atmosphere. After the first four legs of the DSDP had failed to recover core representing a continuous Cretaceous-Tertiary transition, Worsley (1971) elaborated on these ideas and proposed that a late Cretaceous progressive shallowing of the carbonate compensation depth (CCD) into the photic zone might have affected the calcareous plankton directly. It has since been shown that carbonate deposition in at least some deep ocean basins was continuous across the Cretaceous-Tertiary boundary and that preservation of carbonate microfossils even improves between late Maastrichtian and earliest Danian (Thierstein, in press).

CLAY MINERALOGY

There are few data available on mineralogical changes in the carbonates and deep-sea clays across the Cretaceous-Tertiary boundary. Christensen *et al.* (1973) reported that increases in detrital quartz, silt, and trace-element concentrations in the boundary clay (Fiskeler) in Denmark were correlated with a decrease in carbonate contents, suggesting deposition of locally derived detritus in a starved, anoxic basin. Chamley and Robert (1979) in a study of DSDP sites in the Atlantic Ocean found that the stratigraphically and geographically variable supply of dominantly primary detrital clays through the late Cretaceous was suddenly terminated in the Paleocene. The trend, they concluded, was inconsistent with a latest Cretaceous cooling, which had been proposed based on oxygen isotopic data.

TRACE-ELEMENT GEOCHEMISTRY

The recent discovery of a strong increase in the concentration of iridium in the Cretaceous-Tertiary boundary clays (Alvarez *et al.*, 1980; Ganapathy, 1980; Smit and Hertogen, 1980) lends strong support to Urey's (1973) proposal of a meteorite impact as cause for the observed mass-extinctions. Noble element concentrations in differentiated crustal material are much lower than in undifferentiated extraterrestrial materials: measured iridium concentrations in Holocene deep-sea sediments average 0.3 part per billion (ppb); in manganese nodules, owing to low bulk accumulation rates, they average 8.5 ppb; and in iron meteorites they go up to 500 ppb (Crocket *et al.*, 1973).

Alvarez *et al.* (1980) explained the observed increase in iridium in the noncarbonate fraction from an average of 0.3 ppb in the uppermost Cretaceous limestones at Gubbio (Italy) to a maximum of 8.4 ppb in the boundary clay as a consequence of an asteroid impact. With isotopic data they ruled out the possibility of a supernova or other galactic cause for the extinction event (Russell, 1979; Napier and Clube, 1979) and argued convincingly for a solar-system source of their impacting object. The mass-extinction in their scenario would be caused by a thick veil of dust thrown up into the stratosphere by the impact, which then, through shielding of sunlight for a few years, would inhibit photosynthesis. There are several aspects of this proposal, however, that appear to require further analysis and scrutiny (see also Chapter 5). Light attenuation, for instance, would have to exceed 99 percent for months to prohibit survival of calcareous phytoplankton (Blankley, 1971). To achieve this attenuation, mass concentrations of about 500 $\mu g\ m^{-3}$ over a stratospheric dust layer of 10-km thickness would have to be maintained (Cadle and Grams, 1975), assuming average grain-size distributions similar to those observed in the stratosphere during recent large volcanic explosions (e.g., Mossop, 1964). Additional uncertainties revolve around the possibility of noble-element enrichment processes within the global geochemical cycle, such as in the biosphere, in soils, or in hard grounds.

Excessive heat production during a meteorite impact and cyanide poisoning from a cometary collision have also been proposed recently (Emiliani, 1980; Hsü, 1980).

Future studies of other boundary sections, of the geochemical cycle of noble elements, and of rocks associated with known meteorite impacts may well yield conclusive evidence for the proposed model.

STABLE ISOTOPES

Some of the most important information known about climates in the late Mesozoic is derived from comparatively few studies on the oxygen isotopic composition of fossil carbonate shells. These studies were concentrated on an evaluation of major stratigraphic trends and paleolatitudinal gradients (Frakes, 1979). Recent results show that oxygen isotopic gradients preserved in late Cretaceous microfossils were very weak, both vertically and latitudinally, when compared with those of the Holocene (Table 8.1). The deepest benthic foraminifera measurements available for the Cretaceous, however, are from a paleodepth of only 2.5 km. Truly abyssal waters, representative of polar winter temperatures, may well have been considerably colder than the 7 to 10°C measured at intermediate depths. The possibility of lowered temperature gradients suggests a less-vigorous deep-water circulation than observed today, which may well have led to an amplification of salinity gradients in surface waters. The present-day range of average annual salinity anomalies from the expected latitudinal values is 11 ‰ between the eastern Mediterranean, a significant deep-water contributor, and the Panama Basin, which is unaffected by river runoff (Dietrich *et al.*, 1975). This difference is equivalent to a range in the oxygen isotopic composition of about 6.5 ‰ (Craig and Gordon, 1965). Because oceanic circulation in the Cretaceous was probably less intense than today's, regional variation can be expected in the isotopic composition of oceanic surface waters, as evaporation and precipitation were at least as large as at present. In particular, the latitudinal surface salinity gradients may have been considerably larger than today's. These gradients could have been maintained by increased moisture transport from low to high latitudes, which at the same time would have provided an efficient heat-transport mechanism. Recent sensitivity studies of global climate models suggest an increase in latitudinal moisture transport with increasing atmospheric CO_2 content (Manabe and Stouffer, 1980). Higher atmospheric CO_2 contents may have been caused by the lowered solubility of gases in warmer ocean waters (Plass, 1972).

The possibility of considerable latitudinal salinity fractionation of oceanic surface waters in the late Cretaceous is a basic premise of the brackish-water injection model (Gartner and Keany, 1978). Preliminary oxygen isotope data from the coccolith fraction at DSDP Site 356 (South Atlantic) appear to support a sudden freshening of oceanic surface waters at the Cretaceous-Tertiary boundary (Thierstein and Berger, 1978). The oxygen isotopic shift, however, could not be confirmed at other sites or in other fossil groups. In addition, the original biostratigraphic evidence used by Gartner and Keany (1978) has been reinterpreted as a submarine slump, further weakening the brackish-water injection hypothesis (Thierstein, *in press*).

Recent studies on the carbon isotopic variability of the dissolved carbon dioxide in ocean water as well as in Recent and Neogene carbonate shells indicate that carbon isotopes may be used to trace fertility patterns (Goodney *et al.*, 1980) as well as major changes in the partitioning and cycling of carbon between the main carbon reservoirs. Changing flux rates between the reservoirs leave their traces in the carbonate skeletons of the transport vehicles. Such major changes have been documented in the middle Cretaceous (anoxic events), the Cretaceous-Tertiary boundary (collapse of the oceanic plankton community), and the late Miocene (beginning of Messinian salinity crisis) (Thierstein and Berger, 1978; Scholle and Arthur, 1980; Vincent *et al.*, 1980).

A virtually unexplored field of great paleoclimatic potential is the study of seasonal variability preserved in the stable isotopic composition of carbonate shells of macroinvertebrates.

TABLE 8.1 Oxygen Isotopic Temperature Gradients Preserved in Holocene and Latest Cretaceous Carbonate Microfossils

	Holocene	Latest Cretaceous
Planktonic forams (vertical, 0-200 m)	9°C[a]	5°C[b]
Planktonic/benthic forams (vertical, 0-2500 m)	21°C[c]	5°C[d]
Nannofossils (0-55° paleolatitude)	25°C[e]	13°C[b]

[a]From Berger *et al.* (1978).
[b]From unpublished data by author.
[c]From Vincent *et al.* (in press).
[d]From Savin (1977).
[e]From Goodney *et al.* (1980).

THE ENIGMA OF THE WARM AND STABLE CLIMATES IN THE LATE CRETACEOUS

The Cenozoic cooling trend in high latitudes was not gradual but proceeded in spurts and relapses as documented in the paleobotanical and oxygen isotopic evidence (Figure 8.3). The high-latitude cooling trend is generally thought to be caused by tectonic processes, such as the movements of continents into polar areas, global regression, and mountain building, interrupted by more local events, such as closure or opening of oceanic gateways, the isolation and reconnection of ocean basins, volcanism and the like, with amplification or attenuation by climatic feedback mechanisms, such as changes in albedo, in oceanic circulation, or in the carbon cycle. High-resolution stratigraphy may provide clues to these climatic feedback mechanisms and to the stability of the various climatic states that the Earth has experienced in glacials and interglacials, in the warm and stable late Cretaceous, and during the mid-Cretaceous period of widespread deep-water anoxia.

What were the conditions that made the late Cretaceous climates so stable? Was it climatic stability in itself, which, when disrupted, helped to produce the unprecedented biotic extinctions? The enigma of the properties of the late Cretaceous climatic system may be just as important as that of the disturbance that led to the extinctions. How different were late Cretaceous climates from Pleistocene ones?

From a late Phanerozoic viewpoint, today's oceans are characterized by their small and highly dispersed areal coverage, by their pronounced temperature stratification into a thin, warm surface layer and a large cold deep layer, and by vigorous meridional surface circulation, leading to a strong geographical separation of fertility. The deep-water reservoir is mainly replenished in the cold polar areas, which are characterized by high surface albedo. Net oceanic sensible heat transport is limited to low latitude surface waters (Figure 8.4). A major portion of the poleward heat transport occurs as sensible heat in the atmosphere through the upper part of the subtropical Hadley cells and across the polar front. The latent heat transport (heat stored during evaporation and released during precipitation) is closely tied to the atmospheric circulation patterns and is strongly mirrored in the average latitudinal surface salinities in tropical and subtropical areas (Figure 8.4).

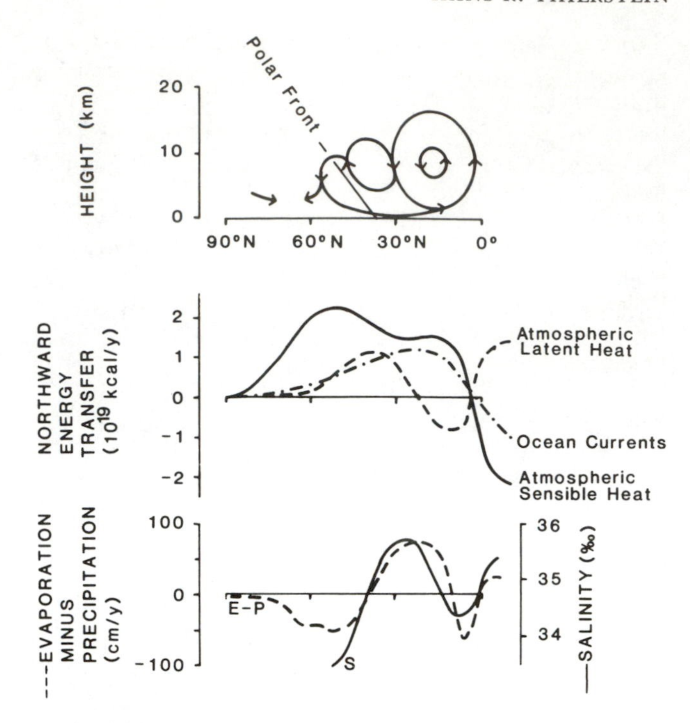

FIGURE 8.4 Schematic representation of vertical, meridional atmospheric circulation, annual latitudinal heat flux, evaporation/precipitation pattern, and surface salinities for the northern hemisphere (after Frakes, 1979; Sellers, 1965; Dietrich *et al.*, 1975).

In the late Cretaceous, on the other hand, the overall climatic circulation patterns must have been considerably different for the following reasons. Lower surface albedo, on a global scale, because of higher sea level and the absence of ice in high latitudes, must have resulted in a more efficient global heat absorption. If there was indeed no significant temperature decrease below 2500-m water depth or in polar latitudes, the density of oceanic waters would have been determined to a much larger degree by salinity, rather than temperature, suggesting the possibility of a deep-water circulation reversal and thus an increase of the oceanic sensible heat transport into high latitudes. High-latitude upwelling of saline, but relatively warm water, should have been countered efficiently, however, by negative feedback through latent and sensible heat loss and increase in salinity. Could the resulting density increase have been balanced by dilution with precipitation and runoff surplus? A significantly longer residence time in the deep-water reservoir compared to today appears unlikely, as

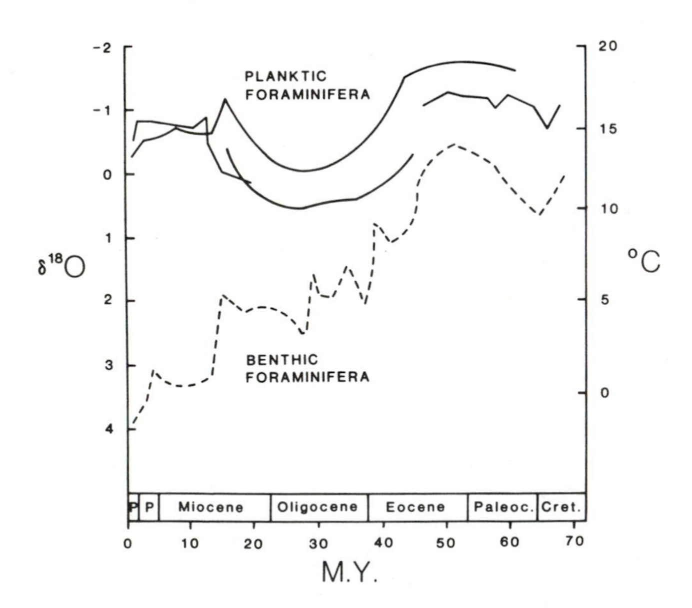

FIGURE 8.3 Isotopic paleotemperature data for latest Cretaceous and Cenozoic planktonic and benthic foraminifera, illustrating the post-Cretaceous irregular cooling trend in high latitudes (after Savin, 1977).

there is no evidence of latest Cretaceous deep-water anoxia or greatly lowered fertility, as judged from calcium carbonate sedimentation rates and sedimentology. Did high-latitude upwelling occur locally or seasonally? Increased seasonality in high latitudes should have derived from higher heat retention due to decreased albedo (no ice), particularly during summers, when the total daily radiation is higher on the poles than anywhere else. Could evidence for seasonality be derived from the annual variability in the stable isotopic composition of high-latitude macroinvertebrate shells and be used to test Wolfe's (1978) claim for a significantly lowered inclination of the Earth's rotational axis?

A number of open questions concerning late Cretaceous climates can be answered by careful study of available sediments and fossils and by using available techniques, whereas others will have to be addressed by way of analogy and inference from the longer-range Phanerozoic climatic history of the Earth.

REFERENCES

Alvarez, W., and D. W. Vann (1979). Comments and replies on "Biostratigraphy and magnetostratigraphy of Paleocene terrestrial deposits, San Juan Basin, New Mexico," *Geology 7*, 66-71.

Alvarez, L. W., W. Alvarez, F. Asaro, and H. V. Michel (1980). Extraterrestrial cause for the Cretaceous-Tertiary extinctions: Experiment and theory, *Science 208*, 1095-1108.

Beckmann, J. P. (1960). Distribution of benthonic foraminifera at the Cretaceous-Tertiary boundary of Trinidad (West Indies), *Report 21st Geol. Congr., Norden, part V*, 57-69.

Berger, W. H., J. S. Killingley, and E. Vincent (1978). Stable isotopes in deep-sea carbonates: Box core ERDC-92, west equatorial Pacific, *Oceanol. Acta 1*, 203-216.

Birkelund, T. (1979). The last Maastrichtian ammonites, in *Cretaceous/Tertiary Boundary Events Symposium. I. The Maastrichtian and Danian of Denmark*, T. Birkelund and R. G. Bromley, eds., U. of Copenhagen, Copenhagen, Denmark, pp. 51-57.

Blankley, W. F. (1971). Auxotrophic and heterotrophic growth and calcification in coccolithophorids, Ph.D. Thesis, U. of California, San Diego, 186 pp.

Bramlette, M. N. (1965). Massive extinctions in biota at the end of Mesozoic time, *Science 148*, 1696-1699.

Butler, R. F., E. J. Lindsay, L. L. Jacobs, and N. M. Johnson (1977). Magnetostratigraphy of the Cretaceous-Tertiary boundary in the San Juan Basin, New Mexico, *Nature 267*, 318-323.

Cadle, R. D., and G. W. Grams (1975). Stratospheric aerosol particles and their optical properties, *Rev. Geophys. Space Phys. 13*, 475-501.

Chamely, H., and C. Robert (1979). Late Cretaceous to early Paleogene environmental evolution expressed by the Atlantic clay sedimentation, in *Cretaceous/Tertiary Boundary Events Symposium. II. Proceedings*, W. K. Christensen and T. Birkelund, eds., U. of Copenhagen, Copenhagen, Denmark, pp. 71-77.

Christensen, L., S. Fregerslev, A. Simonsen, and J. Thiede (1973). Sedimentology and depositional environment of lower Danian fish clay from Stevns Klint, Denmark, *Bull. Geol. Soc. Denmark 22*, 193-212.

Craig, H., and L. I. Gordon (1965). Deuterium and oxygen-18 variations in the ocean and the marine atmosphere, in *Stable Isotopes in Oceanographic Studies and Paleotemperatures, Spoleto, July 26-30, 1965*, Cons. Naz. Richerche, Lab. Geol. Nucleare, Pisa, pp. 9-130.

Crocket, J. H., J. D. MacDougall, and R. C. Harriss (1973). Gold, palladium and iridium in marine sediments, *Geochim. Cosmochim. Acta 37*, 2547-2556.

Danis, G. P. L, J. A. Foster, and D. A. Russell (in press). The paleomagnetic record and the Cretaceous-Tertiary boundary in North America, *Syllogeus*.

Dietrich, G., K. Kalle, W. Krans, and G. Siedler (1975). *Allgemeine Meereskunde*, Gebr. Borntraeger, Berlin, 593 pp.

Emiliani, C. (1980). Death and renovation at the end of the Mesozoic, *EOS: Trans. Am. Geophys. Union 61*, 505-506.

Frakes, L. A. (1979). *Climates Throughout Geologic Time*, Elsevier, Amsterdam, 310 pp.

Ganapathy, R. (1980). A major meteorite impact on the earth 65 million years ago: Evidence from the Cretaceous-Tertiary boundary clay, *Science 209*, 921-923.

Gartner, S., and J. Keany (1978). The terminal Cretaceous event: A geologic problem with an oceanographic solution, *Geology 6*, 708-712.

Gartner, S., and J. P. McGuirk (1979). Terminal Cretaceous extinction: Scenario for a catastrophe, *Science 206*, 1272-1276.

Goodney, D. E., S. V. Margolis, W. C. Dudley, P. Kroopnick, and D. F. Williams (1980). Oxygen and carbon isotopes of Recent calcareous nannofossils as paleooceanographic indicators, *Mar. Micropaleontol. 5*, 31-42.

Hakansson, E., and E. Thomsen (1979). Distribution and types of bryozoan communities at the boundary in Denmark, in *Cretaceous/Tertiary Boundary Events Symposium. I. The Maastrichtian and Danian of Denmark*, T. Birkelund and R. G. Bromley, eds., U. of Copenhagen, Copenhagen, Denmark, pp. 78–91.

Hsü, K. J. (1980). Terrestrial catastrophe caused by cometary impact at the end of the Cretaceous, *Nature 285*, 201-203.

Kauffman, E. G. (1979). The ecology and biogeography of the Cretaceous-Tertiary extinction event, in *Cretaceous/Tertiary Boundary Events Symposium. II. Proceedings*, W. K. Christensen and T. Birkelund, eds., U. of Copenhagen, Copenhagen, Denmark, pp. 29-37.

Kent, D. V. (1977). An estimate of the duration of the faunal change at the Cretaceous-Tertiary boundary, *Geology 5*, 769-771.

Krassilov, V. A. (1978). Late Cretaceous gymnosperms from Sakhalin and the terminal Cretaceous event, *Paleontology 21*, 893-905.

Lerbekmo, J. F., M. E. Evans, and H. Baadsgaard (1979). Magnetostratigraphy, biostratigraphy and geochronology of Cretaceous-Tertiary boundary sediments, Red Deer Valley, *Nature 279*, 26-30.

Luterbacher, H. P., and I. Premoli Silva (1964). Biostratigrafia del limite Cretaceo-Terziario nell' Appennino Centrale, *Riv. Ital. Paleontol. LXX*, 67-128.

Manabe, S., and R. J. Stouffer (1980). Sensitivity of a global climate model an increase of CO_2 concentration in the atmosphere, *J. Geophys. Res. 85*, 5529-5554.

Mossop, S. C. (1964). Volcanic dust collected at an altitude of 20 km, *Nature 203*, 824-827.

Napier, W. M., and S. V. M. Clube (1979). A theory of terrestrial catastrophism, *Nature 282*, 455-459.

Newell, N. D. (1962). Paleontological gaps and geochronology, *J. Paleontol. 36*, 592-610.

Okada, H., and H. R. Thierstein (1979). Calcareous nannoplankton—Leg 43, Deep Sea Drilling Project, in *Initial Reports Deep Sea Drilling Project 43*, U.S. Government Printing Office, Washington, D.C., pp. 507-573.

Pitman, W. C., III (1978). Relationship between eustacy and stratigraphic sequences of passive margins, *Geol. Soc. Am. Bull. 89*, 1389-1403.

Plass, G. N. (1972). Relationship between atmospheric carbon dioxide amount and properties of the sea, *Environ. Sci. Technol. 6*, 736-740.

Russell, D. A. (1979). The enigma of the extinction of the dinosaurs, *Ann. Rev. Earth Planet. Sci. 7*, 163-182.

Savin, S. M. (1977). The history of the Earth's surface temperature during the past 100 million years, *Ann. Rev. Earth Planet. Sci. 5*, 318-355.

Schindewolf, O. H. (1954). Über die möglichen Ursachen der grossen erdgeschichtlichen Faunenschnitte, *Neues Jahrb. Geol. Palaeontol. Monatsh. 10*, 457-565.

Scholle, P. A., and M. A. Arthur (1980). Carbon isotope fluctuations in Cretaceous pelagic limestones: Potential stratigraphic and petroleum exploration tool, *Am. Assoc. Petrol. Geol. Bull. 64*, 67-87.

Sellers, W. D. (1965). *Physical Climatology*. U. of Chicago Press, Chicago, 272 pp.

Smit, J., and J. Hertogen (1980). An extraterrestrial event at the Cretaceous-Tertiary boundary, *Nature 285*, 198-200.

Tappan, H. (1968). Primary production, isotope extinctions and the atmosphere, *Palaeogeogr. Palaeoclimatol. Palaeoecol. 4*, 187-210.

Thierstein, H. R. (in press). Late Cretaceous calcereous nannoplankton and the change at the Cretaceous-Tertiary boundary, in *The Deep Sea Drilling Project: A Decade of Progress*, R. G. Douglas and E. L. Winterer, eds., Soc. Econ. Paleontol. Mineral. Spec. Publ. 30.

Thierstein, H. R., and W. H. Berger (1978). Injection events in ocean history, *Nature 276*, 461-466.

Thierstein, H. R., and H. Okada (1979). The Cretaceous/Tertiary boundary event in the North Atlantic, in *Initial Reports Deep Sea Drilling Project 43*, U.S. Government Printing Office, Washington, D.C., pp. 601-616.

Urey, H. C. (1973). Cometary collisions and geological periods, *Nature 242*, 32-33.

Vincent, E., J. S. Killingley, and W. H. Berger (1980). The magnetic Epoch-6 carbon shift: a change in the oceans' $^{13}C/^{12}C$ ratio 6.2 million years ago, *Mar. Micropaleontol. 5*, 185-203.

Vincent, E., J. S. Killingley, and W. H. Berger (in press). Stable isotope composition of benthic foraminifera from the equatorial Pacific, *Nature*.

Wolfe, J. (1978). A paleobotanical interpretation of Tertiary climates in the northern hemisphere, *Am. Sci. 66*, 694-703.

Worsley, T. R. (1971). Terminal Cretaceous events, *Nature 230*, 318-320.

9 Long-Term Climatic Oscillations Recorded in Stratigraphy

ALFRED G. FISCHER
Princeton University

INTRODUCTION

The marine stratigraphic record reveals cyclic changes of various sorts, including periodic interruptions of deposition, change in the sedimentary constituents supplied, change in faunas and floras, and change in the nature of the depositional environment. Many of these changes are too general in character and in distribution to be attributed to local causes: they seem to reflect global changes in climate and their effects on the marine system.

These phenomena are discussed here in sequence of increasing period. Bedding phenomena visible at the outcrop level appear to correspond to climatic changes induced by the Earth's orbital perturbations—in the 20,000-500,000-yr range—the same forces that drove the glacial advances and retreats of the Pleistocene. Broader phenomena that must generally be synthesized from regional or global data suggest a possible climatic cycle in the 30-36 million years (m.y.) realm. This in turn appears to ride on an extremely long cycle (not necessarily of fixed period) that brought on alternation of "icehouse" and "greenhouse" climates: in the last 700 m.y., the Earth seems to have completed two and started on a third of these cycles.

CHANGES IN THE 20,000-500,000-YEAR (MILANKOVITCH) RANGE

21,000- and 43,000-Year Cycles

Sediments when viewed at the level of a roadcut or hillside are characterized by stratification. This is generally attributed to random fluctuations in the supply of sediment to, and removal of sediment from, a given depositional site. Such processes might be expected to produce a fairly random aggregation of thicker and thinner strata, yet many sedimentary "formations" show rather striking uniformity of bedding thickness. This is particularly true of many limestones, in which thicker beds of biogenically formed carbonate alternate with thin interbeds of shale, recording a simple oscillation cycle, as recognized by Gilbert (1895, 1900), Schwarzacher (1975), Fischer (1980, 1981), and others.

Gilbert (1895) attributed rhythmic bedding in the Cretaceous of Colorado to climatic influence of the axial precession having a period of about 21,000 yr. These variations in insolation were first worked out quantitatively by Milankovitch (1941) and have been revised by Berger (1980).

The timing of such rhythms may be approached in two different ways, from "below" or from "above." In varved sequences, in which beds are composed of presumed annual laminations, the duration of a bed in years should equal the number of varves within it. The alternative is to take a radiometrically well-dated interval, such as a stage or an epoch, and to divide its length by the number of beds found within it.

The first of these methods was applied to the Green River Oil Shale (Eocene) by Bradley (1929). Bradley did not actually count the varves in a bed—he determined the mean thickness of varves from thin sections and the mean thickness of beds from measurements on outcrop and found the ratio to be 21,000 : 1. Whereas this work needs independent verification, it seems to have confirmed Gilbert's hypothesis in principle.

The second method has been applied to Cretaceous pelagic and hemipelagic limestones by Arthur (1979a) and by Fischer (1980, 1981). Various problems arise with this approach. One is that few rhythmic bedding sequences and continuous exposures span the length of a stage, so that is becomes necessary in most sequences to extrapolate. Another is that in rival time scales—Obradovich and Cobban (1975) versus Van Hinte (1976)—some stages differ by a factor of 3, so that it becomes necessary to average the results from several stages. A series of 11 Cretaceous sequences from Colorado, France, and Italy (Fischer 1980, 1981) yielded raw averages ranging from 10,000 to 87,000 yr per bed. The eight shorter ones yielded a mean of 17,125 on the Obradovich-Cobban scale and 26,375 on the Van Hinte scale, for a combined mean of 21,750. Of the remaining three, one is poorly dated and the other two seem to lie in the vicinity of 50,000 yr and might be related to the 43,000-yr cycle in obliquity (Milankovitch, 1941; Berger, 1980).

Thus the existence of the 21,000-yr rhythm—and therewith of a precessional influence on Cretaceous sedimentary regimes—appears to be moderately well established. The case for a sedimentary record of the cycle in tilt, on the other hand, is not strong except in the deep-sea record (Arthur, 1979b).

100,000-Year Cycle

Simple bedding rhythms of the type discussed above tend to occur in sets, and while there is considerable variation in the number per set, statistical averages out of any one sequence usually yield a mean number of about five (Schwarzacher, 1975). This ratio holds for the Precambrian-Cambrian boundary beds in Morocco (Monninger, 1979); for Carboniferous limestones in Ireland, Triassic limestones in the Alps, Triassic lake deposits in New Jersey, and Jurassic limestones in southern Germany (Schwarzacher, 1975); and for five of the eight Cretaceous sequences studied (Fischer, 1980, 1981). If the Cretaceous sequences cited above are of precessional origin, then the "Schwarzacher bundles," which they compose, would seem to have a timing of about 100,000 yr. Furthermore, by analogy, it appears reasonable to interpret this bedding pattern, characteristic of various parts of the Phanerozoic, as a record of precessional cycles grouped into 100,000-yr sets.

This 100,000-yr rhythm is the strongest of the glacial rhythm signals in the Pleistocene marine record (Hays *et al.*, 1976), and is attributed there to the orbital quasi-rhythm in eccentricity (Milankovitch, 1941; Berger, 1980). It is extremely tempting to consider the Schwarzacher bundles as the product of the precession coupled with eccentricity. Indeed, the precession can influence climate only by way of orbital eccentricity, so that a bundling of precessional beds into larger sets is virtually demanded by theory.

Cycles at the 500,000-Year Level

An example of long rhythms in stratal sequences is provided by the Permo-Carboniferous megacycles of Kansas (Moore *et al.*, 1951; Heckel, 1977), in which terrigenous deposits at base and top separate a marine sequence characterized by a perculiarly patterned alternation of shales and limestones. Some 25 of these megacycles characterize the 10 m.y. of Missourian-Virgilian time, a mean duration of 400,000 yr.

In various other sequences, such as the Precambrian-Cambrian boundary beds in Morocco (Monninger, 1979), the Triassic lake deposits of New Jersey (Van Houten, 1964), and the Cretaceous and Eocene in central Italy, Schwarzacher bundles are in turn grouped into sets of four to six, representing about 0.5 m.y. each.

This cycle too appears to have a match in orbital perturbations, namely in a longer cycle of eccentricity, which emerges from Berger's (1980) calculations. This cycle has recently been recognized in the Pleistocene record by Briskin and Berggren (1975).

Multiple Pathways of Expression

Whereas the sediments in which these bedding patterns have been found are mainly limestone sequences, they include different depositional regimes, in which rhythmicity is induced by different factors. In the alpine Triassic, for example, the cause is a variation in sea level, leading to repeated emergence and submergence of carbonate banks (Fischer, 1964). In the late Cretaceous of central Italy, the setting is one of deep water throughout, and the rhythms reflect a change from carbonate deposition to clay deposition—either because the carbonate supply was reduced or because of carbonate dissolution on the seafloor (Arthur and Fischer, 1977). In the Cretaceous of Kansas, and in the Mid-Cretaceous (Aptian-Albian) of Italy and of the present ocean floor, some of the rhythmicity was produced by changes in oxygen content of bottom waters: the depositional sites oscillated between aerobic and anaerobic conditions (Arthur and Fischer, 1977; Arthur, 1979b; Fischer, 1980). In the Triassic lake deposits of the Newark rift (Van Houten, 1964), the rhythmicity resulted from changes in lake level and in the chemistry of the lake. The only common denominator for all of these changes is climate—climatic fluctuation so severe as to change sea level (presumably by growth of glaciers), the chemistry and behavior of the oceans, and the salinity of lakes.

Conclusions and Problems Regarding Cyclicity at the Milankovitch Level

The data summarized above have led me to conclude that climatic oscillations driven by the Earth's orbital perturbations have not been limited to the Pleistocene but have affected the Earth's climates through Phanerozoic time—the last 600 m.y. We have barely begun to recognize their record in the sediments and are far from having adequate descriptions, let alone understanding. There are suggestions that the nature of this record has changed with time. In the Late Pleistocene record, for example, the eccentricity signal is strongest, the obliquity signal next, and the precessional signal weakest. In the Late Carboniferous cyclothems of Kansas—another glacial time—eccentricity cycles—in particular, the 400,000-yr cycle—seem again to dominate the picture. At nonglacial times the precessional signal seems the strongest. Are there definite time changes in the kinds of rhythms—signals of changes in orbital character or of changes in the Earth's response to constant signals? Is there a solar factor in addition? We do not know the answers. What the record tells us is that different parts of the Earth recorded the climatic changes in different ways, and this in turn should serve to develop some understanding of the functioning of the Earth. Puzzling, for example, are the sharp sea-level changes suggested by the record for times generally thought to have been free of polar ice. Was there perhaps mountain glaciation on scales far beyond that of today? Such questions call for further studies.

CHANGES AT THE 30-MILLION-YEAR AND 300-MILLION-YEAR LEVELS

30-Million-Year Cycle

While historical geologists since Lyell have given lip service to the principle of uniformitarianism, in which the Earth is viewed as having developed in a gradual and steady manner, a majority of stratigraphers and tectonicists, going back to Cuvier and d'Orbigny, including Chamberlin, Grabau, and Umbgrove, have been impressed with the segmentation of geologic history into episodes. Some of these changes appear to be rhythmic, and one of the rhythms represented lies at the 30-36-m.y. level.

Dorman (1968) suggested a 30-m.y. cycle in global temperatures, based on oxygen isotope analyses of Cenozoic mollusks. Damon (1971) analyzed the record of marine transgressions and regressions from the continents and concluded that Phanerozoic sea level rose and fell with a periodicity of 36 ± 11 m.y. and that this bore some correspondence to periodicities in global mountain building and in regional plutonism.

Fischer and Arthur (1977) suggested that the Mesozoic-Cenozoic part of Earth history is logically subdivided not into four periods as currently practiced but into seven, with a mean duration of 32 m.y., corresponding essentially to Grabau's (1940) seven pulses: the Triassic, Liassic, "Jurassic," Comanchean, Gulfian or "Cretaceous," Paleogene, and Neogene. Each of these corresponds to an expansion of organic diversity in the pelagic marine realm (development of polytaxy) followed by a decline to an "oligotaxic" state. This pattern appears in global counts of coexisting genera and species as well as in the structure of marine communities, in which the polytaxic state is characterized by the development of superpredators, while the crash leading to oligotaxy is accompanied by the spread of opportunistic generalists (Figure 9.1). Fischer and Arthur tentatively recognized some reflections of this cycle in marine temperature regimes (Figure 9.2), in the oxygenation of the oceans, in carbon isotope ratios, in the ups and downs of the calcite compensation depth, in the development of submarine unconformities, and in other factors.

Their overall conclusion was that oceanic structure and behavior have changed on a time scale of about 32 m.y., responding to some change in general climate: in polytaxic episodes the high latitudes were warmer and the temperature of the ocean mass as a whole was higher than during oligotaxy. However, these fluctuations ride on a much longer oscillation, which will be discussed below. During the last 100 m.y. the Earth has passed through three polytaxic episodes—that of the Late Cretaceous, that of the Eocene, and that of the Miocene—delimited by three oligotaxic ones—the Maastrichtian-Danian boundary crisis, the Oligocene crash, and the current decline. During this time we have experienced the "climatic deterioration" long recognized by the terrestrial paleobotanists (Dorf, 1970). Each episode of polytaxy has been merely a step back "up" in what has been a general "downward" trend toward colder high latitudes and colder ocean masses—a trend that finally culminated in the glacial episode in which we find ourselves now.

The history of pelagic diversity in the Paleozoic offers some support for the existence of the 30-m.y. cycle through the Paleozoic, but the precision of Paleozoic data remains marginal. For that matter, the existence of this cycle in the Mesozoic-Cenozoic is still a matter of debate; Hallam (Chapter 17), for example, finds no convincing evidence for the postulated Mid-Jurassic break, and some of the polytoxic episodes recognized by Fischer and Arthur are split by minor reductions in faunal diversity.

The general causes for the postulated 30-m.y. cycle remain uncertain. The pattern suggests that it was a minor modulation of the long (300-m.y.) greenhouse-icehouse cycle discussed below. I am therefore inclined to think that it, too, was engendered by changes (lesser ones) in atmospheric carbon dioxide pressure and that it, too, expresses imbalances between the rates at which carbon dioxide is added to the outer Earth by volcanism and withdrawn from it by weathering and sedimentation (see discussion below). Indeed, just as the long-range cycle is here attributed to first-order changes in volcanism and in sea level, so the 30-m.y. cycle seems to match shorter fluctuations in these factors (Damon, 1971).

300-Million-Year Level

In the last 700 m.y. the Earth has undergone three major episodes of glaciation, during which ice caps not only covered one or both of the polar regions but extended at times more than

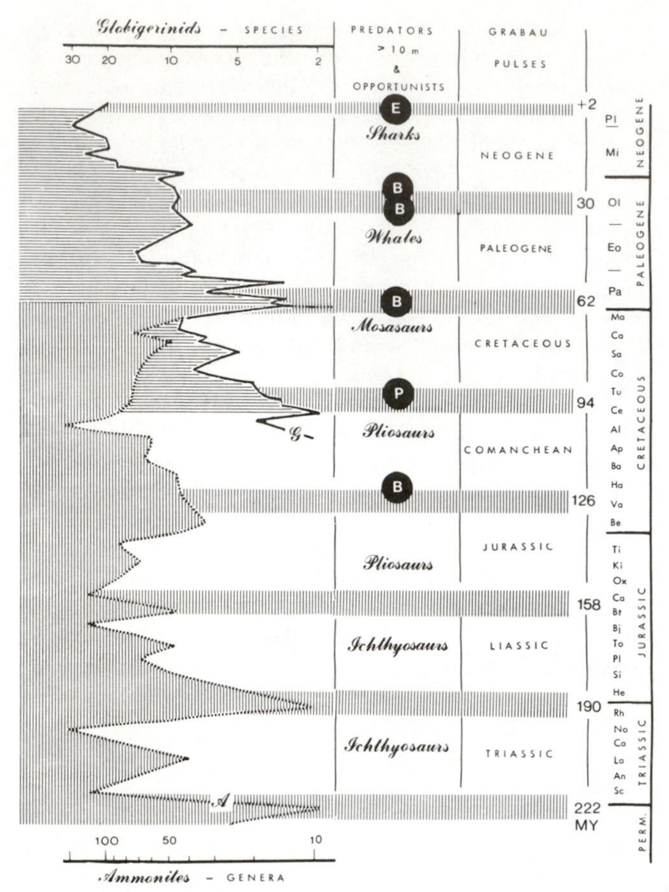

FIGURE 9.1 Pelagic diversity, superpredators, and blooms of opportunists over last 220 m.y. On left, changes in global diversity. Genera of ammonities (A) and species of planktic (globigerinacean) foraminifera (G), plotted logarithmically. Episodes of increasing diversity are separated by biotic crises of varying magnitude. Crises of moderate and high intensity recur at intervals of approximately 32 m.y. (shaded bands, defining seven cyclic episodes or pulses of diversification: polytaxy). These essentially coincide with transgressive pulses of Grabau (right). Each polytaxic pulse brought superpredators exceeding 10 m in length, a role that has been successively filled by ichthyosaurs, pliosaurs, mosasaurs, whales, and sharks, as shown in middle. Superpredators are known only from stages opposite the names. Mid-Triassic ichthyosaurs: Cymbospondylus and Shastasaurus; Toarcian ichthyosaur: Stenopterygius; Oxfordian pliosaur: Stretosaurus; Albian pliosaur: Kronosaurus; Campanian-Maastrichtian mosasaurs: Hainosaurus and others; Eocene whale: Basilosaurus; Mio-Pliocene shark: Carcharodon megalodon. Biotic crises are accompanied by local mass-occurrences of single pelagic species, rare in normal biotas. These are interpreted as blooms of opportunists and have been plotted in black circles. B, Braarudosphaera, a coccolithophorid; P, Pithonella, a problematicum; E, Ethmodiscus rex, a giant diatom. From Fischer and Arthur (1977).

halfway to the equator (see Chapter 6 for summary and literature). The times of these first-order glaciations (Figure 9.3) are Late Precambrian [about 750-650 m.y. ago (Ma)], Late Carboniferous-early Permian (340-255 Ma), and Pleistocene-Recent, having commenced about 2 Ma and stretching on into the unknown future. Another glacial episode, of lesser vigor and short duration, was associated with the end of the Ordovician, about 435 Ma.

In between, the world seems to have lacked polar ice caps. The Paleogene of the Arctic region, for example, contains a warm temperate to subtropical forest assemblage including large trees, remains of amphibians, a wide range of reptiles, and, among the mammals, horses and monkeys (Koch, 1963; West and Dawson, 1978). Also, the marine molluscan fauna of western Greenland is distinctly subtropical (Kollmann, 1979).

This paleobotanical evidence for a once very different world is corroborated by the marine record: The present ocean's warm waters are confined to a thin surficial layer in the lower

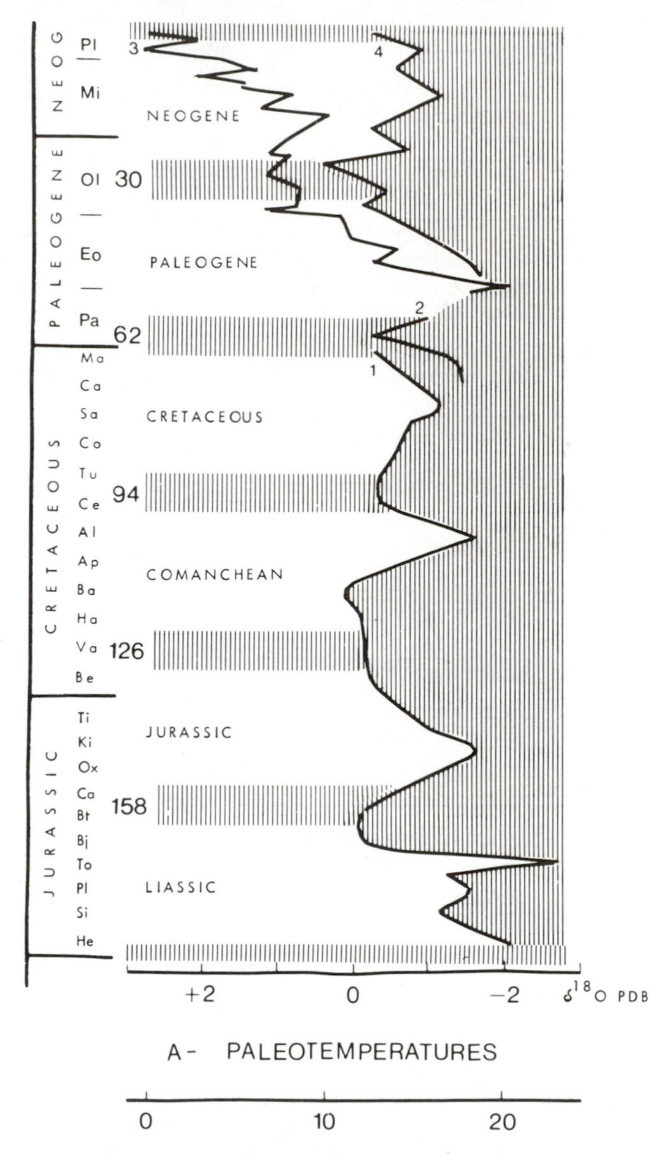

FIGURE 9.2 Paleotemperatures derived from oxygen isotope ratios in calcitic fossil skeletons, assuming constant oxygen isotope ratios in seawater. 1, belemnites, northwestern Europe, uncorrected lat. 45-55°; 2, planktonic foraminifera, South Atlantic, uncorrected lat. 30-32°; 3, planktonic foraminifera, South Pacific, uncorrected lat. 47-52°; 4, planktonic foraminifera, tropical Pacific, uncorrected lat. 7-19°. From Fischer and Arthur (1977).

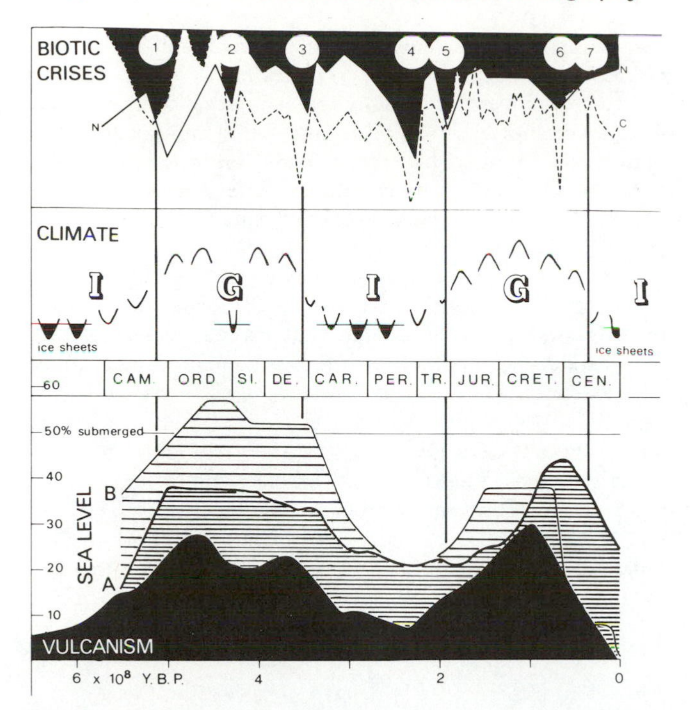

FIGURE 9.3 Relation of inferred climates to secular patterns in volcanism, sea level, and organic diversity. Volcanism: emplacement of plutons in North America, after Engel and Engel (1964). Sea level: A, first-order eustatic curve of Vail *et al.* (1977); B, compromise between North American and Russian records, constructed from Hallam (1977); the scale at left refers to this curve. Biotic record: N, Stehli *et al.*'s (1969) curve of disappearance of animal families; C, net gain-and-loss curve of Cutbill and Funnel (1967), overlap shaded. Inferred climatic states from Fischer (1981); minor oscillations (which may bring about growth of ice sheets, shaded) after Fischer and Arthur (1977). Diagram modified from Fischer (1981).

and middle latitudes. Bottom waters are close to the freezing point throughout the oceans, and the mean temperature of the oceanic water masses lies at about 3°C. While we do not have reliable measurements of paleotemperatures—oceanic or otherwise—for the ice ages of the Paleozoic and Precambrian, it seems likely that their oceans had a temperature and structure rather similar to that of today.

In contrast, studies of oxygen isotopes from Cretaceous oceanic deposits (e.g., Douglas and Savin, 1975) show that the temperatures of the bulk of the Cretaceous ocean masses lay in the vicinity of 15°C, i.e., that the mean temperature of the oceans averaged some 10° above that of the present ones. This suggests that bottom waters were not formed as today, by chilled polar surface waters mixed with a strong dash of meltwater from the polar ice caps. Several alternatives appear possible. Either lightly cooled surface waters descended in the high latitudes, to form a bottom water of simple origin, or, alternatively, bottom waters generated in the paraequatorial dry belts (horse latitudes) to form a deep warm layer, as yet not sampled. A likely compromise is that bottom waters arose from a mixture of both of these processes.

In short, it appears that the Cretaceous and Paleogene periods had a markedly different climate and ocean: tropical temperatures were much as they are today, while the temperature gradients toward the poles (and, in the ocean, toward the bottom) were very much lower. This implies a more uniform distribution of energy received from the Sun. The most appealing mechanism for this is that of a climatic "greenhouse" (Budyko, 1977; Manabe and Wetherald, 1975), in which an enrichment of atmospheric carbon dioxide inhibits radiation losses of energy into space. Temperatures in the tropical seas would not rise appreciably, because of increased evaporation, but the water content of the atmosphere would increase, and the transport of heat to the higher latitudes would occur largely as latent heat of evaporation, released in the high latitudes by heavy rainfall.

Fischer (1981) has contrasted these climatic states of the Earth as the "icehouse state" and the "greenhouse state" (Figure 9.3). I view the history of the last 700 m.y. as a passage through two great icehouse-greenhouse cycles and into the beginning of a third. Associated phenomena include first-order changes in sea level, mean ocean temperatures, and oceanic aeration, possibly linked to changes in volcanicity and plate motions, as explained below. The transitions from one state to the other appear to be punctuated by the four major biotic crises. A special explanation, in this scheme, has to be invented for the glaciation and biotic crisis at the end of Ordovician time.

Associated Phenomena

Volcanism Long-term secular variations in global volcanism are not easily apprehended. Of the various volcanic processes, those associated with the generation of the oceanic lithosphere rank largest, but their pre-Jurassic record is lost by the recycling of the oceanic crust back into the mantle. On the continents, the great andesitic volcanic piles formed at convergent plate boundaries are largely lost to erosion. The best record of long-term volcanism is probably that of the granitic plutons that form the substructure of these belts (Fischer, 1981).

If we may take the emplacement of such plutons as indicative of volcanic activity in general, and if North America is representative, then a plot of the rate of "granite" emplacement in North America, through geologic time, should serve as an index to worldwide volcanicity. Such a plot, by Engel and Engel (1964) is shown at the bottom of Figure 9.3. It is essentially bimodal, showing one broad (and bifid) peak that matches the inferred greenhouse state of the Ordovician-Devonian and another, sharper peak that matches the Mesozoic greenhouse interval.

Sea-Level Change The eustatic curves by Vail *et al.* (1977) show the fluctuations of sea level relative to the continents. While this demonstrates a lively history of sea-level oscillations, its most generalized version—Vail's first-order curve or Hallam's (1977) curve (here modified by Fischer)—coincide with the inferred succession of icehouse and greenhouse states: the three major glaciations occur within the three lows of the curve, while the greenhouse states correspond to the highs. While glaciation itself drives sea level through oscillations because of withdrawal of water from the hydrosphere into ice caps, the longer persistence of these first-order lows shows them to have been a condition for the growth of ice sheets. The latter merely contributed feedback. For the reasons why high sea levels should coincide with greenhouse states, see below.

Oceanic Aeration The ocean of our day is moderately well oxygenated throughout, so that animal life, dependent on aerobic respiration, is possible on almost all bottoms and throughout almost all of the water mass. Exceptions are encountered only in localized areas such as the Black Sea, the Cariaco Trench, and the Santa Barbara Basin and, seasonally, in certain tropical belts of upwelling. Because of this, bottom sediments are almost everywhere stripped of organic matter by scavengers and bacteria and are plowed and mixed in the process (Fischer and Arthur, 1977). Accumulation of petroleum source beds is at a minimum.

In contrast, during much of the past, black, organic-rich, finely laminated sediment was widely deposited on marine bottoms deprived of free oxygen. In my experience, widespread anaerobism of this sort peaked in the Ordovician to Devonian interval and again in the Jurassic and Cretaceous (Fischer and Arthur, 1977; Jenkyns, 1980). These times correspond to the greenhouse states. Berry and Wilde's (1979) alternative explanation, that the Paleozoic black shales are holdovers from the poorly oxygenated atmosphere of the Precambrian, fails to explain the earlier (Lower Cambrian) spread of highly oxygenated (red) fossiliferous marine sediments, as well as the recurrence of anoxia in the Mesozoic. Within these episodes, there was a waxing and waning of anoxia, on what Fischer and Arthur (1977) interpreted as a 30-m.y. cycle, as well as on the yet smaller scales of the Milankovitch cycles.

The causes for these variations in oceanic aeration are not resolved. We may think of the ocean in analogy to an organism that digests food: In anaerobic periods, the sea has indigestion. This may be brought on in one of two ways: either by a surfeit of organic matter supplied to it, which overwhelms its digestive capacity, or by a breakdown in the digestive system itself.

Fischer and Arthur (1977) ascribed mainly to the latter cause and linked aeration to the cooling of oceans, which (a) permits a given volume of water to absorb more oxygen while in contact with the atmosphere and (b) favors a vigorous marine circulation, shortening the residence time of water in the depths, between times of recharge at the surface. This explanation thus offers indirect evidence for a warming of seas between times of massive glaciation. On the other hand, the greenhouse state is likely to have increased organic production on the lands, owing to more plant growth in high latitudes, to more widespread rainfall, and to the greater availability of carbon dioxide. It seems likely that during greenhouse states the supply of organic matter from the lands to the oceans may have been greater than it is today. I have therefore come to believe that both factors worked together to promote oceanic anaerobism.

Punctuation by Biotic Crises

Two curves at the top of Figure 9.3 illustrate changes in faunal diversity revealed by the fossil record. N is Stehli *et al.*'s (1969) compilation of the disappearance of animal families; C is Cutbill and Funnel's (1967) analysis, depicting net gain and loss in invertebrate diversity. Six times of large-scale disappearance of taxa—first-order biotic crises, numbered 1-6—stand out in both curves. Of these, numbers 1, 3, and 5 coincide rather well with the boundaries between the climatic states suggested in the middle of the diagram. Number 2 coincides with the brief plunge into glaciation that occurred in the middle of the Early Paleozoic greenhouse. In a previous paper (Fischer, 1981) I sought to relate crises 4 and 6 to the climatic transitions as well, but this does not appear feasible. Both occur too soon, 4 before the breakup of Pangea, 6 before the greenhouse came to an end, as evidenced by the warm nature of the Arctic in Paleogene times (see above). Also, there is now strong evidence for an extraterrestrial origin of crisis 6 (Alvarez *et al.*, 1980). The transition to the Late Cenozoic icehouse state is marked, instead, by the Late Eocene-Oligocene biotic crisis, which does not show on Stehli *et al.*'s curve, is a second-order crisis in Cutbill's and Funnell's compilation, and was strongly developed in the marine realm (Fischer and Arthur, 1977).

Possible Causes

Elsewhere (Fischer, 1981), I have suggested that the fluctuations in atmospheric carbon dioxide content result from fluctuations in the rate of supply (from volcanism) and in the rate of withdrawal (by the linked processes of weathering and sedimentation) (Budyko, 1977; Holland, 1978). Both are linked to a major cyclic process—mantle convection. The hypothesis can only be outlined here.

While there is much uncertainty about the manner in which the lithospheric plates are driven, thermal convection of the mantle is generally taken to be the ultimate cause (e.g., Morgan, 1972). There is no reason to suppose that this process runs at a constant rate, and Fischer (1981) has proposed that the historical pattern in which episodes of continental dispersion and episodes of continental aggregation succeed each other results from a cyclicity in mantle convection in the following manner.

In one phase, the mantle is in a quiescent state, having few and slowly turning cells, which tend to sweep the continental masses together into one pangea. As a result of few cells, the total length of midoceanic ridge is relatively short; because of slow convection and slow spreading rates, the ridge is relatively narrow. It follows (Russell, 1968; Hays and Pitman, 1973) that the displacement of waters from the ocean basins by the ridge is small and that the continents stand high and dry. Contributions of carbon dioxide from the interior to the ocean-atmosphere system are at a low, because both basaltic volcanism in rift zones (which brings carbon dioxide from the mantle) and andesitic volcanism from subduction zones (which recycle lithospheric carbon back to the atmosphere) are minimal. At the same time, the large area of the lands implies that uptake of carbon dioxide by weathering is at a maximum. As a result, a high atmospheric content of carbon dioxide, inherited from a former state, cannot be maintained. Carbon dioxide pressure will drop until the rate of weathering slows and the rate of carbon dioxide withdrawal approaches the rate of volcanic addition. This low balance results in the development of the icehouse state. In this, the growth of a glacial armor over the lands further reduces weathering, providing a negative feedback that may help to explain why the Earth has never turned into a complete iceball.

Mantle convection increases, by development of more cells and of more vigorous convection. New rifts develop, and old ones spread more rapidly. As marine ridges grow in length as well as in width, and as continents spread out by rifting, sea-water is displaced, flooding the continents. The increased volcanism in rift belts and in belts of plate convergence raises the output of carbon dioxide to the atmosphere-hydrosphere system. At the same time, flooding of the continents cuts carbon dioxide losses to the lithosphere by weathering. The net result is that atmospheric CO_2 pressure must rise, until the weathering rates, increased thereby, once again withdraw carbon dioxide at rates matching the volcanic addition. This high balance results in the greenhouse state.

The late Ordovician ice age, coming in the middle of a greenhouse episode, at a time of sea-level highs, and associated with a major biotic crisis, seems altogether exceptional and asks for special explanation. One possibility that comes to mind is that of a greenhouse that overshot, producing a cloud cover so dense as to reflect enough of the solar radiation to cool the Earth. The alternative is to find a geologically transient sink for carbon dioxide.

Conclusions and Problems

Global climates have alternated between states susceptible to widespread glaciation (icehouse states) and greenhouse conditions. Two such cycles have been completed in the 600 m.y. of Phanerozoic time. The reasons for them are not firmly established. While several authors (cf., Pearson, 1978) have suggested a tie to changes in insolation, related to the cycle, of galactic rotation, an apparent correlation with sea-level changes and volcanicity suggests an internal cause. This is here sought in hypothetical cycles of mantle convection, which drive sea levels and atmospheric carbon dioxide content by independent pathways linked by a feedback mechanism (weathering). Whereas the Phanerozoic record suggests a length of about 300 m.y. per cycle, a rigorous periodicity throughout Earth history is not implied, inasmuch as mantle behavior must change in a cooling Earth. Nevertheless, it seems likely that we are in the early part of an icehouse state.

Riding on this long cycle are a family of smaller climatic fluctuations, of which one seems to have a periodicity of perhaps slightly more than 30 m.y. In this one, too, the climatic effect may depend on changes in carbon dioxide, but the mechanisms remain obscure.

CONCLUSIONS

In summary, I suggest that the stratigraphic record holds evidence of a wide range of global changes in climatic state. Largest among these are the 150-m.y.(?) alternations between the major greenhouse and icehouse states. We know only the latter, and the traditional attempts to reconstruct the Mesozoic or the mid-Paleozoic world along strictly actualistic lines are grossly inadequate.

Upon this great cycle rides a smaller one, having a period somewhere around 30 m.y. to 36 m.y., which mimics the large one on a smaller scale and which is recorded in its effects on life. The oligotaxic times that happen to coincide with the turnover in the large cycle, from its greenhouse phase to its icehouse phase and vice versa, are particularly pronounced as some of the world's great biotic crises.

Smaller pre-Pleistocene climatic oscillations are seen at the 500,000-yr level, at the 100,000-yr level, at the 50,000-yr level, and at the 20,000-yr level, in round numbers. These appear to match similar periods in ice flux within the Pleistocene, which have been attributed to climatic effects of the Earth's orbital perturbations. These rhythmic events are recorded in a wide variety of sedimentary settings and record a multitude of pathways by means of which climatic change became expressed in sediments. We have only begun to recognize them and are far from any understanding of them.

In the normal course of development, we could expect to slide more deeply into the icehouse state for some millions of years to come, with continued gradual loss of species. The burning of fossil fuels may instead provide a brief brush with the greenhouse state within a generation. That would be a brief passage only, limited by the amounts of fossil fuels available. The effects on climate, I must leave to others more qualified. However, I believe that a full greenhouse of the kinds that existed in the mid-Paleozoic and Mesozoic would require the complete melting of the ice caps and the warming of the oceans as a whole. That would produce a major biotic crisis of the sort that brought about the partial or complete collapse of some of the world's organic communities in Devonian, Triassic, and Oligocene time. That event seems unlikely to occur for another 70 m.y., but even a brief brush with greenhouse conditions may upset the accustomed structure and behavior of atmosphere and oceans. It might thus have marked effects on the biosphere and on human life and history.

REFERENCES

Alvarez, L. W., W. Alvarez, F. Asaro, and W. V. Michel (1980). Extraterrestrial cause of the Cretaceous-Tertiary extinctions, *Science 208*, 1095-1108.

Arthur, M. A. (1979a). Sedimentologic and geochemical studies of Cretaceous and Paleogene pelagic sedimentary rocks: The Gubbio sequence, Dissertation, Princeton U., Part I, 174 pp.

Arthur, M. A., (1979b). North Atlantic Cretaceous black shales: The record at Site 398 and a brief comparison with other occurrences, in *Initial Reports of the Deep Sea Drilling Project Vol. 47, pt. 2*, U.S. Government Printing Office, Washington, D.C.

Arthur, M. A., and A. G. Fischer (1977). Upper Cretaceous-Paleocene magnetic stratigraphy at Gubbio, Italy. I. Lithostratigraphy and sedimentology, *Geol. Soc. Am. Bull. 88*, 367-371.

Berger, A. L. (1980). The Milankovitch astronomical theory of paleoclimates: A modern review, *Vistas in Astronomy 24*, 103-122.

Berry, W. B. N., and P. Wilde (1979). Progressive ventilation of the oceans—an explanation for the distribution of the Lower Paleozoic black shales, *Am. J. Sci. 278*, 257-275.

Bradley, W. H. (1929). The varves and climate of the Green River epoch, *U.S. Geol. Surv. Prof. Pap. 645*, 108 pp.

Briskin, M., and W. A. Berggren (1975). Pleistocene stratigraphy and quantitative oceanography of tropical core V-16 205, *Micropaleontol. Spec. Publ. 1*, 167.

Budyko, M. I. (1977). *Climatic Changes*, American Geophysical Union, Washington, D.C., 261 pp.

Cutbill, J. L., and B. M. Funnel (1967). Computer analysis of the fossil record, in *The Fossil Record*, W. B. Harlan *et al.*, eds., Geol. Soc. London, pp. 791-820.

Damon, P. E. (1971). The relationship between late Cenozoic volcanism and tectonism and orogenic-epeirogenic periodicity, in *Late Cenozoic Glacial Ages*, K. K. Turekian, ed., Yale U. Press, pp. 15-36.

Dorf, E. (1970). Paleobotanical evidence of Mesozoic and Cenozoic climatic changes, in *Proceeding of the North American Paleontological Convention, Vol. 2*, E. I. Yochelson, ed., Allen and Unwin, London, pp. 323-346.

Dorman, F. H. (1968). Some Australian oxygen isotope temperatures and a theory for a 30-million-year world temperature cycle, *J. Geol. 76*, 297-313.

Douglas, R. G., and S. M. Savin (1975). Oxygen and carbon isotope analyses of Cretaceous and Tertiary microfossils from Shatsky Rise and other sites in the North Pacific Ocean, in *Initial Reports of Deep Sea Drilling Project, Vol. 32*, U.S. Government Printing Office, Washington, D.C., pp. 509-521.

Engel, A. E. J., and C. G. Engel (1964). Continental accretion and the evolution of North America, in *Advancing Frontiers in Geology and Geophysics*, A. P. Subramaniam and S. Balakrishna, eds., Indian Geophysical Union, pp. 17-37e.

Fischer, A. G. (1964). The Lofer cyclothems of the alpine Triassic, *Kansas State Geol. Surv. Bull. 169*, Vol. 1, pp. 107-150.

Fischer, A. G. (1980). Gilbert—bedding rhythms and geochronology, in *The Scientific Ideas of G. K. Gilbert*, E. I. Yochelson, ed., Geol. Soc. Am. Spec. Pap. 183, pp. 93-104.

Fischer, A. G. (1981). Climatic oscillations in the biosphere, in *Biotic Crises in Ecological and Evolutionary Time*, M. Nitecki, ed., Academic, New York, pp. 103-131.

Fischer, A. G., and M. A. Arthur (1977). Secular variations in the pelagic realm, in *Deep Water Carbonate Environments*, H. E. Cook and P. Enos, eds., Soc. Econ. Paleontol. Mineral. Spec. Publ. 25, pp. 18-50.

Gilbert, G. K. (1895). Sedimentary measurement of geologic time, *J. Geol. 3*, 121-127.

Gilbert, G. K. (1900). Rhythms and geologic time, *Am. Assoc. Adv. Sci. Proc. 49*, 1-19.

Grabau, A. W. (1940). *The Rhythm of the Ages*, Henry Vetch Publ. Co., Peking, 561 pp.

Hallam, A. (1977). Secular changes in marine inundation of USSR and North America through the Phanerozoic, *Nature 269*, 769-772.

Hays, J. D., and W. C. Pitman III (1973). Lithospheric motion, sea level changes and climatic and ecological consequences, *Nature 246*, 18-22.

Hays, J. D., J. Imbrie, and N. J. Shackleton (1976). Variations in the Earth's orbit: Pacemaker of the ice ages, *Science 194*, 1121-1132.

Heckel P. H. (1977). Origin of phosphatic black shale facies in Pennsylvanian cyclothems of midcontinent North America, *Am. Assoc. Petrol. Geol. Bull. 61*, 1045-1068.

Holland, H. D. (1978). *The Chemistry of Oceans and Atmospheres*, Wiley, New York, 351 pp.

Jenkyns, H. C. (1980). Cretaceous anoxic events: From continents to oceans, *J. Geol. Soc. 137*, 171-188.

Koch, D. E. (1963). Fossil plants from the lower Paleocene of Agatdalen (Angmartusuk) area, central Nuqssuaq Peninsula, northwest Greenland, *Medd. Groenl. 172*, 1-120.

Kollmann, H. (1979). Distribution patterns and evolution of gastropods around the Cretaceous-Tertiary boundary, in *Cretaceous-Tertiary Boundary Events*, Vol. II Proceedings, W. K. Christensen and T. Birkelund, eds., U. of Copenhagen, Copenhagen, Denmark.

Manabe, S., and R. T. Wetherald (1975). The effect of doubling the CO_2 concentration on the climate of a general circulation model, *J. Atmos. Sci. 32*, 3-15.

Milankovitch, M. (1941). *Kanon der Erdbestrahlung und seine Anwendung auf das Eiszeitenproblem*, Ed. Spec. Acad. Roy. Serbe, Belgrade, 633 pp., English translation, U.S. Department of Commerce.

Monninger, W. (1979). The section of Tiout (Precambrian/Cambrian boundary beds, Anti-Atlas, Morocco): An environmental model, *Arb. Palaeontol. Inst. Wuerzburg (Germany)*, 289 pp.

Moore, R. C., J. C. Frye, J. M. Jewett, W. Lee, and H. O'Connor (1951). The Kansas rock column, *Kansas Geol. Surv. Bull. 89*, 132 pp.

Morgan, W. J. (1972). Plate motions and deep mantle convection, in *Studies in Earth and Space Sciences*, R. Shagam, ed., Geol. Soc. Am. Mem. 132, 7-22.

Obradovich, J. D., and W. A. Cobban (1975). A time scale for the Late Cretaceous of the Western Interior of North America, in *The Cretaceous System in the Western Interior of North America*, W. G. E. Caldwell, ed., Geol. Assoc. Canada Spec. Pap. 13, pp. 31-54.

Pearson, R. (1978). *Climate and Evolution*, Academic, New York, 274 pp.

Russell, K. L. (1968). Oceanic ridges and eustatic changes in sea level, *Nature 218*, 861-862.

Schwarzacher, W. (1975). *Sedimentation Models and Quantitative Stratigraphy*, Developments in Stratigraphy, Vol. 19, Elsevier, New York, 377 pp.

Stehli, F. G., R. G. Douglas, and N. D. Newell (1969). Generation and maintenance of gradients in toxonomic diversity, *Science 164*, 947-949.

Vail, P. R., R. M. Mitchum Jr., and S. Thompson (1977). Seismic stratigraphy and global changes in sea level, part 4, in *Seismic Stratigraphy*, C. E. Peyton, ed., Am. Assoc. Petrol. Geol. Mem. 26, pp. 83-97.

Van Hinte, J. E. (1976). A Cretaceous time scale, *Am. Assoc. Petrol. Geol. Bull. 60*, 498-516.

Van Houten, F. B. (1964). Cyclic lacustrine sedimentation, Upper Triassic Lockatong Formation, central New Jersey and adjacent Pennsylvania, *Kansas Geol. Surv. Bull. 169*, 497-531.

West, R. M., and M. R. Dawson (1978). Vertebrate paleontology and Cenozoic history of the North Atlantic region, *Polarforschung 48*, 103-119.

10 Climatic Significance of Lake and Evaporite Deposits

HANS P. EUGSTER
The Johns Hopkins University

INTRODUCTION

Lake deposits and evaporites are sensitive indicators of local climatic conditions and changes in climate—lakes because they are ephemeral and owe their existence to a delicate balance between hydrology, sedimentation, and tectonics and evaporites because they represent extreme stages of desiccation. The very aspects that make these deposits valuable for our purposes are also responsible for some of the difficulties encountered in extracting climatic inferences. Few lake deposits are large enough to have attracted the general attention of geologists, notable exceptions being Devonian Lake Orcadie (Old Red Sandstone), Triassic Lake Lockatong, and Eocene Lakes Gosiute and Uinta (Green River Formation). The lacustrine record in the geologic column has yet to be tapped, although a fine summary of paleolimnology in North America has been provided (Bradley, 1963).

Evaporite deposits, especially of marine derivation, have been studied extensively, if not exhaustively, in part because of their economic significance (see recent summaries by Dean and Schreiber, 1978; Holser, 1979). Evaporite minerals respond readily to postdepositional changes. Because depositional and postdepositional processes form a continuum, mineral textures and sedimentary structures must be interpreted with great care.

LAKE DEPOSITS

The study of ancient lake deposits has been helped immeasurably by limnologists, sedimentologists, and geochemists, who make observations on active lakes, and by geologists, who have concerned themselves with Pleistocene lakes and their Holocene residues. Paleoclimatic information may be contained in the nature of the basin itself as well as in the material that filled it. Every lake deposit is testimony to at least two important environmental changes, one initiating lacustrine deposition and the other terminating it. Climatic factors may be involved in either or both events. For instance, glacial lakes often form during a warming trend, and they can be recognized from the shape of the basins and the fact that many are dammed by moraines. Tectonic events unrelated to overall climatic change may lead to steepened hydraulic gradients and increased precipitation in the watershed and hence earlier filling of the basin. Choking with sediment probably is the most common death of lakes, but lakes can also dry up. In that case, a clear record is usually left in the sediment in the form of an evaporitic terminal stage.

Reliable reconstruction of depositional environments of lake sediments, a prerequisite for extracting climatic inferences, depends on a whole array of geologic, geochemical, and geophysical tools. Most of our experiences have been gained from Holocene-Pleistocene settings (see e.g., Gilbert, 1890; Hansen, 1961; Jones, 1965; Neev and Emery, 1967; Anderson and Kirkland, 1969; Brunskill, 1969; Ludlam, 1969; Irion, 1970, 1973; Müller *et al.*, 1972; Degens *et al.*, 1973; Eardley *et al.*, 1973; Dean and Gorham, 1976; Eugster and Hardie, 1976, 1978; Friedman *et al.*, 1976; Hardie *et al.*, 1978; Jones and Bowser, 1978; Kelts and Hsü, 1978; Lerman, 1978; Matter and Tucker, 1978; Müller and Wagner, 1978; Smith, 1978, 1979; Stoessel and Hay, 1978; Stoffers and Hecky, 1978; Cerling, 1979; Yuretich, 1979; Bradbury and Whiteside, 1980; Eugster, 1980; McKenzie *et al.*, 1981; Sims *et al.*, 1981; Spencer *et al.*, 1981a, 1981b; Smith *et al.*, in press). Once the stratigraphic time lines have been established for a particular basin, using traditional approaches as well as radiogenic dating, tephrochronology, paleomagnetics, amino acid racemization, and other methods, facies maps are constructed. Information on sediment type, sedimentary structures, mineralogy, isotopic compositions, biota (such as ostracodes, diatoms, algae, molluscs, and fish), and pore-water chemistry are combined in a detailed reconstruction of the depositional environments and their evolution in space and time. Comparisons with adjacent basins may then make it possible to separate local changes due to hydrology or tectonics from regional climatic trends.

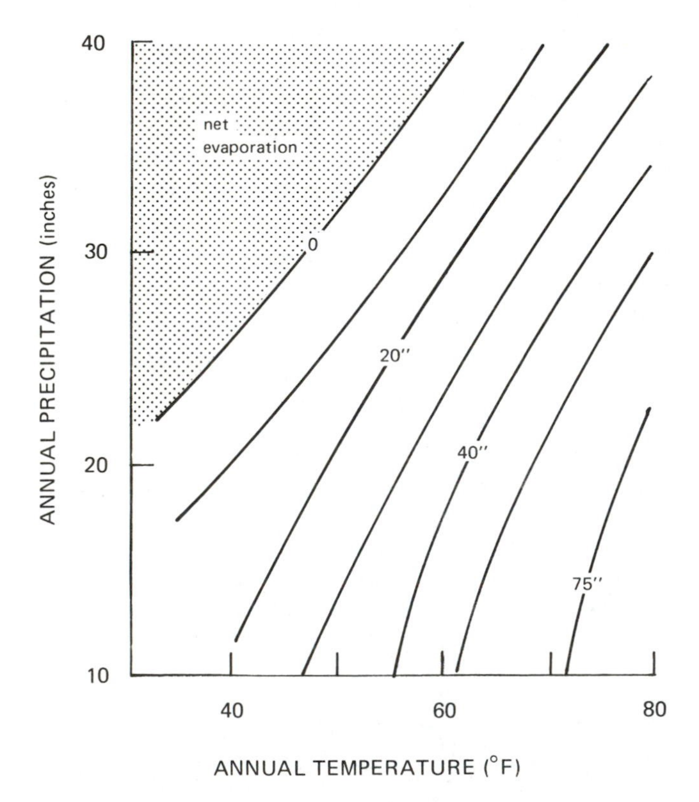

FIGURE 10.1 Correlation between annual precipitation, net evaporation (evaporation from lake surface-precipitation on lake surface), and mean annual temperature for closed-basin lakes. From Langbein (1961).

Such studies are difficult at best and are most likely to succeed in Holocene-Pleistocene basins, where one moment in time can be studied directly. For older deposits there often exists doubt whether they are in fact lacustrine or not. Such deposits can usually be identified as such from lithologies, sedimentary structures, and fossil content. Associated rock units formed immediately preceding or contemporaneously with lake deposits may also be good indicators. For instance, the Green River Formation is defined as a lens of lacustrine deposits following and interbedded with the unquestionably continental and largely fluviatile Wasatch formation. Among evaporites, lacustrine deposits can most readily be identified if their chemistry was alkaline, an evolutionary path not accessible to seawater derivatives (Hardie and Eugster, 1970).

The mere existence of lakes in a particular locality contains little climatic information, because lakes can exist from arctic to tropical environments. Futhermore, many lakes are formed by nonclimatic events, including tectonic graben lakes, oxbow lakes (channel migration), volcanic lakes, and landslide lakes. Most useful will be glacial lakes and evaporitic lakes. We will focus on the latter; glacial deposits are treated elsewhere (Chapter 6). Geologic and geochemical work on saline lakes has been summarized recently (Eugster and Hardie, 1978; Hardie *et al.*, 1978; Eugster, 1980).

SALINE LAKES

Saline lakes form in hydrologically closed basins under conditions where evaporation exceeds inflow. They respond to climatic changes, and hence their deposits contain excellent records of such changes. Most published studies have dealt with Pleistocene-Holocene examples such as Searles Lake (Smith, 1979), Lake Bonneville-Great Salt Lake (Eardley *et al.*, 1973), or Lake Lisan-Dead Sea (Begin *et al.*, 1974).

A classic study of climatic parameters that define closed-basin lakes was presented by Langbein (1961). Figure 10.1 shows the interrelations between temperature, precipitation, and net evaporation. In response to changes in the inflow-evaporation balance, closed lakes go through cycles with respect to lake level, areal extent, volume, and salinity; such cycles are reflected in their deposits in a variety of ways. Transgressive-regressive events may be recognized by sedimentary structures and changes in lithologies. Good examples are the depositional cycles described by Eugster and Hardie (1975) from the Wilkins Peak Member of the Green River formation. Transgression is indicated by flat-pebble conglomerates that are usually followed by oil shales, indicating a higher lake stand. Oil shales gradually give way to mudflat facies rocks in response to falling lake levels. Extreme regression is indicated by evaporite accumulation. In reading this record we must be conscious of the many ways in which the inflow-evaporation balance can be effected: tectonic changes, hydrology, and climate all playing their part.

In fresh lakes, the mineralogy of the sediments is usually inherited from the bedrock lithologies of the watershed (Jones and Bowser, 1978). In contrast, progressive desiccation of a closed basin brings with it profound changes in the chemistry of the waters and hence also the chemical sediments (Eugster and Jones, 1979). Some of these changes may be useful as indicators of the onset of arid conditions even when evaporite deposition never took place. Perhaps the most useful minerals are the alkaline earth carbonates calcite, aragonite, and dolomite. In general, because $CaCO_3$ precipitation is related to evaporative concentration, increasing salinity increases the Mg/Ca ratio of the water. In response, the principal carbonate precipitate may change from low-Mg calcite to high-Mg calcite and eventually aragonite. Good examples have been described by Müller from a variety of Holocene and Pleistocene lakes (for example, Lake Balaton, Müller and Wagner, 1978). Great Salt Lake, Utah, near the Holocene-Pleistocene boundary experienced a drop of 100 m in less than 2000 yr, a drop recorded in the carbonate mineralogy as well as in the oxygen isotopes of the carbonates (McKenzie *et al.*, 1981; Spencer *et al.*, 1981a, 1981b).

An excellent record of arid conditions may also be left in the authigenic mineral assemblages, because such minerals form by reaction of intestitial brine with volcanogenic or chemical sediments. Best known are the zeolite sequences of alkaline lake deposits (Surdam and Sheppard, 1978). Increasing salinity is reflected in the sequence smectite-zeolite-anaclime-K-feldspar, with the dominant zeolites being phillipsite, clinoptilolite, erionite, and chabazite. Such sequences may be displayed laterally or vertically. Because high-pH brines are necessary, zeolites do not form in Cl or SO_4-dominated bases. Saline, alkaline conditions are also indicated by the presence of Magadi-type cherts, that is, cherts that have formed from a sodium silicate precursor, such as magadiite.

The climatic information to be extracted from saline lake deposits must be calibrated in comparison with active salt lakes. Considering such diverse settings as Great Salt Lake (Utah), Death Valley (California), Dead Sea (Asia), Lake Chad (Africa), Lake Uyuni (Bolivia), Lake Vanda (Antarctica), Tso-Kar Lake (Lhadak), it becomes clear that saline lakes do not give independent information on latitude, temperature, precipitation, and other climatic parameters, but they do tell us that in these localities evaporation exceeds inflow. Evaporation is experienced at the lake level, but inflow is a complex product of hydrologic processes in the entire watershed. Mean parameters, such as mean annual humidity, are not safe guides, because a rainy season can be compensated for by an intense dry period. Also, we realize that salt lakes do not occur in the most extreme deserts for lack of inflow. Ideal settings are down-faulted valley flows in the rain shadow of mountains that act as snow catch.

There is no doubt that the most valuable climatic information contained in lake deposits relates to short-term temporal changes and to local differences between adjacent basins. This is most clearly documented in the Eocene Green River deposits of Utah, Colorado, and Wyoming.

THE GREEN RIVER FORMATION

Paleolimnology, paleoclimatology, and the Green River formation are inextricably linked, as demonstrated through the work of Bradley (1929, 1931, 1948, 1963), who combined a wealth of information to deduce the climate of the region at the time of deposition. Bradley was forced to oversimplify and generalize in deducing a Green River climate. Paleotopographic arguments fixed the basin floors at 1000 ft. Using the pine-spruce distribution of the Southern Appalachians, mountain crests were fixed at a minimum elevation of 6000-8000 ft. Several lines of evidence yielded mean annual temperatures of 67°F. From Langbein's (1961) formulas, Bradley calculated an annual precipitation of 34 in. at the lake level for a just-full stage and an average of 38 in. for the whole watershed. During times of overflow, precipitation must have been higher. During evaporite deposition (Wilkins Peak time), mean annual precipitation was estimated at 24 in. The large and varied flora, which contained cypress and palm trees (see, for example, Brown, 1934, MacGinitie, 1969), persuaded Bradley to call the overall climate warm-temperate like that of the present Gulf Coast states.

Recent work on the Green River formation has led to somewhat different conclusions (Eugster and Hardie, 1975, 1978; Surdam and Stanley, 1979, 1980). We have focused more on the differences between basins, on the temporal changes, and on the complexity of the environments. In modifying Bradley's conclusions, we have relied heavily on sedimentological arguments and on our experience with active continental evaporites.

The Eocene Green River formation was deposited in at least three basins, which were occupied by Lakes Uinta and Gosiute and Fossil Lake (see Figure 10.2). The Uinta and Green River (Gosiute) Basins, separated by the Uinta Mountains, were very large, with wide, flat floors. Their principal sediments are oil shales, which are rocks rich in carbonates and kerogen. Although it is clear that Lake Gosiute at times drained into the Uinta Basin (Bradley, 1963; Surdam and Stanley, 1979, 1980), we believe that there is good sedimentological evidence that both basins were hydrologically closed throughout most of their lives. Shallow, oil-shale producing lakes were surrounded by playa fringes, where carbonate mud was produced, modified, transported, and redeposited (see Smoot, 1978). Lake levels fluctuated extensively, and waters varied from brackish to saline. The basins south of the Uinta Mountains contained lakes for a much longer time (at least 12 million years) than did the Green River Basin (at least 5 million years). In fact, Lake Gosiute not only was of shorter duration, but during part of its life it was more ephemeral in nature. At least 25 separate episodes have been recorded during Wilkins Peak time, where the lake essentially dried up and became a salt pan. Salt deposition also occurred in Lake Uinta to the south, but less frequently and on a more localized scale. During much of its existence, Lake Uinta was a permanent lake with oil shale one of its principal deposits. Although the stratigraphic correlations and age dates are still in doubt (O'Neill *et al.*, 1981), it seems clear from the summary of Wolfbauer (1971) that evaporites frequently were deposited in the north while oil shales formed in the

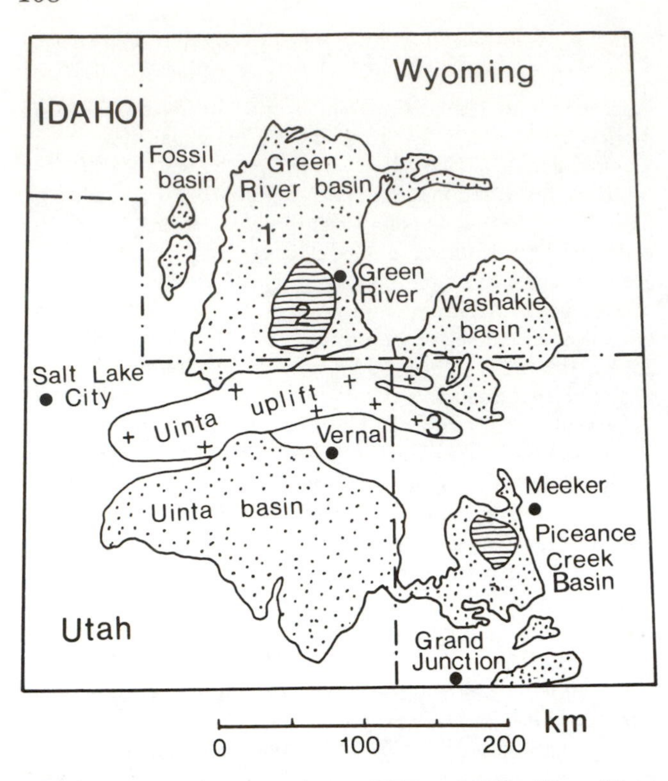

FIGURE 10.2 Sedimentary deposits of Eocene Lakes Uinta (Uinta and Piceance Creek Basins) and Gosiute (Green River and Washakie Basins) and Fossil Lake (Fossil Basin). 1, Green River Formation; 2, bedded salts; 3, basement. From Eugster and Hardie (1978).

south. Such local differences in the evaporation-inflow balance are not surprising and may be due to a number of factors. However, O'Neill *et al.* (1981) conclude that the "high stands of Lakes Gosiute and Uinta were essentially synchronous and therefore were likely controlled by climatic rather than tectonic factors." Surdam and Stanley (1980), in a comparison of the drainage basins and sedimentation histories of the two basins, take the opposite position: "Basin filling and enlargement of the drainage system were probably a consequence of tectonic activity and stability of the basins and adjacent uplifts, although climatic conditions that increased sediment yield and runoff in the hydrographic basins also could have hastened their filling. However, it is difficult to explain patterns of evaporite minerals, oil shale, mudstone, and sandstone formed in Lakes Gosiute and Uinta if climate was the dominant factor."

While Lakes Gosiute and Uinta went through their evolutionary changes, the much smaller Fossil Lake to the west (Figure 10.2) was caught between the ridges of the Thrust Belt. Fed perhaps by a precursor of the Snake River, this lake was more like a tub, narrow, long, and probably deeper (most of the time). Its sediments are the classic varve limestones often taken as prototype Green River rocks, which contain the famous, exquisitely preserved fish fossils. We now know that these rocks owe their existence to special circumstances that have to do as much with hydrology and sedimentation as with climate. No sizable accumulations of evaporites or oil shales occurred, probably because the lake was deeper and sediment-choked. Yet there is a clear record of several periods of desiccation or near desiccation, with trees or bushes growing on the bed of the lake.

This brief summary of three contrasting basins should make it obvious how difficult it is to speak of a Green River climate. Nevertheless, I do not agree that the present Gulf Coast climate provides a valid analog. The overall conditions were much drier, and I believe that the significance of the subtropical flora has been overemphasized. The nature of the sediments indicates a dry, continental climate with evaporation exceeding inflow most of the time. Bradley was aware of the evidence indicating such an environment; in speaking of *Ephedra* pollen (Bradley, 1963, pp. 639-640) he notes that "If its Eocene forebears lived in the same parched environment as the *Ephedra* of the Rocky Mountain region of today, it is odd that it should show up in the sediments of the earliest stage (Luman) of Gosiute Lake. It also occurs, as would be expected, in the deposits of the later saline stage. The presence of *Ephedra* pollen agrees with an observation by R. W. Brown (1934, pp. 280-281) that certain thick, coraceous leaves in the Green River flora indicate hot, dry summers like those of southern California."

Southeastern California probably provides a much better climatic comparison. Though none are very large, the region contains many hydrologically closed basins, with waters ranging from brackish to very saline. Palm trees grow wherever there is enough water.

MARINE EVAPORITES

Marine evaporites have traditionally been useful climatic indicators because of their bulk and obvious cyclicity. Although they occur throughout the geologic column from the Precambrian onward, there are definite centers of evaporite depositions in time and space (see Kozary *et al.*, 1968), for example, during the Devonian in North America, the Permian and the Triassic in Europe, and the Jurassic in North America and Miocene in Europe. As has been pointed out often, no large-scale evaporites are forming at present anywhere on Earth. There seems to be no lack of evaporitic environments, but the necessary depositional conditions do not coincide. Again, the presence of evaporites is due as much to tectonic as climatic factors.

Onset of evaporitic conditions can often be recognized from underlying and laterally contiguous sediments. Biogenic limestones and reef deposits are common precursors, indicating that tropical to subtropical climates prevailed. Other organism-rich sediments such as oil shales record the increasing preservation, if not productivity, of organic matter in penesaline environments. Laterally associated with evaporite deposition, continental sediments such as red beds, arkoses, and conglomerates may accumulate in down-faulted basins.

Although there must be a much better correlation between marine evaporites and latitude than there is between continental evaporites and latitude, there is no intrinsic reason why marine evaporites could not form at high latitudes, except for the fact that evaporation rates are so much lower. It should be possible to recognize such deposits by the presence of mirabilite, $Na_2SO_4 \cdot 10H_2O$, a typical precipitate formed by chilling of brines having the composition of concentrated seawater such as those of the Great Salt Lake of Utah.

For the present discussion it is useful to separate marine evaporites into three groups: (1) those dominated by gypsum and anhydrite, (2) those dominated by halite, and (3) potash-bearing deposits. By far the largest number of fossil deposits belong to the gypsum-anhydrite group, and these deposits are best understood because there are good Holocene analogs. Sabkha settings such as those of the Persian Gulf (Purser, 1973) are usually invoked for these deposits, which often consist of CaMg carbonate-gypsum laminites. Typical examples and criteria for recognizing them have been described from the Miocene of Sicily (Hardie and Eugster, 1971), and a depositional model has been suggested by Eugster and Hardie (1978). To produce such deposits, seawater has to be concentrated at least three but not more than ten times. As pointed out by Kinsman (1976), the average humidity must vary from 98 to 76 percent. Transgressive-regressive cycles are common at this stage, in response to the interplay between climatic and tectonic forces. Excellent examples have been described from the Permian of northern Italy (Bosellini and Hardie, 1973) and the Permian of Texas and New Mexico (Anderson *et al.*, 1972). Subsidence is essential for continued deposition in shallow basins like those envisaged for group 1 evaporites, but each step down may lead to transgression across the bar that protects the basin from the open sea. Similarly, subsidence must be slow enough so that sedimentation is able to keep pace and the basin never becomes deep.

In contrast, there is increasing evidence that the thick halite-dominated accumulation (group 2) so characteristic of the saline giants (Hsü, 1972) probably formed in deep basins (see Gill, 1977; Hardie *et al.*, 1978; Briggs *et al.*, 1980; Harvie *et al.*, 1980). Brine bodies several hundreds of meters deep may accumulate as a consequence of increased subsidence coupled with more intense evaporation. Such a body, though of different composition, exists now in the Dead Sea. An appropriate hydrologic model to account for the sequence of progressively evaporated mineral assemblages has been proposed by Harvie *et al.* (1980). Evaporation for halite sequences to form must be from 10- to 20-fold seawater concentration. This may or may not signal increasing aridity, depending on whether or not a hydrologic system for preconcentration is available.

Most marine evaporites stop at the halite stage, and the final concentration products (group 3 evaporites) are either not formed or not preserved. These products are the K-Mg salts, so important for economic reasons and also as climatic indicators. Minerals such as polyhalite [$K_2MgCa_2(SO_4)_4 \cdot 2H_2O$], epsomite ($MgSO_4 \cdot 7H_2O$), carnallite ($KMgCl_3 \cdot 6H_2O$), and kieserite ($MgSO_4 \cdot H_2O$) require such extreme desiccation that it is difficult to envisage an appropriate near-marine climatic setting. This has been pointed out by Kinsman (1976), who attempted to correlate mean annual humidity with stage and intensity of evaporation. The activity of H_2O, aH_2O, of a brine is a direct measure of the humidity of the air in equilibrium with such a brine, as

$$aH_2O = \frac{P_{H_2O}\,\text{brine}}{P_{H_2O}\,\text{water}} \quad \text{and} \quad RH = 100\,aH_2O,$$

where P_{H_2O} is the water-vapor pressure and RH is the relative humidity in percent. Calculated aH_2O values for seawater concentrates and minerals with which they are in equilibrium are shown in Figure 10.3 (see Eugster *et al.*, 1980). Potash deposits do not form until aH_2O falls below 0.7, the Mg sulfates epsomite and kieserite require values below 0.6, and carnallite 0.45, that is, an average humidity of less than 45 percent is required for carnallite to accumulate, a condition difficult to imagine in a marine setting.

For the deposition of group 3 evaporites we favor a depositional model, termed the playa model. To reach this stage, the deep brine basin filled up with evaporites; that is, subsidence did not keep pace with sedimentation. In the resulting flat, broad basin, now under the influence of extreme aridity, fractional dissolution and recycling of the most soluble salts will occur, as observed in continental playas (Eugster and Jones, 1979). In consequence, the most soluble minerals will accumulate in the hydrographic center as a bedded deposit. These minerals are quite hygroscopic, and they can be preserved only

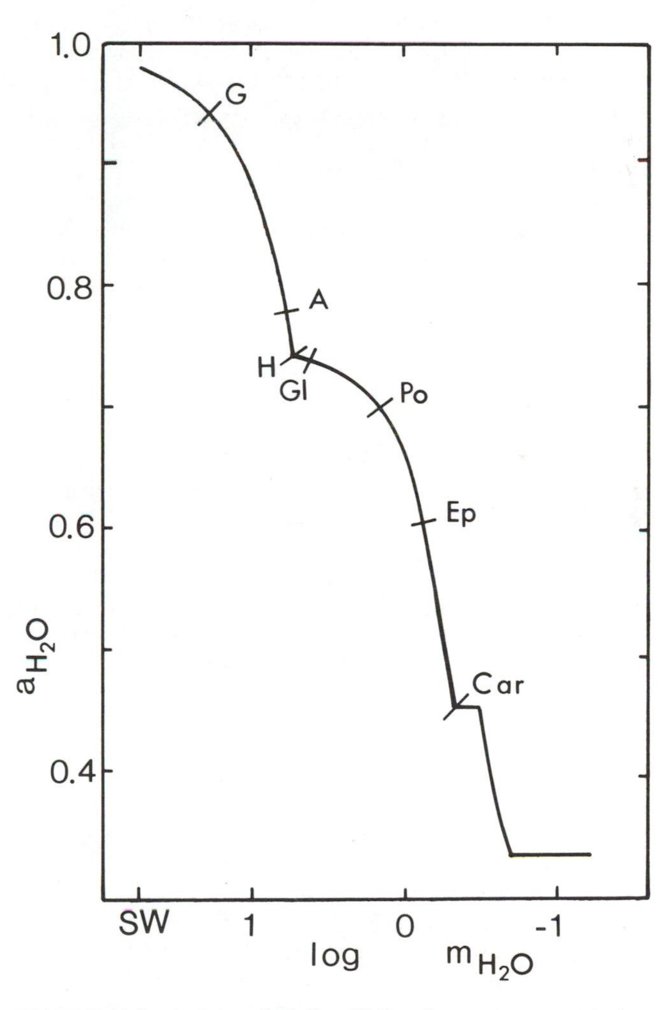

FIGURE 10.3 Activity of H_2O, aH_2O, of seawater concentrates. Evaporation begins with 55.5 moles of seawater (SW) and proceeds to the right. First appearances of precipitates are noted. G, gypsum; A, anhydrite; H, halite; Gl, glauberite; Po, polyhalite; Ep, epsomite; Car, carnallite. From Eugster *et al.* (1980).

by extreme conditions of aridity. The appropriate situation can perhaps be created by combining the hydrologic setting of the Persian Gulf with the climate of the coast of Chile.

There is one final group of evaporites, often considered to be of marine parentage, that require still further evaporation. These are the deposits that contain tachyhydrite ($CaMgCl_4 \cdot 6H_2O$) (see, for example, Wardlaw, 1972; Szatmari *et al.*, 1979). However, as pointed out by Hardie (1979), such deposits may in fact not be directly related to seawater. Nevertheless, where they did form and were preserved, climatic conditions must have been unimaginably severe.

CONCLUSIONS

Lake deposits are sensitive indicators of paleoclimatic changes. Particularly useful are glacial and saline lakes. The latter are not related to latitude and form when evaporation exceeds inflow, recording a complex balance between climate, hydrology, and tectonics. Changes in depositional conditions are indicated by lithologies, sedimentary structures, and mineralogy and often result in transgressive-regressive cycles. Local differences may override general climatic conditions. This is illustrated by a reinterpretation of the Eocene Green River climate—a climate exemplified by southeastern California is preferred to that of the Gulf Coast states.

Climatic interpretation of marine evaporites depends on the special hydrologic requirements. Three stages are distinguished, with each stage dominated in turn by Ca sulfates, halite, and K-Mg salts. The bulk of marine evaporites belong to stage 1. A sabkha setting is a good depositional model. Seawater can be concentrated up to tenfold before halite precipitates. Thick halite deposits may accumulate in a deep stratified basin such as that suggested by Harvie *et al.* (1980). For the final stage, we prefer the playa model, which requires relative humidities of less than 60-45 percent. Final products are hygroscopic, and it is not easy to reconcile their requirements for extreme aridity with a marine setting.

This brief summary should make it clear how difficult it is to derive unequivocal information about paleoclimates from sedimentary deposits, because of the delicate interplay between climatic, tectonic, and hydrologic forces. In order to arrive at valid climatic conclusions, all the evidence must be reviewed, including paleontologic and sedimentologic aspects. We must calibrate and sharpen our tools on the sediments of currently active lakes and evaporitic environments.

REFERENCES

Anderson, R. Y., and D. W. Kirkland (1969). Paleoecology of an early Pleistocene lake on the high plains of Texas, *Geol. Soc. Am. Mem. 113*, 211 pp.

Anderson, R. Y., W. E. Dean, D. W. Kirkland, and H. I. Snide, (1972). Permian Castile varved evaporite sequence, West Texas and New Mexico, *Geol. Soc. Am. Bull. 83*, 59-86.

Begin, A. B., A. Ehrlich, and Y. Nathan (1974). Lake Lisan, the Pleistocene precursor of the Dead Sea, *Geol. Surv. Israel Bull. 63*, 30 pp.

Bosellini, A., and L. A. Hardie (1973). Depositional theme of a marginal marine evaporite, *Sedimentology 20*, 5-27.

Bradbury, J. P., and M. C. Whiteside (1980). Paleolimnology of two lakes in the Klutlan glacier region, Yukon Territory, Canada, *Quat. Res. 4*, 149-168.

Bradley, W. H. (1929). The varves and climate of the Green River epoch, *U.S. Geol. Surv. Prof. Pap. 158-E*, 87-110.

Bradley, W. H. (1931). Origin and microfossils of the oil shale of the Green River Formation of Colorado and Utah, *U.S. Geol. Surv. Prof. Pap. 168*, 58 pp.

Bradley, W. H. (1948). Limnology and the Eocene lakes of the Rocky Mountain region, *Geol. Soc. Am. Bull. 59*, 635-648.

Bradley, W. H. (1963). Paleolimnology, in *Limnology in North America*, D. G. Frey, ed., U. of Wisconsin Press, Madison, Wisc., pp. 621-652.

Briggs, L. I., D. Gill, D. F. Briggs, and R. D. Elmore (1980). Transition from open marine to evaporite deposition in the Silurian Michigan Basin, in *Hypersaline Brines and Evaporitic Environments*, A. Nissenbaum, ed., Developments in Sedimentology 28, Elsevier, Amsterdam, pp. 253-270.

Brown, R. W. (1934). The recognizable species of the Green River flora, *U.S. Geol. Surv. Prof. Pap. 185-C*, pp. 45-68.

Brunskill, G. J. (1969). Fayetteville Green Lake, New York, II. Precipitation and sedimentation of calcite in a meromictic lake with laminated sediments, *Limnol. Oceanogr. 14*, 830-847.

Cerling, T. E. (1979). Paleochemistry of Plio-Pleistocene Lake Turkana, Kenya, *Paleogeogr. Paleoclimatol. Paleoecol. 27*, 247-285.

Dean, W. E., and E. Gorham (1976). Major chemical and mineral components of profundal surface sediments in Minnesota lakes, *Limnol. Oceanogr. 21*, 259-284.

Dean, W. E., and B. C. Schreiber, eds. (1978). *Marine Evaporites*, Soc. Econ. Paleontol. Mineral. Short Course Notes, Oklahoma City, 188 pp.

Degens, E. T., R. P. Von Herzen, H. K. Wong, and H. W. Tannasch (1973). Structure, chemistry and biology of an East Africa Rift Lake, *Geol. Rundsch. 62*, 245-276.

Eardley, A. J., R. T. Shuey, V. Gvosdetsky, W. P. Nash, M. D. Picard, D. C. Grey, and G. J. Kukla (1973). Lake cycles in the Bonneville basin, Utah, *Geol. Soc. Am. Bull. 84*, 211-216.

Eugster, H. P. (1980). Lake Magadi, Kenya, and its precursors, in *Hypersaline Brines and Evaporite Environments*, A. Nissenbaum, ed., Developments in Sedimentology 28, Elsevier, Amsterdam, pp. 195-232.

Eugster, H. P., and L. A. Hardie (1975). Sedimentation in an ancient playa-lake complex: The Wilkins Peak member of the Green River Formation of Wyoming, *Geol. Soc. Am. Bull. 86*, 319-334.

Eugster, H. P., and L. A. Hardie (1976). Some further thoughts on the depositional environment of the Solfifera Series of Sicily, *Mem. Geol. Soc. Italy 16*, 29-38.

Eugster, H. P., and L. A. Hardie (1978). Saline lakes, in *Lakes—Chemistry, Geology, Physics*, A. Lerman, ed., Springer-Verlag, New York, pp. 237-293.

Eugster, H. P., and B. F. Jones (1979). Behavior of major solutes during closed-basin brine evolution, *Am. J. Sci. 279*, 609-631.

Eugster, H. P., C. E. Harvie, and J. H. Weare (1980). Mineral equilibria in the six-component sea water system, Na-K-Mg-Ca-SO_4-Cl-H_2O at 25°C, *Geochim. Cosmochim. Acta 44*, 1335-1347.

Friedman, I., G. I. Smith, and K. Hardcastle (1976). Studies of Quaternary saline lakes—II. Isotopic and compositional changes during desiccation of the brines in Owens Lake, California 1969-71, *Geochim. Cosmochim. Acta 40*, 501-511.

Gilbert, G. K. (1890). Lake Bonneville, *U.S. Geol. Surv. Monogr. 1*, 438 pp.

Gill, D. (1977). Saline A-1 sabkha cycles and the late Silurian paleogeography of the Michigan Basin, *J. Sed. Petrol. 47*, 979-1017.

Hansen, K. (1961). Lake types and lake sediments, *Verh. Internat. Verein Limnol. 14*, 121-145.

Hardie, L. A. (1979). Evaporites: Marine or non-marine? *Geol. Soc. Am. Abstr. Programs 1979*, 368.

Hardie, L. A., and H. P. Eugster (1970). The evolution of closed-basin brines, *Mineral. Soc. Am. Spec. Publ. 3*, 273-290.

Hardie, L. A., and H. P. Eugster (1971). The depositional environment of marine evaporites: A case for shallow, clastic accumulation, *Sedimentology 16*, 187-220.

Hardie, L. A., J. P. Smoot, and H. P. Eugster (1978). Saline lakes and their deposits: A sedimentological approach, *Int. Assoc. Sedimentol. Spec. Publ. 2*, 7-41.

Harvie, C. E., J. H. Weare, L. A. Hardie, and H. P. Eugster (1980). Evaporation of sea water: Calculated mineral sequences, *Science 208*, 498-500.

Holser, W. T. (1979). Mineralogy of evaporites, in *Marine Minerals*, R. B. Burns, ed., Min. Soc. Am. Short Course Notes 6, pp. 211-294.

Hsü, K. H. (1972). Origin of saline giants: A critical review after the discovery of the Mediterranean evaporite, *Earth Sci. Rev. 8*, 371-396.

Irion, G. (1970). Mineralogisch-sedimentpetrographische und geochemische untersuchungen am Tuz Gölö (Salzee), Türkei, *Chem. Erde* 29, 163-226.

Irion, G. (1973). Die anatolischen Salzeen, ihr Chemismus und die Entstehung ihrer chemischen Sedimente, *Arch. Hydrobiol. 71*, 517-557.

Jones, B. F. (1965). The hydrology and mineralogy of Deep Springs Lake, Inyo County, California, *U.S. Geol. Surv. Prof. Pap. 502-A*, 56 pp.

Jones, B. F., and C. J. Bowser (1978). The mineralogy and related chemistry of lake sediments, in *Lakes—Chemistry, Geology, Physics*, A. Lerman, ed., Springer-Verlag, New York, pp. 179-227.

Kelts, K., and K. J. Hsü (1978). Freshwater carbonate sedimentation, in *Lakes—Chemistry, Geology, Physics*, A. Lerman, ed., Springer-Verlag, New York, pp. 295-323.

Kinsman, D. J. J. (1976). Evaporites: Relative humidity control of primary mineral facies, *J. Sed. Petrol. 46*, 273-279.

Kozary, M. T., J. C. Dunlap, and W. E. Humphrey (1968). Incidence of saline deposits in geologic time, in *Saline Deposits*, R. B. Mattox, ed., Geol. Soc. Am. Spec. Pap. 88, pp. 43-57.

Langbein, W. B. (1961). The salinity and hydrology of closed lakes, *U.S. Geol. Surv. Prof. Pap. 412*, 1-20.

Lerman, A., ed. (1978). *Lakes—Chemistry, Geology, Physics*, Springer-Verlag, New York, 363 pp.

Ludlum, S. D. (1969). Fayetteville Green Lake, New York III. The laminated sediments, *Limnol. Oceanogr. 14*, 848-861.

MacGinitie, H. D. (1969). The Eocene Green River flora of northwestern Colorado and northeastern Utah, *U. Calif. Publ. Geol. Sci. 83*, 203 pp.

Matter, A., and M. E. Tucker, eds. (1978). *Modern and Ancient Lake Sediments*, Int. Assoc. Sedimentol. Spec. Publ. 2, 290 pp.

McKenzie, T., K. Kelts, J. Pika, H. P. Eugster, R. C. Spencer, B. F. Jones, M. J. Baedeker, S. L. Rettig, M. B. Goldhaber, and C. J. Bowser (1981). Oxygen-18 composition of Great Salt Lake sediments: An indicator of lake level fluctuations (abstract), American Association of Petroleum Geologists Meeting, San Francisco.

Müller, G., and F. Wagner (1978). Holocene carbonate evolution in Lake Balaton (Hungary): A response to climate and impact of man, in *Modern and Ancient Lake Sediments*, A. Matter and M. E. Tucker, eds., Internat. Assoc. Sedimentol. Spec. Publ. 2, pp. 57-81.

Müller, G., G. Irion, and U. Förstner (1972). Formation and diagenesis of inorganic Ca-Mg Carbonates in the lacustrine environment, *Naturwissenschaften 59*, 158-164.

Neev, D., and K. O. Emery (1967). The Dead Sea—depositional processes and environments of evaporites, *Israel Geol. Surv. Bull. 41*, 147 pp.

O'Neill, W. A., J. F. Sutter, and K. O. Stanley (1981). Sedimentation history of rich oil-shale sequences in Eocene Lakes Gosiute and Uinta, Wyoming and Utah, *Geol. Soc. Am., Rocky Mountain Sec. Abstr. Programs.*

Purser, B. H., ed. (1973). *The Persian Gulf*, Springer-Verlag, New York, 471 pp.

Sims, T. D., D. P. Adam, and M. J. Rymer (1981). Late Pleistocene stratigraphy and palynology of Clear Lake, Lake County, California, *U.S. Geol. Surv. Prof. Pap. 1041.*

Smith, G. I., ed. (1978). Climate variation and its effect on our land and water, *U.S. Geol. Surv. Circ. 776-B*, 52 pp.

Smith, G. I. (1979). Subsurface stratigraphy and geochemistry of late Quaternary evaporites, Searles Lake, California, *U.S. Geol. Surv. Prof. Pap. 1043*, 130 pp.

Smith, G. I., V. J. Barczak, G. F. Moulton, and J. C. Liddicoat (in press). Core KM-3, a surface-to-bedrock record of late Cenozoic sedimentation in Searles Valley, California, *U.S. Geol. Surv. Prof. Pap.*

Smoot, J. P. (1978). Origin of the carbonate sediments in the Wilkins Peak Member of the lacustrine Green River Formation (Eocene), Wyoming, U.S.A., in *Modern and Ancient Lake Sediments*, A. Matter and M. E. Tucker, eds., Internat. Assoc. Sedimentol. Spec. Publ. 2, pp. 109-127.

Spencer, R. C., H. P. Eugster, K. Kelb, J. McKenzie, B. F. Jones, M. J. Baedeker, S. L. Rettig, M. B. Goldhaber, and E. J. Bowser (1981a). Late Pleistocene and Holocene sedimentary history of Great Salt Lake, Utah (abstract), American Association of Petroleum Geologists meeting, San Francisco.

Spencer, R. C., H. P. Eugster, B. F. Jones, M. J. Baedeker, S. L. Rettig, M. B. Goldhaber, C. J. Bowser, K. Keltig, and J. McKenzie (1981b). Mineralogy and pore fluid geochemistry of Great Salt Lake, Utah (abstract), American Association of Petroleum Geologists meeting, San Francisco.

Stoessel, R. K., and R. L. Hay (1978). The geochemical origin of sepiolite and kerolite at Amboseli, Kenya, *Contrib. Mineral. Petrol. 65*, 255-267.

Stoffers, P., and R. E. Hecky (1978). Late Pleistocene-Holocene evolution of the Kivu-Tanjanyika basins, in *Modern and Ancient Lake Sediments*, A. Matter and M.E. Tucker, eds., Internat. Assoc. Sedimentol. Spec. Publ. 2, pp. 43-44.

Surdam, R. C., and R. A. Sheppard (1978). Zeolites in saline, alkaline-lake deposits, in *Natural Zeolites, Occurrence, Properties, Use*, L. B. Sand and F. A. Mumpton, ed., Pergamon, Oxford, pp. 145-174.

Surdam, R. C., and K. O. Stanley (1979). Lacustrine sedimentation during the culminating phase of Eocene Lake Gosiute, Wyoming, *Geol. Soc. Am. Bull. 90*, 93-110.

Surdam, R. C., and K. O. Stanley (1980). Effects of changes in drainage-basin boundaries on sedimentation in Eocene Lakes Gosiute and Uinta of Wyoming, Utah, and Colorado, *Geology 8*, 135-139.

Szatmari, P., R. S. Carvalho, and I. A. Simoes (1979). A comparison of evaporites facies in the late Paleozoic Amazon and the Middle Cretaceous South Atlantic salt basins, *Econ. Geol. 74*, 432-447.

Wardlaw, N. C. (1972). Unusual marine evaporites with salts of calcium and magnesium chloride in Cretaceous basins of Sergipe, Brazil, *Econ. Geol. 67*, 156-168.

Wolfbauer, C. A. (1971). Geologic framework of the Green River Formation in Wyoming, *Contrib. Geol. U. Wyoming 10*, 3-8.

Yuretich, R. F. (1979). Modern sediments and sedimentary processes in Lake Rudolf (Lake Turkana) eastern Rift Valley, Kenya, *Sedimentology 26*, 313-331.

Ancient Soils and Ancient Climates

11

FRANKLYN B. VAN HOUTEN
Princeton University

INTRODUCTION

The essence of this paper is a consideration of how ancient soils may reflect long-term changes in climate and in the composition of the atmosphere. Detailed field studies and laboratory analyses of ancient nonmarine deposits commonly reveal remnants of fossil soils. These paleosols, together with transported clay- and iron-rich sediments derived directly from them, provide clues for reconstructing ancient climates, because they were formed under a limited range of temperature and humidity.

Basic inquiries that direct such an investigation focus on (1) the relation of climate to a global temperature gradient and (2) patterns of precipitation controlled by the distribution of landmasses and their mountain belts, as well as by (3) the global distribution of drifting continents and (4) the modifying effect of the stand of global sea level and flooding of the continents. Paleosols amplify other geologic information about trends in climate change through Earth history and about those conditions that produced widespread coal swamps and continental glaciation. The data also reflect the role of major regressions of the seas and impose critical constraints on speculation about Precambrian climate and atmosphere.

METHOD

Procedure

The discipline of paleopedology (Yaalon, 1971) is founded on the well-documented proposition that widespread soils in the geologic past, as in the present (White, 1979), reflect atmospheric temperature and humidity, the former varying mostly with latitude, the latter asymmetrically east to west. These, in turn, are conditioned and augmented by effects of local relief and biota, of stability of the landmass (time factor), and of the distribution of land and sea.

Analysis of an ancient weathered profile necessary to establish its particular pedologic character (Table 11.1) requires data concerning the petrography of the parent rock, mineralogic and chemical composition of the profile, and possible diagenetic alteration. Such a study deals with those downward-mobilizing processes that oxidize plant debris, lower soil-water pH, dissolve minerals and remove soluble cations, produce new clay minerals (Singer, 1980), and concentrate insoluble ferric and aluminum oxides in the upper horizon of the soil profile. Current procedures and modes of interpretation involved in paleosol analyses are well illustrated in studies by McPherson (1979) and Retallack (1976, 1977).

TABLE 11.1 Criteria Diagnostic of Pedogenic Profiles

- Relative thin (0.5-3 m), but extensive, tabular
- Transitional lower boundary, sharp upper boundary
- Rooted, burrowed to bioturbated
- Wavy to disturbed bedding, or obliterated
- Vein network
- Patterned shrinkage-swelling features (gilgai)
- Clay coating (cutans) on grains and fragments
- Color mottling (gleying)—gray-blue, violet, red-brown colors independent of stratification
- Calcareous nodules to calcrete—translocation of calcium carbonate
- Fe_2O_3 crusts—ferricrete
- SiO_2 crusts—silcrete; corroded quartz grains
- Concentration of Al_2O_3 and Fe_2O_3 in upper horizon and mobilization of Fe oxides
- Clay minerals modified toward low-silica kaolin
- Depletion of cations, except in calcrete concentration

Paleosols comprise those profiles of weathering developed on rocks or sediments that were exposed long enough to be modified by soil-forming processes. For the present review, potentially meaningful ones are limited largely to lateritic products of hot and humid weathering (Thomas, 1974) composed essentially of ferric and aluminous oxides (sequioxides) and high-aluminum kaolinite clay and to calcretes produced by less intense dry-climate weathering and composed of ferric oxides, clay minerals, and calcium carbonate. Today laterite and calcrete predominate on relatively stable landmasses within the intertropical climate zone (Figure 11.1; Van Houten, 1961, 1973). Here the temperature is less important than differences in amount and seasonal distribution of rainfall in determining the kind of soil produced.

Laterite

Lateritic soils or latosols (Millot, 1970; Paton and Williams, 1972), both aluminous bauxite and ferruginous laterite, form under conditions of prolonged stability and intense weathering, accompanied by essentially no erosion or aggradation. They are favored by relatively uniform maritime conditions on windward sides of continents within the humid intertropical, or tropical forest, zone. Sequioxides and kaolinitic products of laterization can be transported to local depressions, as in a karst terrane, or to depositional basins where distinctive aluminum-rich claystones and oolitic ironstone may accumulate (Millot, 1970).

Calcrete

Carbonate-enriched calcrete (caliche) and calcareous red-earth soils reflect warm, seasonally dry climate and commonly develop on drier leeward sides of continents. In as much as calcrete requires less-intense weathering than laterite, it can form in periods of thousands of years. Estimates suggest that calcrete profiles develop at rates of 10-50 mm/10^3 yr with a maximum of about 1000 mm/10^3 yr. In the geologic record calcretes are commonly associated with reddish-brown detrital deposits containing more silica-rich clay minerals such as illite, montmorillonite, and chlorite. Used prudently these redbeds (Millot, 1970) can provide suggestive evidence of warm, dry climate.

LIMITATIONS

Ancient soils as clues to ancient climate have limitations. Even though laterites and calcretes commonly developed on uplands and well-drained nonmarine deposits in the past, terranes of this sort constitute only a minor part of the geologic record. Moreover, to provide useful information, ancient soils, wherever they formed, must have been buried relatively rapidly to prevent subsequent chemical modification or loss by erosion. Conversely, soils could develop on nonmarine deposits only where accumulation was interrupted or was slow enough to permit development of a distinctive profile. Yet, commonly in such a situation pedological processes were arrested before completion, or the upper diagnostic part of the profile was eroded. In addition, many ancient soils cannot be dated accurately as to time of origin, and after burial some were altered by diagenetic effects that obscure primary climatic indicators. The use of Recent laterite and calcrete as a model is limited, particularly because modern soils have developed during a time of considerable tectonic activity, a low stand of sea level, and assembled continental blocks (Figure 11.1). These factors reduce the global extent of maritime conditions and cratonic stability that favored the formation of laterite in the past.

GEOLOGIC RECORD

Phanerozoic Time

The available data plotted on currently reasonable paleocontinental drift reconstructions (Smith and Briden, 1977; Scotese *et al.*, 1979) and transferred to a time-latitude chart (Figure 11.1) provide a basis for evaluating paleoclimatic information supplied by ancient laterite and calcrete. The record of their temporal and geographic distribution reflects only large-scale motion and change in the atmosphere, hydrosphere, and lithosphere. During the Phanerozoic interplay of global climate zones, latitudinal position of wandering continents, and the longitudinal extent of orogenic belts, and of assembled cratonic blocks, the role of global climate change probably was minimal.

As today, laterites in the past lay mostly within the intertropical zones in east-coast locations (Valeton, 1972; Bardossy, 1973). They were widespread during siderolithic periods (Millot, 1970) when extensive, relatively stable areas of the continents were subjected to prolonged weathering in hot and humid climate, producing ferric oxides, kaolinitic clay, beds of lignite and coal, and quartz-rich sand. Commonly, development of laterite was accompanied by the nearby accumulation of transported bauxite or kaolinite and oolitic ironstones.

A general northward drift of most of the major landmasses during Phanerozoic time (Figure 11.1) produced a northward shift of the broad belt of laterite on Laurasian continents and

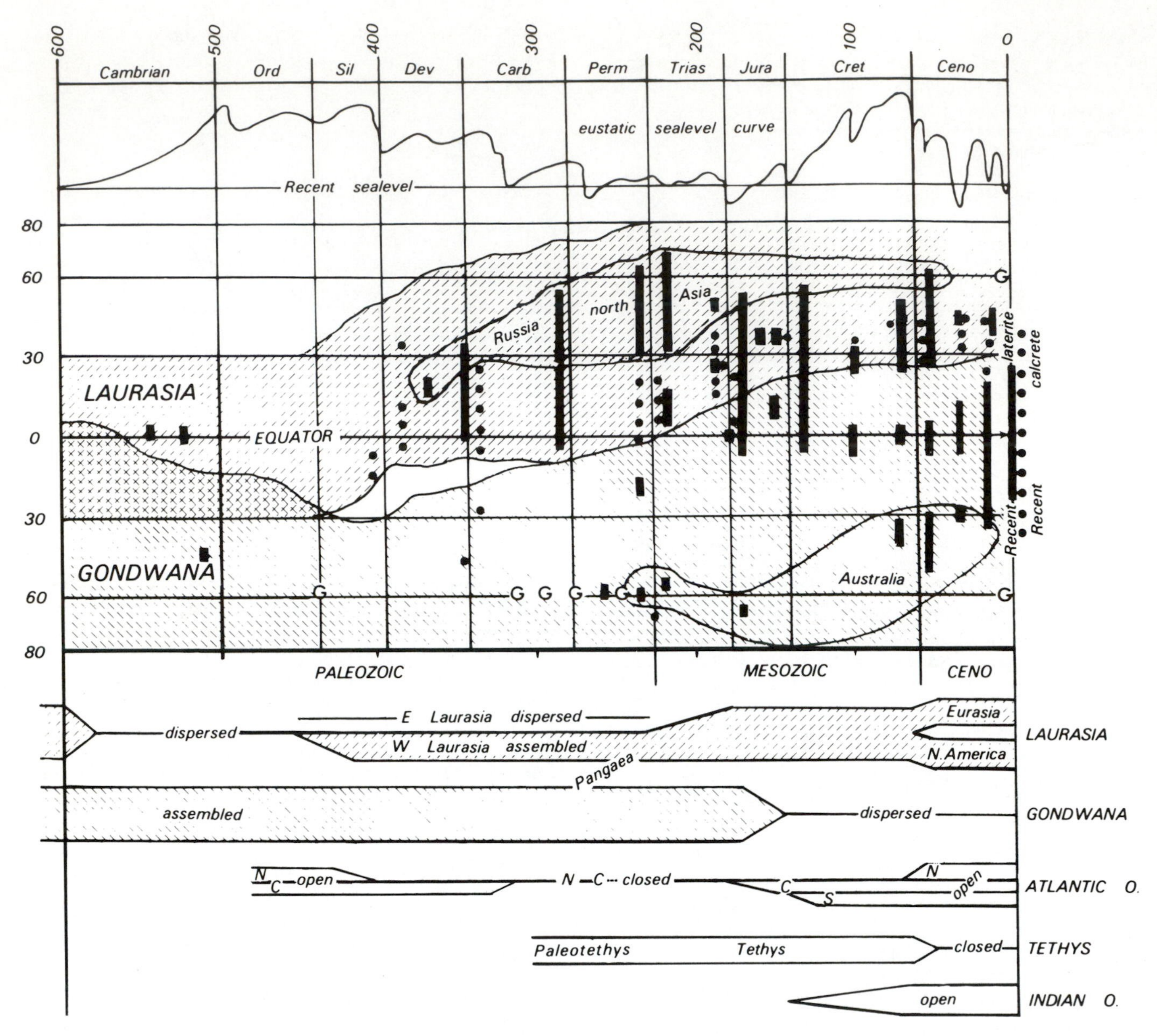

FIGURE 11.1 Time-latitude chart of distribution of Phanerozoic laterite (bars) and calcrete (dots) on cratonic blocks of Laurasia (left diagonal pattern) and Gondwana (right diagonal pattern). Eustatic sea-level curve after Vail *et al.* (1977). Lower part of chart diagrams trends of assemblage and dispersal of blocks of Laurasia and Gondwana and of opening and closing of ocean basins. Based on paleocontinental reconstructions by Scotese *et al.* (1979) and Smith and Briden (1977). G, continental glaciation; N, northern; C, central; and S, south Atlantic.

led to low-latitude development on northern blocks of Gondwana in Mesozoic and Cenozoic time. This trend was modified by a late Mesozoic-Cenozoic southward expansion of the laterite belt when Australia broke away from Antarctica. In middle and late Paleozoic time favorable lateritic conditions were widespread on dispersed Laurasian blocks in middle and high latitudes but occurred on assembled Laurasian blocks only in low latitudes. Similarly, in Mesozoic and early Cenozoic time laterites developed in high northern latitudes toward leeward coasts (Figure 11.2) when the open Atlantic, Tethys, and Caribbean permitted latitudinal oceanic circulation among continental fragments.

The limited resolution of the laterite data is illustrated by the fact that a reconstructed 10° northward drift of Africa and Australia between Eocene and Miocene time is not reflected in the known distribution pattern of laterite. Early Permian to Cenozoic laterite in southern latitudes higher than 30° S are known only in southeastern Australia (Figures 11.1-11.3). The Permian and Triassic paleosols have been interpreted to be products of humid tropical to subtropical intervals within and following the extensive Gondwana glaciation (Loughnan *et al.*, 1974; Loughnan, 1975). Development of oolitic ironstone in Australia in late Permian and early Jurassic time also suggests periods of warm, moist climate. Nevertheless, Retallack

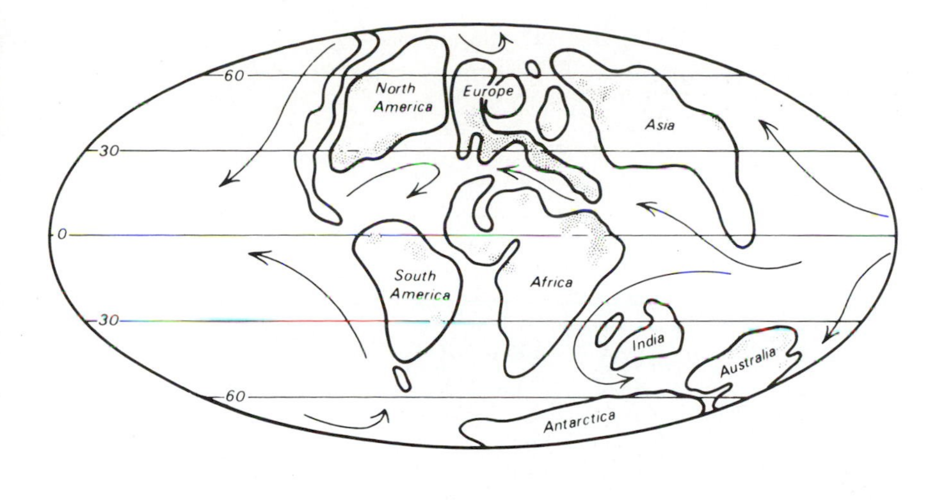

FIGURE 11.2 Cretaceous distribution of laterite (shaded) on landmasses and estimated major oceanic surface currents. Adapted from Smith and Briden (1977).

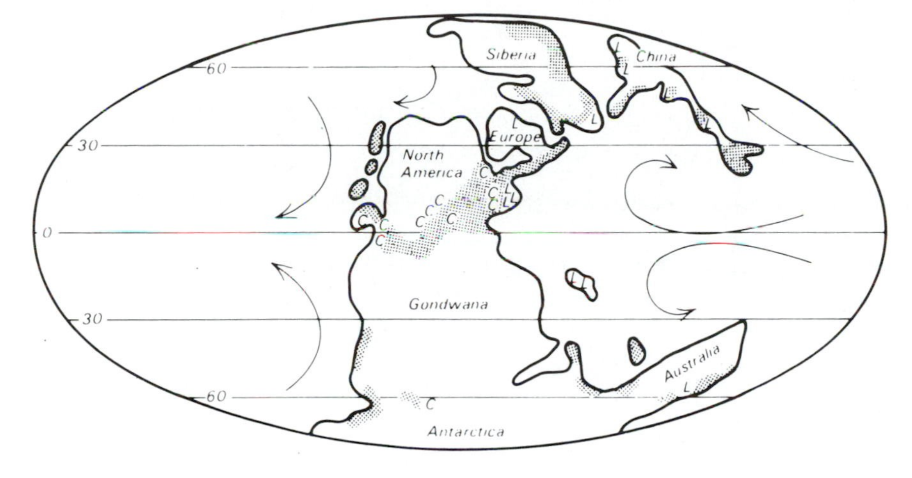

FIGURE 11.3 Late Permian and Early Triassic distribution of laterite (L) and calcrete (C) on landmasses and estimated major oceanic surface currents. Mountains shaded. Adapted from Scotese *et al.* (1979) and Smith and Briden (1977).

(1977) claims that the early Triassic profiles are podzols and that "the climate in southeastern Australia seems to have remained cool temperate from the later Permian to the later Triassic." Late Paleozoic and early Mesozoic laterite in high northern latitudes accumulated in Russia and northern Asia-Siberia when they were the principal landmasses in the far north (Figures 11.1 and 11.3). According to Strakhov (1969, p. 214, Figure 94), these paleosols developed in moist, tropical to subtropical climate.

Almost all of the ancient calcretes recorded formed on intermittently aggraded redbeds in unstable (mobile) belts, commonly within the intertropical zone and on the leeward side of a landmass. All but one lay between latitudes 45° N and 45° S, and most of these are preserved on Laurasian continental blocks. During late Permian and Triassic time, longitudinal assemblage of Laurasia and Gondwana prevented latitudinal oceanic circulation and led to a preponderance of calcrete over laterite in the intertropical zone of North America and Europe (Figures 11.1 and 11.3). This reconstructed low-latitude belt of dryness was rapidly succeeded by relatively humid Early Jurassic climate and the widespread development of laterite and oolitic ironstone in Eurasia and northeastern Gondwana.

Precambrian

A Proterozoic record of laterite and oolitic ironstone, calcrete, and nonmarine redbeds (Chandler, 1980) and aluminous sandstone (Young, 1973) on the Canadian Shield is augmented by redbeds as old as 2000 million years (Ma) in southern Africa (Button and Tyler, 1979) and 2380 Ma in India (Windley, 1977). These data imply an oxygen-deficient atmosphere since about 2300 Ma, and a warm, humid to semiarid climate when the ferric oxide-bearing sediments accumulated. Detailed analysis of this record is complicated by the fact that Canada apparently drifted widely into high and low latitudes during Proterozoic time (Irving, 1979). Nevertheless, both the extensive iron formations and various redbeds that accumulated between 2300 and 1800 Ma lay in relatively low paleolatitudes (Donaldson *et al.*, 1973).

Accumulation of Proterozoic products of oxidative weathering overlapped the widespread blooming of cherty and banded-iron formations by as much as 500 m.y. (Figure 11.4). A specific association of the two, about 2000 Ma, comprises a redbed with weathered andesite pebbles that lies below the Labrador cherty iron formation (Dimroth, 1976). Clearly,

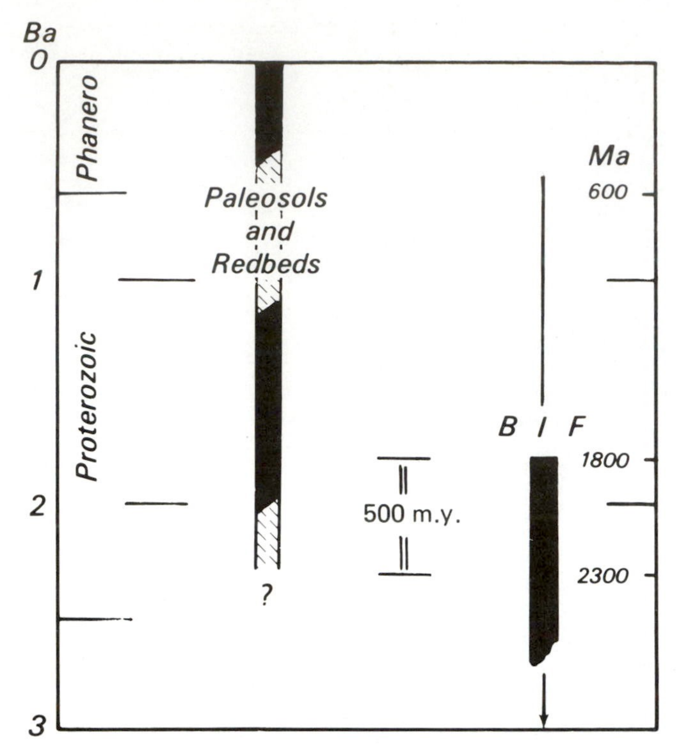

FIGURE 11.4 Geologic range of paleosols, calcrete, and redbeds compared with that of Proterozoic banded (cherty) iron formations (BIF). Black bar, range of effectual record; Ba, billion years ago; Ma, million years ago.

references to an oxygen-deficient atmosphere after 2000-2300 Ma (Button and Tyler, 1979; Chapter 5) are speculations that ignore the facts of the geologic record. In contrast, a marked abundance of CO_2 in the atmosphere and hydrosphere during this interval (Strakhov, 1969) may have been an important factor in the genesis of the unique Proterozoic iron formations.

Evidence of chemical weathering older than 2300 Ma in Canada (2200 Ma in southern Africa; Button, 1979; Button and Tyler, 1979) includes metamorphosed aluminum-rich weathering crusts as old as 3500 Ma (Serdyuchenko, 1968). Commonly, iron has been leached from these profiles, indicating at least local reducing conditions. The evidence does suggest an atmosphere with a limited supply of oxygen (Pienaar, 1963). Nevertheless, one of the gray paleosols is virtually identical to some modern soils formed in temperate, humid climate (Gay and Grandstaff, 1979). Others may be products of oxidative weathering that were altered diagenetically by reducing paleogroundwater. In lieu of substantial evidence, gray profiles alone afford no compelling indication that the early Proterozoic atmosphere was significantly different from that of later time.

CONCLUSIONS

Pedologic effects of ancient climate are more boldly stamped on nonmarine sedimentary rocks than generally acknowledged. Recent detailed studies, such as those by Allen (1974), McPherson (1979), and Retallack (1976, 1977), reveal the amount of useful information that can be gleaned. With careful work the amount of available data should be increased significantly.

There is need for a wider understanding of the criteria for identifying ancient soils (Table 11.1), and for better-trained students who can use the techniques of modern soil science in detailed analyses of nonmarine deposits. Problems involved in this practice include identifying a soil type from a partially eroded profile or recognizing ferruginous and kaolinitic profiles produced by processes other than laterization and distinguishing between the effect of intense warm-climate weathering and of prolonged weathering under more moderate conditions and regional stability.

Generally, the available record of paleosols is more directly related to episodes of widespread nonmarine sedimentation than it is to a global control of climate. This situation, in turn, was closely controlled by tectonic development of appropriate basins. In addition, development and distribution of laterite and calcrete reviewed here imply that the paleolatitudinal position and interrelation of landmasses, and the latitudinal and longitudinal oceanic circulation through open gateways (Chapter 12) during Phanerozoic history of drifting continents, were the principal forcing factors in long-range climate change. These conditions were augmented by effects of the stand of global sea level and extent of flooding of the continents, as well as by the development of continental glaciation or of ice-free polar regions. In contrast, ancient soils contribute little to an understanding of short-term climate-controlled processes or periodicities like those exhibited in the Pleistocene record.

Throughout Phanerozoic time the intertropical belt apparently persisted in essentially steady state but shifted somewhat northward, except during Permo-Triassic time when a meridional assemblage of major landmasses (Pangaea) was a barrier to latitudinal oceanic circulation. During episodes of widespread regression, as in Early Cretaceous time, nonmarine sequences of stable landmasses commonly preserved a record of hot and humid climate in successive ferruginized profiles, as well as in deposits of transported laterites and of kaolinite associated with oolitic ironstones.

Laterite and oolitic ironstones in the early Paleozoic and Proterozoic record demonstrate that development of these products of intense weathering were not uniquely dependent on the presence of vascular plants. In fact, an absence of abundant plant debris may have fostered the development of oxidized soils in an atmosphere with a somewhat diminished oxygen content.

Data and interpretations presented here suggest that both the Earth's atmosphere and its global climate have varied but little during the last 2000 million years. As in Phanerozoic time, Proterozoic paleosols and redbeds reflect conditions favoring preservation of nonmarine deposits rather than variation in oxygen content of the atmosphere. Correlative banded-iron formations may reflect some aspect of a progressive change in the atmosphere. Speculation about this condition is constrained, however, by evidence of an oxidizing atmosphere since about 2300 Ma. In fact, accumulation of banded iron formations may not have been directly related to the oxygen content of the atmosphere and oceans. A significant role of a reducing atmosphere has not been demonstrated (Dimroth and Kimberly, 1976). At present, hypotheses proposing an oxygen-deficient atmosphere in early Proterozoic time are not rigorously convincing.

ACKNOWLEDGMENTS

Research for this report was supported by National Science Foundation grant EAR 77-06007 and funds provided by the Department of Geological and Geophysical Sciences, Princeton University. J. F. Hubert, J. C. Lorenze, J. G. McPherson, and D. L. Woodrow supplied useful information and helpful suggestions.

REFERENCES

Allen, J. R. L. (1974). Studies in fluviatile sedimentation: Implications of pedogenic carbonate units, Lower Old Red Sandstone, Anglo-Welch outcrops, *Geol. J. 9*, 181-208.

Bardossy, G. (1973). Bauxite formation and plate tectonics, *Acta Geol. Acad. Sci. Hung. 3*, 141-154.

Button, A. (1979). Early Proterozoic weathering profile on the 2200 m.y. old Hekepoort Basalt, Pretoria Group, South Africa, *U. Wits. Econ. Geol. Res. Unit Info. Circ. 133*, 19 pp.

Button, A., and N. Tyler (1979). Precambrian paleoweathering and erosion surfaces in Southern Africa: Review of their character and economic significance, *U. Wits. Eco. Geol. Res. Unit Info. Circ. 135*, 37 pp.

Chandler, F. W. (1980). Proterozoic redbed sequences of Canada, *Geol. Surv. Canada 311*, 53 pp.

Dimroth, E. (1976). Aspects of sedimentary petrology of cherty iron-formations, in *Handbook of Stratabound and Stratiform Ore Deposits, Vol. 7*, Elsevier, Amsterdam, pp. 203-254.

Dimroth, E., and M. M. Kimberley (1976). Precambrian atmospheric oxygen: Evidence in sedimentary distribution of carbon, sulfur, uranium and iron, *Can. J. Earth Sci. 13*, 1161-1185.

Donaldson, J. A., E. Irving, J. C. McGlynn, and J. K. Park (1973). Drift of the Canadian Shield, in *Implications of Continental Drift to the Earth Sciences, Vol. 1*, Academic, London, pp. 3-17.

Gay, A. L., and D. E. Grandstaff (1979). Precambrian paleosols at Elliot Lake, Ontario, Canada (abstract), *Geol. Assoc. Canada—Mineral. Assoc. Canada, Joint Ann. Mtg.*, Abstracts, 52.

Irving, E. (1979). Paleopoles and paleolatitudes of North America and speculations about displaced terrains, *Can. J. Earth Sci. 16*, 669-694.

Loughnan, F. C. (1975). Laterites and flint clays in the early Permian of the Sydney Basin, Australia, and their paleoclimatic implications, *J. Sed. Petrol. 45*, 591-598.

Loughnan, F. C., R. Goldberg, and W. N. Holland (1974). Kaolinitic clay rocks in the Triassic Banks wall sandstone of the western Blue Mountains, New South Wales, *J. Geol. Soc. Australia 21*, 392-402.

McPherson, J. G. (1979). Calcrete (caliche) paleosols in fluvial redbeds of the Aztec Siltstone (Upper Devonian), southern Victoria Land, Antarctica, *Sed. Geol. 22*, 267-285.

Millot, G. (1970). *Geology of Clays*, Springer-Verlag, New York, 429 pp.

Paton, T. R., and M. A. J. Williams (1972). The concept of laterite, *Ann. Assoc. Am. Geogr. 62*, 42-56.

Pienaar, P. J. (1963). Stratigraphy, petrology, and genesis of the Elliot Lake Group, Blind River, Ontario, including the uraniferous conglomerate, *Geol. Surv. Canada 83*, 140 pp.

Retallack, G. J. (1976). Triassic paleosols in the Upper Narrabeen Group of New South Wales. Pt. I. Features of the paleosols, *J. Geol. Soc. Australia 23*, 383-400.

Retallack, G. J. (1977). Triassic paleosols in the Upper Narrabeen Group of New South Wales. Pt. II. Classification and reconstruction, *J. Geol. Soc. Australia 24*, 19-36.

Scotese, C. R., R. K. Bambach, C. Barton, R. Van der Voo, and A. M. Ziegler (1979). Paleozoic base maps, *J. Geol. 87*, 217-277.

Serdyuchenko, D. P. (1968). Metamorposed weathering crusts of the Precambrian, their metallogenic and petrographic features, *XXIII Internat. Geol. Congr. 4*, 37-42.

Singer, A. (1980). The paleoclimatic interpretation of clay minerals in soils and weathering profiles, *Earth Sci. Rev. 15*, 303-326.

Smith, A. G., and J. C. Briden (1977). *Mesozoic and Cenozoic Paleocontinental Maps*, Cambridge U. Press, Cambridge, 63 pp.

Strakhov, N. M. (1969). *Principles of Lithologenesis, Vol.2*, Oliver and Boyd, Edinburgh, 609 pp.

Thomas, M. F. (1974). *Tropical Geomorphology*, Wiley, New York, 332 pp.

Vail, P. R., R. M. Mitchum, Jr., and S. Thompson III (1977). Seismic stratigraphy and global changes of sea level, pt. 4, in *Seismic Stratigraphy*, Am. Assoc. Petroleum Geol. Mem. 26, pp. 83-97.

Valeton, I. (1972). *Bauxites*, Elsevier, Amsterdam, 226 pp.

Van Houten, F. B. (1961). Climatic significance of redbeds, in *Descriptive Paleoclimatology*, Interscience, New York, pp. 89-139.

Van Houten, F. B. (1973). The origin of redbeds: A review—1961-1972, *Ann. Rev. Earth Planet. Sci. 1*, 39-61.

White, R. E. (1979). *Introduction to the Principles and Practice of Soil Science*, Wiley, New York, 198 pp.

Windley, B. F. (1977). *The Evolving Continents*, Wiley, London, 385 pp.

Yaalon, D. H., ed. (1971). *Paleopedology—Origin, Nature, and Dating of Paleosols*, Israel U. Press, Jerusalem, 350 pp.

Young, G. M. (1973). Tillites and aluminous quartzites as possible time markers for Middle Precambrian (Aphebian) rocks of North America, *Geol. Assoc. Canada 12*, 97-127.

Role of Ocean Gateways in Climatic Change

12

WILLIAM A. BERGGREN
Woods Hole Oceanographic Institution and Brown University

INTRODUCTION

With its low albedo and large capacity to store heat, the ocean represents one of the most important components of the global climatic system. With a specific heat per unit mass of 4 times that of air, water (i.e., the ocean) has a thermal capacity over 1000 times that of the atmosphere. The energy imbalance from incoming solar radiation between the summer and winter hemispheres and between high and low latitudes results in motions (i.e., transport) in the atmosphere and oceans. The result is a net transport of energy from summer to winter hemispheres and from low to high latitudes. There is little variation at low latitudes in temperature and energy transport. The ocean, with its high thermal capacity, acts as a buffer for such energy changes.

The ocean, at the same time, plays an important role in the meridional transport of energy. The question may then be asked: Has it always been so? The answer is unequivocally no. Present continent-ocean geometry is the result of a long [200 million year (m.y.)] history of changing spatial configurations. The role of gateways (and barriers) is fundamental for recreating the scenario of past circulation histories (and their concomitant effect on paleoclimate). We are only at an early stage in understanding this history, but we can outline the following tentative scenario.

The underlying assumption is the general view that the global climatic cooling characteristic of the Cenozoic is due to natural, earthbound causes, namely the poleward shift of continental masses and the gradual development of a meridional (Cenozoic) rather than latitudinal (Mesozoic) oceanic circulation pattern. Late Neogene (last 15 m.y.) climatic oscillations are, in turn, no doubt the result of the superposition on this global cooling trend of short-term periodic fluctuations in orbital parameters (i.e., the astronomical theory of glaciation, which is less a theory than a mechanism).

The discussion below is presented in two parts. The first deals with a brief summary of the role of the oceans (and their geometry) in heat transfer and thus, ultimately, as a source of climatic change. The second summarizes the main paleogeographic events that have had an influence on oceanic circulation and their effect, where discernible, on climatic evolution.

OCEANS, CLIMATES, AND GATEWAYS

The oceans occupy nearly three quarters of the surface of the Earth and receive over 80 percent of the total solar radiation absorbed at the surface. With a specific heat (0.93 cal g^{-1} °C^{-1}) considerably larger than that of land (0.60-0.19 cal g^{-1} °C^{-1}), seawater yields up its stored heat much more reluctantly than do land areas. With an overturn (i.e., mixing) rate of 500-1500 yr, the oceans may be viewed as vast heat reservoirs that act as a conservative element in maintaining climatic stability.

The Sun is the ultimate source of ocean currents. Uneven absorption of shortwave radiation over small areas on the ocean surface results in density differences that, in turn, set in motion, slow, lateral water-mass movements. Heat energy is transported and distributed accordingly. Deeper waters are warmed by the slower processes of convective mixing. In polar regions such mixing is ultimately complete, whereas elsewhere the thermocline denotes the level to which mixing is effective. In spite of the large amount of radiation that reaches the surface of the ocean, vertical heat transfer is limited to the surficial layers because nearly 90 percent of the heat energy is used in evaporation.

The ocean accounts for nearly a third of the poleward transfer of heat energy [the remaining two thirds is accounted for by atmospheric transport in the form of sensible (50 percent) and latent (15 percent) heat]. This is accomplished primarily in the form of wind-driven surface currents. In a typical ocean basin, surface circulation assumes the form of an anticyclonic gyre centered on a semipermanent (seasonally shifting) area of stable high pressure within the basin. Equatorward the gyre is constrained by the seasonally migrating intertropical convergence zone, poleward by the high-latitude easterlies and longitudinally by landmasses. Gyres commonly consist of four components: a westerly equatorial current, a polar easterly current, a poleward ascending western boundary current, and an equatorial descending eastern boundary current. The intensity and volume of a given current gyre is dependent on wind stress and such geostrophic factors as the Coriolis force, the piling up of waters on the western margins of ocean basins, and the constrictions or deflections by continental geometry.

Heat energy is absorbed by surface waters as a function of latitude. The equatorial zone is the basic source of most of the heat energy in oceanic gyres, although additional radiation is added over the entire extent of a meridional current. In subequatorial belts, westward-moving gyre currents absorb heat at the rate of about 0.1°C for each degree of longitude. Because of this lengthy exposure to high insolation, the maximum surface temperatures are found in the western equatorial extremities of ocean basins. Eastwardly moving currents on the polar sides of gyres lose heat at about the same rate. Meridional currents moving poleward or equatorward cool and warm, respectively, at an average of 0.3°C for each degree of latitude.

At present, the most important oceanic currents from a climatic point of view are the poleward moving extremities of the western boundary currents (such as, North Atlantic Current, Kuroshio, and Brazil Current). These currents transport warm surface waters into regions of low temperatures and thus enhance high-latitude atmospheric heating and evaporation.

By way of contrast, the presence of the Antarctic Circumpolar Current in the southern hemisphere prevents penetration of westerly boundary current beyond 40° S and serves to isolate the Antarctic continent from major sources of precipitation.

This brief summary gives an idea of the role that the oceans play in transporting heat energy and their role as a stabilizing force in maintaining the climatic *status quo*.

At the same time, it will be apparent that the oceans contain the potential for playing a significant role in climatic change given the right set of conditions. The fact that the geologic record provides ample evidence of significantly different climates in the past (when compared with that of today) suggests significant changes in the ocean currents and atmospheric circulation patterns of the past.

Gateways (and barriers) represent the means for transporting (or blocking) heat energy from one area of the globe to another. Identifying these gateways (barriers) and their geometry and refining the time of their inception is of fundamental importance to delineating the scenario of the past climatic history of the Earth.

The major Mesozoic and Cenozoic paleogeographic events (opening or closing of marine gateways) and the associated effect on paleocirculation and paleoclimate are summarized in Figures 12.1-12.4.

TRIASSIC

Global paleography during the Triassic suggests that the major continental landmasses were essentially fused together into a single supercontinent, Pangaea, which was of bipolar extent. Our knowledge of detailed ocean-continent paleogeography during this time, however, is such as to preclude interpretations of the possible role of ocean gateways on climate. The Tethys Sea extended westward as a triangular re-entrant of Panthalassa—the global ocean—between Africa-Arabia to the south and southeast Asia to the north. A westward flowing equatorial current may be deflected into northern and southern high latitudes on encountering the eastern margins of Pangaea. Relative cooling would have occurred at higher latitudes so that these currents would have descended equatorward as cool eastern boundary currents along the western margins of North and South America.

The great abundance and extensive latitudinal distribution of evaporites in the mid-Triassic and the mid-late Triassic development of coral reefs (including the earliest hermatypic hexacorals, at present restricted to shallow warm waters circumscribed by the 30° parallel in the northern and southern hemisphere) along the northern margin of the Tethys Seaway (~ 30° N) attest to warm (i.e., tropical) climates extending to 60° latitude in both hemispheres. The effects of increased continentality (fusion of Pangaea in the Permian) was no doubt an important factor in the climate of the Triassic.

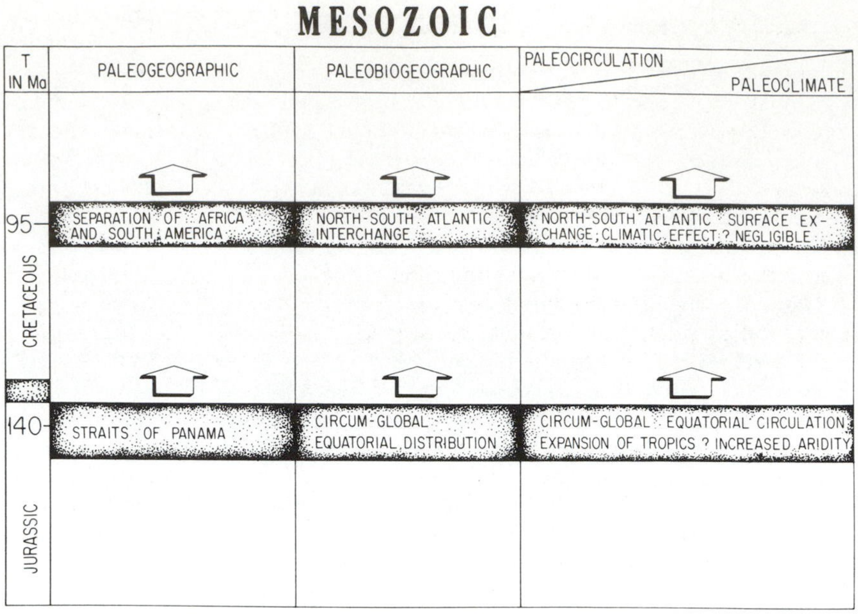

MESOZOIC

T IN Ma	PALEOGEOGRAPHIC	PALEOBIOGEOGRAPHIC	PALEOCIRCULATION / PALEOCLIMATE
CRETACEOUS			
95	SEPARATION OF AFRICA AND SOUTH AMERICA	NORTH-SOUTH ATLANTIC INTERCHANGE	NORTH-SOUTH ATLANTIC SURFACE EXCHANGE; CLIMATIC EFFECT ? NEGLIGIBLE
140	STRAITS OF PANAMA	CIRCUM-GLOBAL EQUATORIAL DISTRIBUTION	CIRCUM-GLOBAL EQUATORIAL CIRCULATION; EXPANSION OF TROPICS ? INCREASED ARIDITY
JURASSIC			

FIGURE 12.1 Mesozoic relationships of paleogeographic, paleobiogeographic, and paleocirculation-paleoclimatic events.

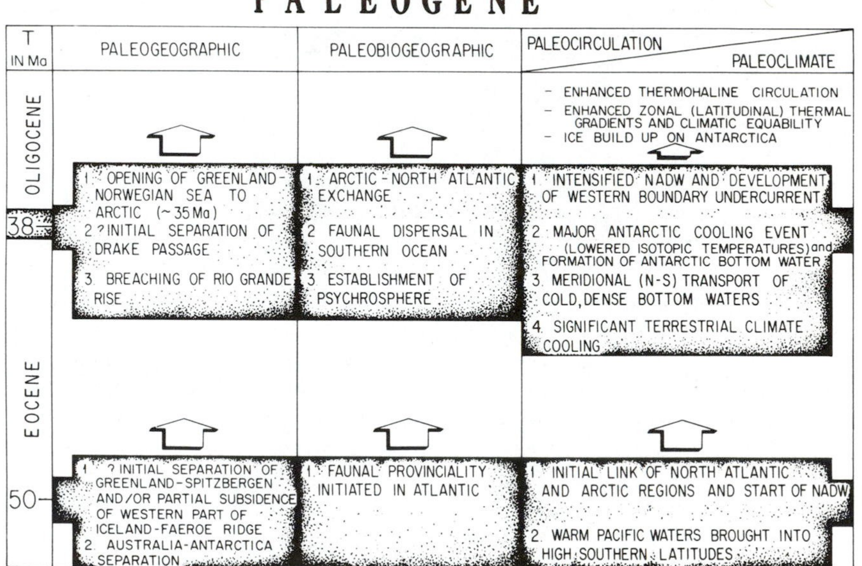

FIGURE 12.2 Paleogene relationships of paleogeographic and paleocirculation-paleoclimatic events.

JURASSIC

Jurassic paleogeography did not differ significantly from the Triassic with one major exception (see below). The latitudinal extension of the continents from pole to pole essentially blocked circumglobal oceanic circulation at high southern latitudes. The Tethys was a large triangular re-entrant of Penthalassa—the global ocean—into the eastern continents from the east. Evidence of distinct temperature-linked gradients is ambiguous during the Jurassic [although floral evidence suggests that a "tropical" belt with cycadlike plants and ferns can be distinguished from "temperate" belts with conifers and ginkgos (Barnard, 1973)]. The essentially latitudinal circulation pattern (deflected into higher latitudes around the polar extremities of Pangaea) would appear to have contributed to the equitable climate of the Jurassic.

A significant circulation event occurred during the late Jurassic—the opening or extension of a seaway through Central America, linking the "Atlantic" and "Pacific" Oceans for the first time in the equatorial region. Coupled with the gradual opening of the North Atlantic by seafloor spreading, this allowed the development of a circumglobal equatorial current system, which has continued, essentially unabated, until relatively recent time [about 3 million years ago (Ma)].

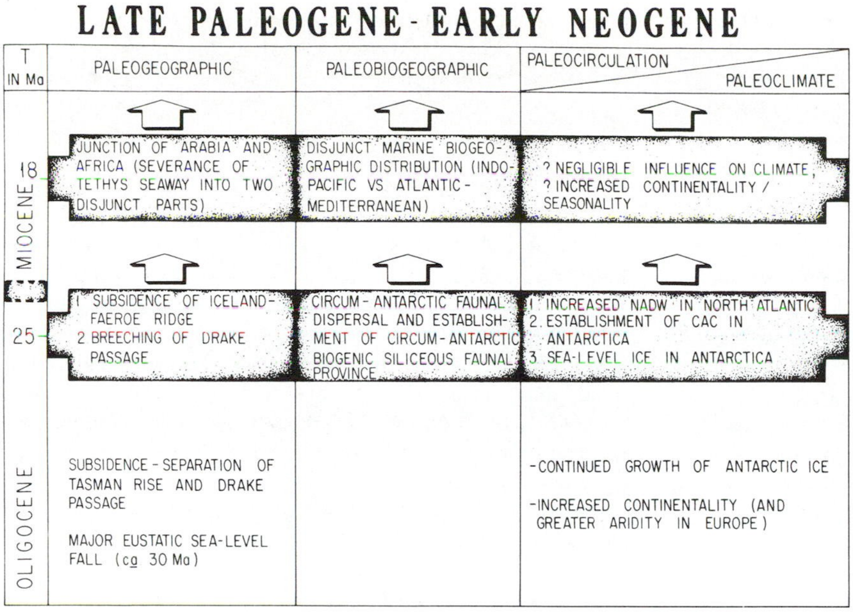

LATE PALEOGENE-EARLY NEOGENE

T IN Ma	PALEOGEOGRAPHIC	PALEOBIOGEOGRAPHIC	PALEOCIRCULATION / PALEOCLIMATE
MIOCENE 18	JUNCTION OF ARABIA AND AFRICA (SEVERANCE OF TETHYS SEAWAY INTO TWO DISJUNCT PARTS)	DISJUNCT MARINE BIOGEOGRAPHIC DISTRIBUTION (INDO-PACIFIC VS ATLANTIC-MEDITERRANEAN)	? NEGLIGIBLE INFLUENCE ON CLIMATE, ? INCREASED CONTINENTALITY / SEASONALITY
25	1 SUBSIDENCE OF ICELAND-FAEROE RIDGE 2 BREECHING OF DRAKE PASSAGE	CIRCUM-ANTARCTIC FAUNAL DISPERSAL AND ESTABLISHMENT OF CIRCUM-ANTARCTIC BIOGENIC SILICEOUS FAUNAL PROVINCE	1 INCREASED NADW IN NORTH ATLANTIC 2. ESTABLISHMENT OF CAC IN ANTARCTICA 3 SEA-LEVEL ICE IN ANTARCTICA
OLIGOCENE	SUBSIDENCE-SEPARATION OF TASMAN RISE AND DRAKE PASSAGE; MAJOR EUSTATIC SEA-LEVEL FALL (ca 30 Ma)		-CONTINUED GROWTH OF ANTARCTIC ICE; -INCREASED CONTINENTALITY (AND GREATER ARIDITY IN EUROPE)

FIGURE 12.3 Late Paleogene-early Neogene relationships of paleogeographic, paleobiogeographic, and paleocirculation-paleoclimatic events.

LATE NEOGENE

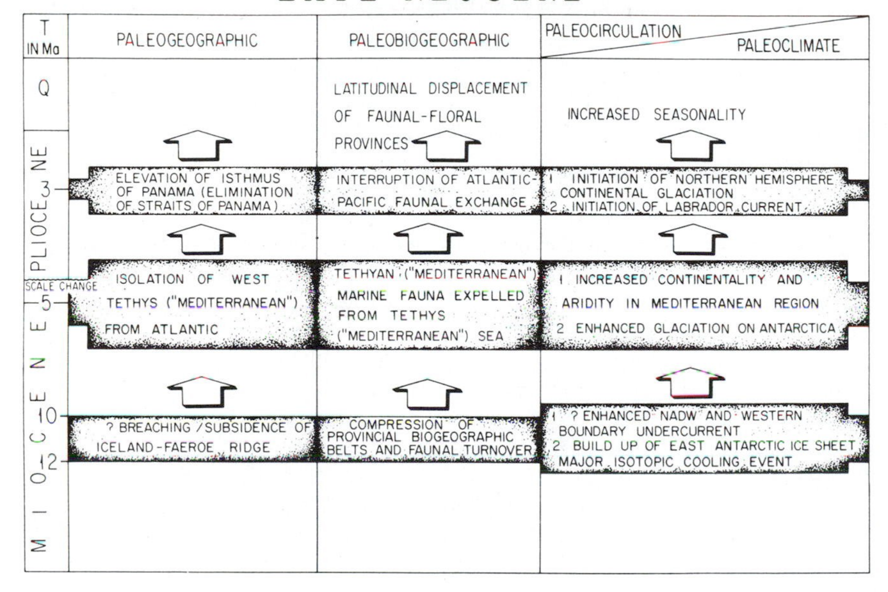

FIGURE 12.4 Late Neogene relationships of paleogeographic, paleobiogeographic, and paleocirculation-paleoclimatic events.

Trans-Atlantic flow of warm, equatorial waters from the Tethys into the equatorial Pacific via the Central America Seaway resulted in a circumglobal equatorial current that served as the main dispersal agent of tropical faunal and floral elements throughout the Late Mesozoic and Cenozoic (Berggren and Hollister, 1974, 1977).

The effect of this event on climate during the late Jurassic is less clear, although the expansion of extensive evaporite deposits into higher latitudes and the concomitant decrease of coal formation suggest increasing aridity. Although there are some discrepancies between data from the rocks (evaporites, corals, bauxites, carbonates), fossils (reefs, microfossils, flora, dinosaurs), and oxygen isotopes, a synthesis of present data suggests warm climates—although on the average the Jurassic may have been slightly cooler than the Triassic (Frakes, 1979). For instance, the latitudinal extent of Jurassic evaporites suggests that it was drier than the Triassic on average. On the other hand, the temporal distribution of evaporites suggests that the early Jurassic, as well as the Late Triassic, was probably cooler and more humid than the middle Triassic.

The significance of the expansion of carbonates into higher latitudes in the Late Jurassic as a climatic indicator is equivocal; it may be a function of eustatic sea-level changes (Hallam, 1975). Although a link between transgressions and an increase in global temperature (resulting in an expansion of the tropical belt and concomitantly, carbonate deposition) appears to be valid in the later Mesozoic and Cenozoic, it is not possible to determine this relationship on the scale treated here.

CRETACEOUS

Data from paleobiogeography and the distribution of rock types suggests that the Cretaceous was a time of high sea levels, shallow epicontinental seas, and warm, dry climates in which tropical conditions comparable with those of today extended under optimum condition beyond 45° N and 70° S latitude (Frakes, 1979). Beyond this, climatic zones were warm to cool-temperate. Mean annual temperatures have been estimated to have been between 10-15°C warmer than the present day, with a temperature gradient about half that of the present.

Oxygen isotope data (Savin, 1977) suggest a general warming trend during the Early Cretaceous (Valanginian-Albian), followed by a gradual but significant cooling in the Late Cretaceous (Campanian-Maestrichtian). Superimposed on this general trend is a shorter-term temperature decrease during the Albian or Cenomanian, followed by a temperature rise in the Turonian and/or Coniacian.

The Early Cretaceous warming may have brought bottom-water temperatures in the central and northwestern Pacific up to an Albian maximum of about 17°C and the Late Cretaceous cooling to a minimum of about 11°C. Concomitant surface temperatures were about 28°C (Albian) and about 19°C (Maestrichtian) (Douglas and Savin, 1975; Frakes, 1979).

The continents were clustered together and continued to extend nearly from pole to pole during the Cretaceous. A single gyre system continued to operate in each hemisphere to about 50-60° latitude, with wide, deep, and sluggish equatorial currents being substantially heated by the long fetch (over 180° of arc) in low latitudes. Poleward transport of this heat in the gyres to the latitude of the descending subtropical circulation cells and the subsequent evaporation nutured the humid, somewhat cooler zones near the poles. Thus the presence of coals at 70° S (New Zealand) and to within a few degrees of the pole in the northern hemisphere (Alaska) may have their explanation in the warm, wide, and deep western boundary-ocean currents that could have flowed unimpeded along the western margins of the Pacific into high latitudes.

The role of ocean pathways in the development of the climatic history of the Cretaceous is difficult to assess at this time simply because they are difficult to identify. One of the major paleogeographic events of the Cretaceous was the separation of Africa and South America and the formation of a permanent marine connection between the North and South Atlantic Oceans during the Turonian Age, about 95 Ma (Reyment and Tait, 1972a, 1972b). However, the existence of barriers to deep-water circulation in the Arctic and Antarctic regions as well as the Rio Grande Rise and Walvis Ridge in the South Atlantic would appear to have nullified any significant contribution to climatic change that the initiation of an oceanic gateway for surface circulation between the North and South Atlantic might have had.

CENOZOIC

Paleogene

The continued fragmentation of the world ocean led to an increasingly inefficient latitudinal thermal energy exchange. Paleobiogeographic and oxygen isotopic studies yield a complementary picture of a long-term global temperature decline, development of a thermally stratified ocean, and enchanced zonal climatic differentiation during the Cenozoic (Berggren and Hollister, 1974, 1977; Frakes, 1979). This climatic decline followed a faunally and florally (Haq and Lohmann, 1976; Haq *et al.*, 1977) and oxygen isotopically (Boersma and Shackleton, 1977; Vergnaud-Grazzini *et al.*, 1978, 1979) recognizable "climatic optimum" in the early Eocene.

The opening of the northeast Atlantic (between Greenland and Norway) began in the early Eocene at about the same time as the separation of Australia and Antarctica (Berggren and Hollister, 1974, 1977). The widespread development of biogenic siliceous deposits in a circumequatorial belt during the Middle Eocene has been linked with the presence of persistent wind-driven (trade winds), diverging ocean-surface currents and associated upwelling, as well as temporally related volcanism in the northeast Atlantic associated with the early phase of rifting and seafloor spreading (Berggren and Hollister, 1974, 1977; Drewry *et al.*, 1974). The high-latitude source of the intensification of oceanic surface and deep-water circulation has been the subject of considerable debate. Although vertebrate faunal evidence suggests elimination of a subaerial barrier in the Norwegian-Greenland Sea at 50 Ma (McKenna, 1972), recent geophysical data suggest a later, early Oligocene date (about 35 Ma) for the separation of Greenland-Spitsbergen (Talwani and Eldholm, 1977). The picture is complicated by the fact that the Greenland-Iceland-Faeroe Ridge may have prevented significant exchange of water from the polar regions to the North Atlantic until the Late Oligocene, about 23 Ma (Tucholke and Vogt, 1979).

An alternate source for the origin of cold, deep water formation—the "big flush"—may lie in the high southern latitude, probably from the Weddell region of the Antarctic (Schnitker, 1980). These cold bottom waters could have moved northward into the North Atlantic through the nascent fracture zones developing in the subsiding Rio Grande Rise.

In the southern hemisphere the northward transit of India diverted the westward flowing Pacific-equatorial current into the counterclockwise subtropical gyre, which penetrated southward, past eastern Australia into the nascent gateway that was opening up between Australia and Antarctica. Isolation of the warm Pacific gyres at these high southerly latitudes led, through adiabatic cooling and evaporation, to precipitation on the Antarctic continent, which eventually led to full continental glacial conditions. This was the beginning of the gradual thermal isolation of the Antarctic continent, which was essentially completed by mid to late Oligocene time (about 25 Ma) with the subsidence of the Tasman Rise and the opening of the Drake Passage and which has continued to this day in the form of the circum-Antarctic current as the gateway has continued to widen during the Neogene.

Oxygen isotope studies indicate a progressive cooling from the Middle Eocene to Early Oligocene time in the sub-Antarctic (Shackleton and Kennett, 1975; Kennett and Shackleton, 1976; L. D. Keigwin, University of Rhode Island, personal communication), North Atlantic (Vergnaud-Grazzini *et al.*, 1978, 1979), South Atlantic (Boersma and Shackleton, 1977), and mid-high latitude Pacific (Douglas and Savin, 1975; Savin *et al.*, 1975). Superimposed on this trend is a pronounced cooling event seen near the Eocene-Oligocene boundary in the Pacific and sub-Antarctic regions, whereas in the Atlantic the cooling appears to be more gradual and of lesser magnitude (Kiegwin, 1980). The relationship of the opening of the Greenland-Norwegian Sea to the Arctic and the possible initial opening of the Drake Passage at about this time (about 38 Ma, J. Sclater, Massachussetts Institute of Technology, personal communication) to the climatic "event" at the Eocene-Oligocene boundary remains unclear because of the lack of precision in dating these various events, but they may be expected to have played a role in enhancing the latitudinal thermal gradients resulting from a greater latitudinal transport of cooler waters originating at high latitudes.

At about the same time, about 40 Ma, the significant generation of cold bottom waters in the Antarctic (and possibly the North Atlantic) resulted in the formation of the psychrospheric fauna, which today lives at temperatures less than 10°C. The appearance of this fauna in the Atlantic Ocean and Tethys Sea at about this time suggests that the Rio Grande Rise has been breached, allowing cold, dense waters to move along a north-south meridional corridor, enhancing the transition from a latitudinal thermospheric circulation to a meridional thermohaline circulation. This transition has been one of the major threshold events in the evolution of the present climate of the world. From this time on, thermally differentiated water masses have moved in a basically meridional pattern, transferring heat across latitudes and establishing a basically latitudinally controlled, zonal climatic gradient of significant proportions.

With the opening of the Drake Passage by mid- to late-Oligocene time the circumpolar current had been established, thermally isolating the Antarctic Continent from the influence of warmer waters to the north (Kennett *et al.*, 1972, 1975). This event appears to be closely linked chronologically with the extension of glaciers on Antarctica to sea level, as shown by drilling in the Ross Sea (Hayes and Frakes, 1975).

Neogene

During the Late Cretaceous and Paleogene, northward movement and rotation of Africa continued to close the Tethys Sea in the east. In the west, right-lateral motion between Africa and Europe narrowed the western junction of the Tethys with the Atlantic between Spain and Morocco. The junction of Arabia and Asia, severing the marine gateway connection between the Indo-Pacific and Atlantic Oceans, occurred during the Early Miocene, about 18 Ma (Van Couvering and Miller, 1971; Berggren, 1972a; Berggren and Hollister, 1974). The transition from a thermospheric to two-layered psychrospheric structure in the Tethys Sea appears to have been effected by Late Eocene time, about 40 Ma (Benson and Sylvester-Bradley, 1971). The effect on climate of the closure of the east-west Tethyan gateway and the subsequent evolution of the Tethys into an eastern (Indo-Pacific Ocean) and western (Mediterranean Sea) part is not clear at this point.

A major decline in isotopic temperature occurs in the Middle Miocene, about 14 Ma (Savin, 1977; Woodruff *et al.*, 1980), and it has been linked with the establishment of the East Antarctic Ice Sheet (Shackleton and Kennett, 1975; Savin, 1977). This cooling event is difficult to link with the development or termination of a structural barrier to flow of water masses because of the disparity in the data that exist on the timing of various events in the North Atlantic. However, the concomitant effects of the severance of the Tethys Seaway and the subsidence of the Iceland-Faeroe Ridge (linking the Arctic polar sea and the Atlantic Ocean) have been linked in a scenario that connects this thermal event with the development of Antarctic glaciation and the evolution of "modern" ocean circulation (Schnitker, 1980). The scenario is as follows:

1. Closure of Tethys in the east (18 Ma) created an evaporation basin that contributed increasing amounts of warm, saline waters to the North Atlantic.
2. Continued fragmentation of Tethys elsewhere (northward drift of Australia separated the equatorial Indian Ocean from the equatorial Pacific) closed off the deep-water connection and reduced the flow of the equatorial current system.
3. Restriction of the Central American Seaway by the gradual emergence of the Isthmus of Panama gradually reduced the flow of the equatorial current between the Atlantic and Pacific. Deflection of the North Equatorial Current northward increased the strength of the Gulf Stream.
4. The breakup of Tethys and its low-latitude circumglobal flow resulted in a temporary increase in surface-water temperatures in the late early Miocene.
5. The gradual subsidence of the Iceland-Faeroe Ridge allowed North Atlantic surface waters into the Norwegian during the early Miocene. By mid-Miocene time subsidence had progressed to the point where reflux of cold bottom water became significant. The Arctic Ocean was then linked with the world ocean as a heat sink.
6. Early North Atlantic deep water, resulting from Norwegian Sea overflow, traversed the length of the Atlantic and inserted itself as an intermediate water mass in the circum-Antarctic current system.
7. Relatively "warm" and saline North Atlantic deep water would rise to the surface near the Antarctic rather than the cold, low-salinity local water. This warm, saline water contained heat that could be converted to latent heat by evaporation. The resulting high evaporation rates supplied moisture

to Antarctica sufficient to push it across the threshold value needed to build up and maintain a stable continental ice cap.

During the latest Miocene (about 5.5-5.0 Ma) the west Tethys Sea was isolated from the Atlantic Ocean. In the Pliocene, marine connections were established and the Mediterranean Sea gradually evolved from a thermohaline to its present thermospheric condition. The effect of the closure of the marine gateway linking the Tethys and Atlantic and the subsequent desiccation of the West Tethys Sea on climate is not clear. In general, the increased continentality would have resulted in enhanced climatic seasonality, although the formation of extensive evaporite deposits (i.e., sabkhas) indicate hot arid conditions as one component of the overall climate. The subsequent opening of the Gilbraltar gateway and the evolution of a thermospheric watermass in the Mediterranean has resulted in an "inland sea" in which there is an excess of surface evaporation over local inflow from rivers and precipitation. The Mediterranean would seem to have a buffering effect on climate, contributing (together with the circum-Mediterranean Alpine chain to the north and the broad, expansive desert areas to the south) to the "Mediterranean climate" to which it gives its name.

Perhaps one of the most significant late Neogene paleogeographic events was the elevation of the Isthmus of Panama, which resulted in the termination of the marine gateway linking the Atlantic and Pacific Oceans, at 3.5 Ma (Berggren and Hollister, 1974). This event had a dramatic effect on oceanic circulation patterns (interruption of the Atlantic to Pacific equatorial current system), paleobiogeography of marine (change from global to disjunct distribution patterns) and terrestrial (re-establishment of north-south migration route—"The Great American Faunal Interchange") organisms. The effect of this barrier on climate is not fully clear, but the proximity in time of its formation to the initiation of northern hemisphere polar glaciation (about 3 Ma; Berggren, 1972b) may indicate a significant role—pumping increased volumes of warm, high-salinity waters (Gulf Stream) into high latitudes where evaporation led to precipitation over the region of eastern Canada and Greenland, cooling, and eventually to the development of the polar ice cap (Luyendyk *et al.*, 1972; Berggren and Hollister, 1974).

During the last 100,000 yr eustatic sea-level fluctuations have led to alternate opening and closing of the Straits of Bab el Mandab in the southern Red Sea and the temporal isolation (and partial evaporation) of the Red Sea (Berggren, 1969). This cyclic process may be expected to have occurred earlier in the Pleistocene as well, but stratigraphic evidence is lacking. What effect these events would have had on climate is not clear, but they may be expected to have been minor and of local extent.

SUMMARY

The history of climate over the past 200 m.y. is one of changing oceanic currents that are in turn the result of changing ocean-continent geometries. During the Mesozoic, the continents of the globe were grouped into a single landmass of essentially bipolar extension. This configuration resulted in a highly efficient, poleward heat transport by large, sluggish gyre systems that were substantially warmed by equatorial transit across nearly half the globe.

The gradual breakup of Pangaea and the consequent dismemberment of Panthalassa into the several connected, but distinct, oceans of today resulted in a decline in the equitable climatic conditions of the Mesozoic. The initial event in the sequence of events leading to the present climatic conditions may have been the mid-Mesozoic (about 140 Ma) development of a circumequatorial, Tethyan seaway that would have had the effect of establishing additional stable gyre systems, thereby diminishing the efficiency of the gyre system in poleward heat transport. As more oceans and gyres have been created and low-latitude seas have become less extensive (through continental drift), global climatic equatability has declined, accelerating significantly during the Cenozoic. In general terms, Mesozoic circulation was latitudinal (and meridional transport of heat energy was relatively efficient), whereas Cenozoic circulation has been predominantly longitudinal (meridional), although meridional heat transport has become increasingly less efficient. The role of gateways has been of fundamental importance in this history of declining climatic equatability because they have served as the conduit through which new current systems are introduced into a new ocean-continent geometry.

ACKNOWLEDGMENTS

I thank J. P. Kennett and A. M. Ziegler for their critical reviews of an earlier draft of the manuscript. This study has been supported by the Submarine Geology and Geophysics Branch of the Oceanography Division of the National Science Foundation, Grant No. OCE-7819769. This is Woods Hole Oceanographic Institution Contribution No. 4759.

REFERENCES

Barnard, P. D. W. (1973). Mesozoic floras, in *Organisms and Continents Through Time*, N. F. Hughes, ed., Spec. Pap. in Palaeontol. 12, Palaeontol. Assoc., London, pp. 175-188.

Benson, R. H., and P. C. Sylvester-Bradley (1971). Deep-sea ostracodes and the transformation of ocean to sea in the Tethys, *Bull. Centre Rech. Paul-SNPA, Suppl.*, 63-91.

Berggren, W. A. (1969). Micropaleontological investigations of Red Sea cores: Summation and synthesis of results, in *Hot Brines and Recent Heavy Metal Deposits in the Red Sea*, E. T. Degens and D. A. Ross, eds., Springer, New York, pp. 329-335.

Berggren, W. A. (1972a). A Cenozoic time-scale—some implications for regional geology and paleobiogeography, *Lethaia* 5, 195-215.

Berggren, W. A. (1972b). Late Pliocene-Pleistocene glaciation, in *Initial Reports of the Deep Sea Drilling Project 12*, U.S. Government Printing Office, Washington, D.C., pp. 953-963.

Berggren, W. A. and C. D. Hollister (1974). Paleogeography, paleobiogeography and the history of circulation in the Atlantic Ocean, in *Studies in Paleooceanography*, W. W. Hay ed., Soc. Econ. Paleontol. Mineral. Spec. Publ. 20, pp. 126-186.

Berggren, W. A., and C. D. Hollister (1977). Plate tectonics and paleocirculation—commotion in the ocean, *Tectonophysics 11*, 11-48.

Boersma, A., and N. J. Shackelton (1977). Tertiary oxygen and carbon isotope stratigraphy, Site 357 (mid latitude South Atlantic), in *Initial Reports of the Deep Sea Drilling Project 39*, U.S. Government Printing Office, Washington, D.C., pp. 911-924.

Douglas, R. G., and S. M. Savin (1975). Oxygen and carbon isotope analyses of Cretaceous and Tertiary microfossils from Shatsky Rise and other sites in the North Pacific Ocean, in *Initial Reports of the Deep Sea Drilling Project 32*, U.S. Government Printing Office, Washington, D.C., pp. 509-520.

Drewry, G. E., A. T. S. Ramsey, and A. G. Smith (1974). Climatically controlled sediments, the geomagnetic field, and trade wind belts in Phanerozoic time, *J. Geol. 82*, 531-553.

Frakes, L. A. (1979). *Climates Throughout Geologic Time*, Elsevier, Amsterdam, 310 pp.

Hallam, A. (1975). *Jurassic Environments*, Cambridge U. Press, Cambridge, 269 pp.

Haq, B. U., and G. P. Lohmann (1976). Early Cenozoic calcareous nannoplankton biogeography of the Atlantic Ocean, *Mar. Micropaleontol. 1*, 119-194.

Haq, B. U., I. Premoli-Silva, and G. P. Lohmann (1977). Calcareous plankton biogeographic evidence for major climatic fluctuations in the early Cenozoic Atlantic Ocean, *J. Geophys. Res. 82*, 3861-3876.

Hayes, D. E., and L. A. Frakes (1975). General synthesis, in *Initial Reports of the Deep Sea Drilling Project 28*, U.S. Government Printing Office, Washington, D.C., pp. 919-942.

Kennett, J. P., and N. J. Shackleton (1976). Oxygen isotopic evidence for the development of the psychrosphere 38 Myr ago, *Nature 200*, 513-515.

Kennett, J. P., R. E. Burns, J. E. Andrews, M. Churkin, Jr., T. A. Davies, P. Dumitrica, A. R. Edwards, J. S. Galehouse, G. H. Packham, and W. J. M. Van der Linden (1972). Australian-Antarctic continental drift, paleocirculation changes and Oligocene deep-sea erosion, *Nature 239*, 51-55.

Kennett, J. P., R. E. Houtz, P. B. Andrews, A. R. Edwards, V. A. Goston, M. Hajos, M. A. Hampton, D. G. Jenkins, S. V. Margolis, A. T. Ovenshine, and K. Perch-Nielsen (1975). Antarctic glaciation and the development of the Circum-Antarctic Current, in *Initial Reports of the Deep Sea Drilling Project 29*, U.S. Government Printing Office, Washington, D.C., pp. 1155-1170.

Kiegwin, L. D., Jr. (1980). Palaeoceanographic change in the Pacific at the Eocene-Oligocene boundary, *Nature 287*, 722-725.

Luyendyk, B. P., D. Forsyth, and J. D. Phillips (1972). Experimental approach to the paleocirculation of the oceanic surface waters, *Geol. Soc. Am. Bull. 83*, 2649-2664.

McKenna, M. (1972). Eocene final separation of the Eurasian and Greenland-North American land masses, *24th Int. Geol. Congr., Montreal, Section 7*, 275-281.

Reyment, R. A., and E. A. Tait (1972a). Faunal evidence for the origin of the South Atlantic, *24th Int. Geol. Congr., Montreal, Section* 7, 316-323.

Reyment, R. A., and E. A. Tait (1972b). Biostratigraphic dating of the early history of the South Atlantic Ocean, *Phil. Trans. R. Soc. London, Ser. B 264*, 55-95.

Savin, S. M. (1977). The history of the Earth's surface temperature during the past 100 million years, *Ann. Rev. Earth Planet. Sci. 5*, 319-355.

Savin, S. M., R. G. Douglas, and F. G. Stehli (1975). Tertiary marine paleotemperatures, *Geol. Soc. Am. Bull. 86*, 1499-1510.

Schnitker, D. (1980). Quaternary deep-sea benthic foraminifers and bottom water masses, *Ann. Rev. Earth Planet. Sci. 8*, 343-370.

Shackleton, N. J., and J. P. Kennett (1975). Paleotemperature history of the Cenozoic and the initiation of Antarctic glaciation: Oxygen and carbon isotope analysis in DSDP sites 277, 279 and 281, in *Initial Reports of the Deep Sea Drilling Project 29*, U.S. Government Printing Office, Washington, D.C., pp. 743-755.

Talwani, M., and O. Eldholm (1977). Evolution of the Norwegian-Greenland Sea, *Geol. Soc. Am. Bull. 88*, 969-999.

Tucholke, B. E., and P. R. Vogt (1979). Western North Atlantic: Sedimentary evolution and aspects of tectonic history, in *Initial Reports of the Deep Sea Drilling Project 43*, U.S. Government Printing Office, Washington, D.C., pp. 791-825.

Van Couvering, J. A., and J. A. Miller (1971). Argon isotope age of the samos *Hipparion* faunas and Late Miocene chronostratigraphy, *Nature 230*, 559-563.

Vergnaud-Grazzini, C., C. Pierre, and R. Le'Tolle (1978). Paleoenvironment of the North-East Atlantic during the Cenozoic: Oxygen and carbon isotope analyses at DSDP sites 398, 400A and 401, *Oceanol. Acta 1*, 381-390.

Vergnaud-Grazzini, C., C. Muller, C. Pierre, R. Le'Tolle, and J. P. Peypouquet (1979). Stable isotopes and Tertiary paleontological paleooceanography in the northeast Atlantic, in *Initial Reports of the Deep Sea Drilling Project 48*, U.S. Government Printing Office, Washington, D.C., pp. 475-491.

Woodruff, F., S. M. Savin, and R. G. Douglas (1980). Biological fractionation of oxygen and carbon isotopes by Recent benthic foraminifera, *Mar. Micropaleontol. 5*, 3-11.

Climatic Acme Events in the Sea and on Land

13

BILAL U. HAQ
Woods Hole Oceanographic Institution

INTRODUCTION

Major climatic events in the geologic past have left a variety of retrievable records in the sediments, both on land and sea. Some marked fluctuations in the climates are evident from widely distributed sedimentological, geochemical, and paleontological features. Here we are mainly concerned with the paleontological record of widespread latitudinal domination of certain floral and faunal elements during times of extreme climatic change. Numerous intervals during which biogeographic patterns show poleward expansions of tropical assemblages, or equatorward extensions of cold, higher-latitude assemblages, have been documented for the Cenozoic. These events have been informally termed climatic *acme events* (indicating peak climatic change) and have been used successfully in the reconstruction of paleoclimate. This paper describes and illustrates the concept of climatic acme events with examples from the Cenozoic marine and terrestrial record and, where available, gives evidence from the oxygen isotope record that corroborates the paleontological conclusions. The usefulness of acme events in biostratigraphic resolution of higher latitudes is also discussed.

The concept of acme events was informally introduced by Haq and Lohmann (1976). The maxima of assemblage migrationary cycles were termed "acme horizons" and defined as the levels in relatively continuous sequences characterized by the maximum abundance of an environmentally sensitive assemblage. If one assumes that times of extreme environmental (climatic, oceanographic) change are, geologically speaking, contemporaneous over large geographic areas, then the acme events will define synchronous horizons. Haq and Lohmann (1976) suggested that the precision with which acme horizons approximate synchronous surfaces depends on our ability unambiguously to identify the acme events (maximum geographic shifts of assemblages). This is affected both by the complexity of the migrationary cycles and by resolution of the method used to delineate them.

The recognition of acme events involves the use of paleobiogeographic data to delineate the cycles of latitudinal migrations of assemblages in stratigraphic sequences over wide geographic areas. The method has been described by Haq and Lohmann (1976) and further discussed by Haq *et al.* (1977) and Haq (1980). It essentially involves the following steps: (1) census of a particular fossil group from relatively continuous sequences in sediment samples in which the fossils are relatively well preserved; (2) standardization and reduction of data by means of an appropriate quantitative analytical technique, which reduces the raw census data into few "factors" (assemblages) that explain a large share of the variance in the data matrix; and (3) delineation of the distribution patterns of the quantitatively defined assemblages on a time-space grid and recognition of individual migrationary events.

CENOZOIC EXAMPLES

In a series of recent studies, the Cenozoic biogeography of calcareous plankton (nannoplankton and planktonic foraminifera) has been delineated for the Atlantic Ocean (Haq and Lohmann, 1976; Haq *et al.*, 1977; Haq, 1980). Numerous latitudinal migrations of assemblages have been recognized in both the Paleogene and the Miocene. Examples of some of the more prominent acme events are included below, and the results are compared to oxygen-isotopic data, where available.

Figure 13.1 shows the nannofloral and planktonic foraminiferal migrationary patterns recognized by Haq *et al.* (1977) in the North Atlantic Paleogene [65-25 million years ago (Ma)]. The most prominent acme events that are evident from both the nannofossil and the foraminiferal migrationary patterns are described below.

1. The shift during the middle Paleocene (60-57 Ma) of both high-latitude nannofloral assemblages to low latitudes and of mid-latitude "low-spired subbotinid" foraminiferal assemblage to low latitudes is in response to a marked cooling at this time.
2. A shift during the latest Paleocene-early Eocene (53-49 Ma) of low-latitude nannofloral assemblages to high latitudes combined with a withdrawal of high-latitude assemblages, combined with a maximum incursion of low-latitude morozovellid foraminiferal assemblage to mid-latitudes, as well as the mid-latitude "low-spired subbotinid" assemblage to higher latitude also takes place as a result of a peak warming.
3. A middle Eocene (46-44 Ma) incursion of higher-latitude assemblages into lower latitudes in response to a cooling, when high-latitude nannofossil assemblages once again return to temperate and tropical areas and the high-latitude globigerinid assemblage makes its first prominent appearance in mid latitudes.
4. A second incursion of the globigerinid assemblage into middle and lower latitudes in the late Eocene-early Oligocene (38-35 Ma) indicates a marked cooling at this time. Nannofossil data are lacking in latest Eocene, but early Oligocene data show a sharp incursion of high-latitude assemblages into low latitudes at 35 Ma.
5. A third incursion of the high-latitude globigerinid assemblage into lower latitude occurs in the middle Oligocene (32-31 Ma). This cooling event is also indicated by the migration of high-latitude nannofossil assemblages into low latitudes at this time. This acme event seems to have been at least equal in intensity to the latest Eocene-early Oligocene event of marked climatic deterioration between 38-35 Ma.

Supporting evidence for most of these paleoclimatic conclusions based on acme events of calcareous plankton comes from other independent sources. On land, the early Eocene acme event, indicating peak warming, manifests itself in a major expansion of tropical-subtropical land plants as far north as 40-45° N latitude (Wolfe, 1978) on the west coast of North America and in the presence of subtropical floral elements as far north as 60° N in the early Eocene in the Gulf of Alaska borderlands area, based on a revised age estimate (see Chapter 16). In general, land plants indicate significantly higher mean annual temperatures in the northern hemisphere and low-latitudinal temperature gradients in the early Eocene (Wolfe, 1978).

The evidence from land plants also corroborates the late Eocene cooling episode: the North American flora indicates that the mean annual temperatures on land were significantly lower than in early Eocene, and, at the same time, mean annual range of temperatures increased dramatically (Wolfe, 1978), supporting a scenario marked by climatic deterioration. Sporomorph evidence from the southeastern United States (Gulf of Mexico and upper Coastal Plains) also suggests a rapid and marked climatic decline near the end of the Eocene, when climates became cooler and drier (Frederiksen, in press). This regime persisted into the early Oligocene.

Oxygen isotopic data from various parts of the ocean also lend support to the marked climatic changes indicated by the Paleogene calcareous plankton temporal and spatial migrations (see Figure 13.2). A decrease in benthic isotopic temperature in the middle Paleocene has been documented by Boersma and Shackleton (1977) from Deep Sea Drilling Project (DSDP) Site 357 on the Rio Grande Rise (South Atlantic) and by Boermsa *et al.* (1979) from DSDP Site 384 (western North Atlantic). At the latter site the planktonic foraminiferal (morozovellid) data also show low isotopic temperature in the middle Paleocene between 61-60 Ma. Buchardt (1978) recorded similarly low temperatures in the middle Paleocene based on oxygen isotope analyses of mollusc shells from northwest Europe.

The latest Paleocene-early Eocene climatic amelioration has been documented in numerous oxygen isotopic studies (Figure 13.2). Shackleton and Kennett (1975) recorded high isotopic temperatures based on both planktonic and benthic foraminifera in southern high-latitude DSDP Site 277 on Campbell Plateau—the temperatures recorded by them were highest in the entire late Paleocene to Recent stratigraphic interval. Buchardt's (1978) molluscan oxygen isotopic curve shows a sharp rise in paleotemperature of shallow marginal seas in northwest Europe in the late Paleocene-early Eocene. Similar trends of isotopic temperature elevation have also been recorded in the late Paleocene-early Eocene sequences at DSDP sites 398 and 401 in the North Atlantic by Vergnaud-Grazzini *et al.* (1978) in both the benthic and planktonic foraminifera.

All the data cited above bear on the conclusion that the latest Paleocene-early Eocene represents the warmest period of the entire Cenozoic in the marine realm. On land, mean annual temperatures were also considerably higher than at present (Wolfe, 1978). This peak climatic amelioration also seems

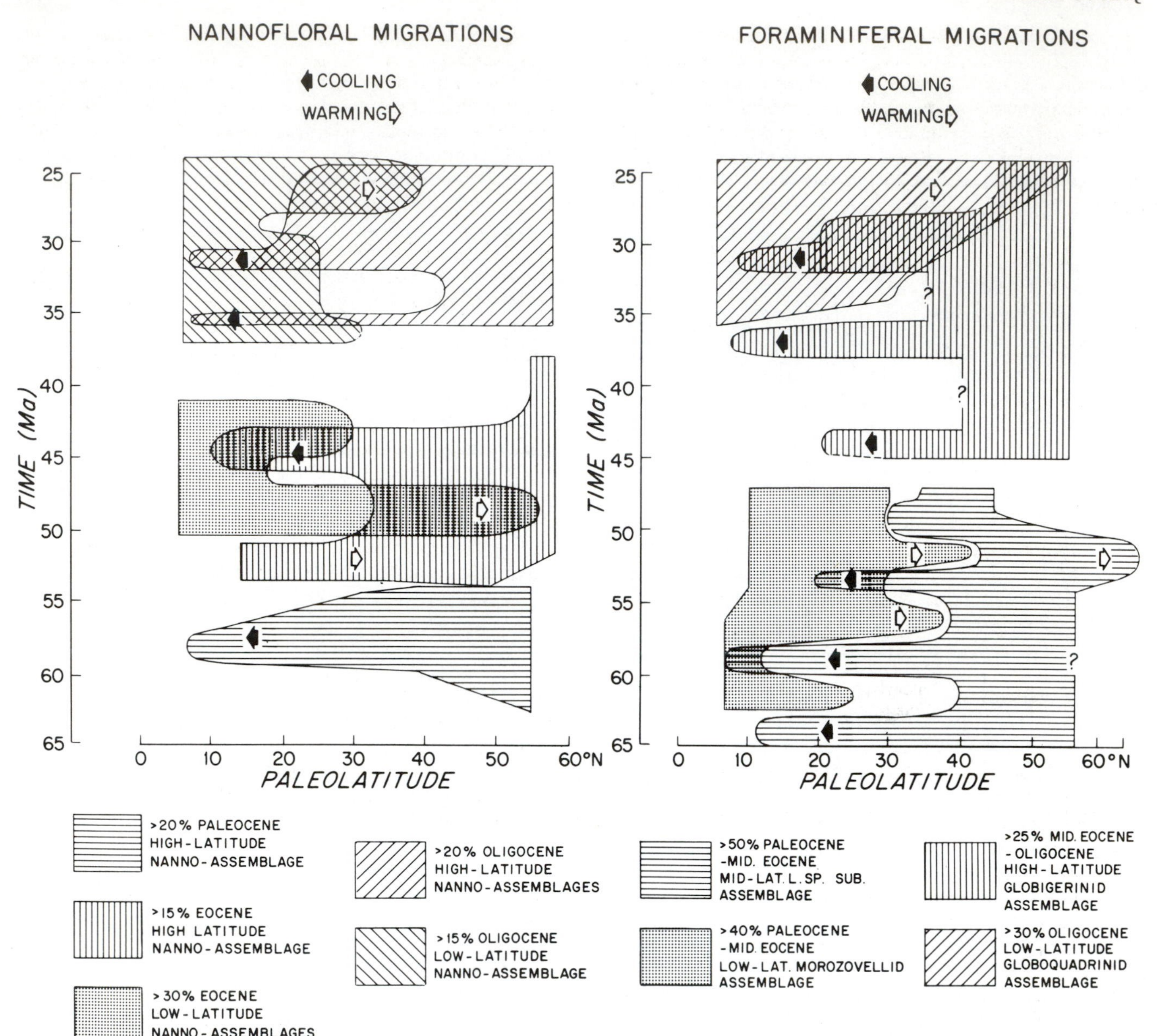

FIGURE 13.1 A summary of the major nannofloral and foraminiferal migrationary patterns in the North Atlantic Ocean through the Early Cenozoic. Migrations toward higher latitudes are interpreted as being caused by climatic warming and toward lower latitudes by climatic cooling. The patterns delineated enclose all samples that contain abundances greater than those indicated in the legend. Arrows indicate the direction of the major shifts of assemblages. Major and minor nannofloral assemblages with similar latitudinal preferences have been combined to obtain composite patterns in some cases. (From Haq *et al.*, 1977, with permission of the American Geophysical Union.)

to have triggered higher evolutionary turnover in phytoplankton and foraminifera, when both nannoplankton and planktonic foraminifera show high evolutionary rates (Berggren, 1969; Haq, 1973), culminating in a peak in pelagic diversity in the middle Eocene (Fischer and Arthur, 1977).

The other nannofloral acme events observed in the Atlantic Ocean are also supported by oxygen isotopic data to varying degrees. The later middle Eocene cooling has been recorded at DSDP site 277 (Shackleton and Kennett, 1975), DSDP site 357 (Boersma and Shackleton, 1977) and in northwest Europe (Buchardt, 1978). The late Eocene-early Oligocene cooling event has been particularly well documented in the oxygen isotopic record. At DSDP site 277 a sharp drop in both planktonic and benthic isotopic temperature is evident (Shackleton and Kennett, 1975); at DSDP sites 398 and 401 (North Atlantic) a marked cooling trend has been recorded (Vergnaud-Grazzini *et al.*, 1978) and the northwest European data show a similar precipitous drop in paleotemperature in the late Eocene-early Oligocene interval (Buchardt, 1978). The slight offset in the timing of this event between DSDP site 277

(Shackleton and Kennett, 1975) and elsewhere is most probably due to the relatively less accurate biochronologic resolution capability because of the high-latitude location of DSDP site 277. The mid-Oligocene cooling event is not so obvious in the isotopic record, which shows relatively low-amplitude variations during most of the Oligocene.

Figure 13.3 summarizes the nannofloral migrations in the North Atlantic in response to climatic acmes during the Miocene (Haq, 1980). Most of the biogeographic "activity" is confined to the mid to high latitudes, which show the most distinct migrationary patterns. This record shows four cooling events, alternating with four warming events of varying in-

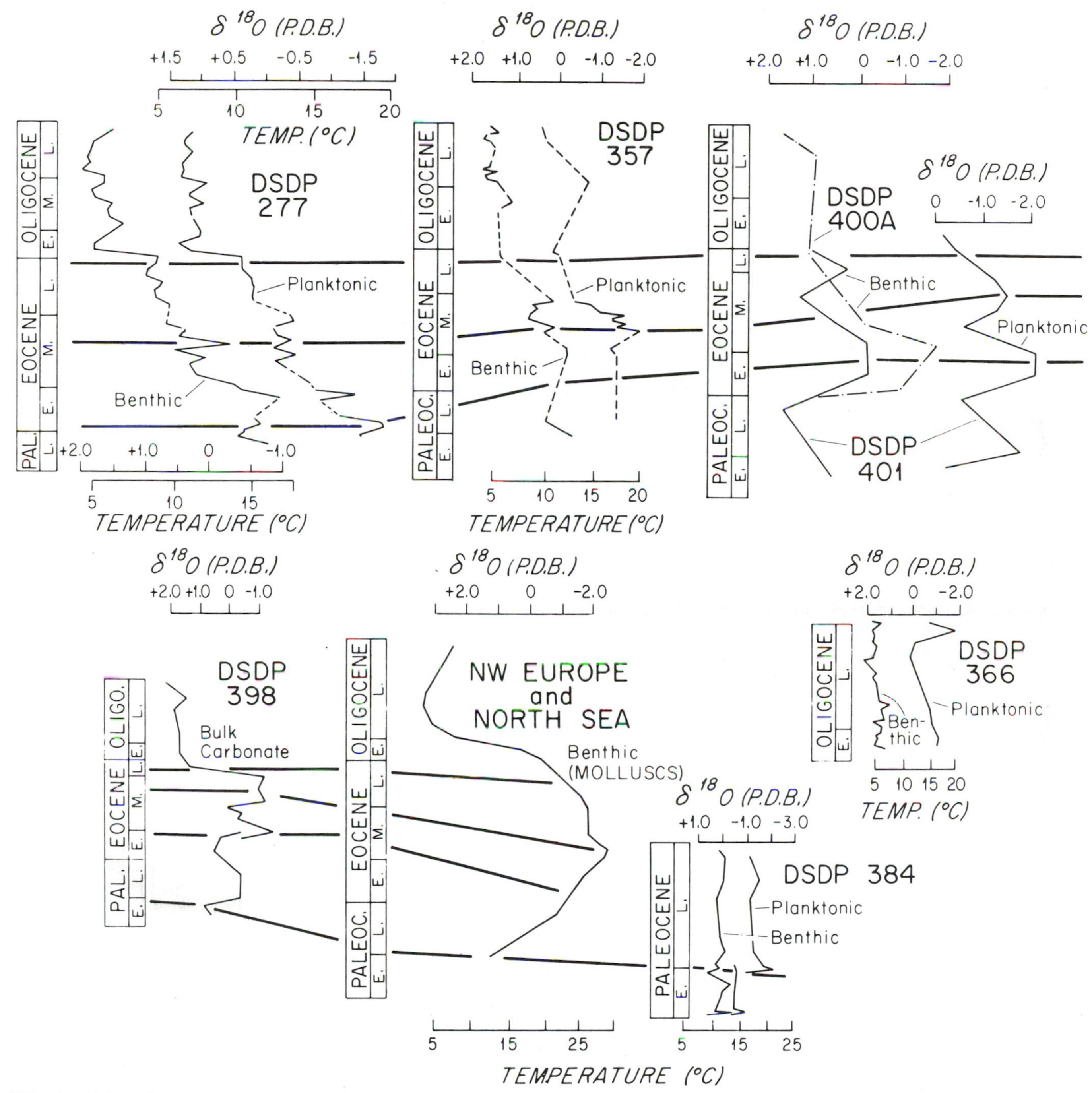

FIGURE 13.2 Paleogene oxygen isotope stratigraphy of DSDP sites from published sources. Planktonic and benthic foraminiferal $\delta^{18}O$ curves from DSDP sites 277 (Shackleton and Kennett, 1975); 357 (Boersma and Shackleton, 1977); and 398, 400, and 401 (Vernaud-Grazzini *et al.*, 1978); generalized mollusc $\delta^{18}O$ curve from northwest Europe (Buchardt, 1978); planktonic and benthic foraminiferal curves from Paleocene of DSDP site 384 (Boersma *et al.*, 1979); and Oligocene planktonic and benthic curves from DSDP site 366 (Boersma and Shackleton, 1978). Lines are drawn through the climatic acme events that have been identified through calcareous plankton migrationary patterns (Haq *et al.*, 1977).

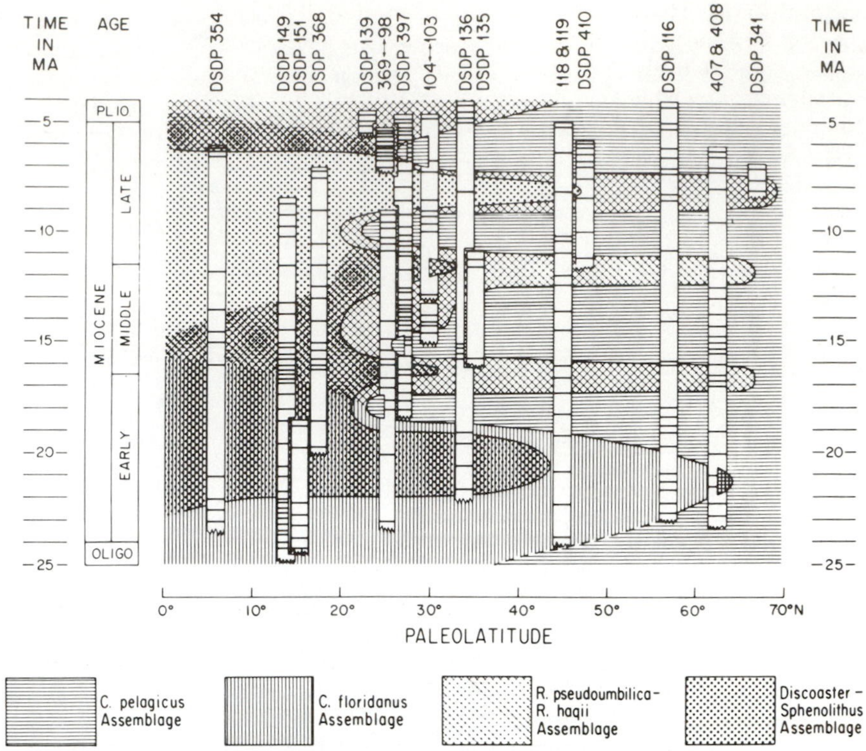

FIGURE 13.3 A summary of nannofloral migrationary patterns in the North Atlantic Ocean through the Miocene. DSDP cores are represented by the appropriate site number, and sample levels in these cores are represented by lines. The Miocene record shows four climatic warming episodes as indicated by low- and mid-latitude assemblage incursions into higher latitudes, and four cooling episodes are indicated by expansion of *Coccolithus pelagicus* (high latitude) assemblage into lower latitudes. (After Haq, 1980.)

tensity and duration. Earliest Miocene is generally cooler, when little or no biogeographic activity is observed and a stress-adapted, cosmopolitan assemblage dominates most latitudes. The first warm acme occurs between 22-20 Ma, with a peak at about 21 Ma. The second warming event occurs between 17-15 Ma, when mid-latitude nannoflora expands into higher latitudes. This is followed by a generally cooler interval up to 12.5 Ma, to be followed by a second incursion of mid-latitude nannofloral element into higher latitudes, indicating a warming between 12.5-11.5 Ma with a peak around 12 Ma. The third expansion of mid-latitude assemblage into higher latitudes occurs between 9-7.5 Ma with a peak around 8 Ma. The late Miocene interval, after this last warming, is characterized by cool climates, and high-latitude nannofloral elements are found as far south as 25° N.

Some of these Miocene acme events also manifest themselves on land, as indicated by the terrestrial plant record from the Pacific Northwest and Alaska (see Chapter 16). A temperature rise in the middle Miocene is indicated between 17-15 Ma, which is followed by a cooling, and then a renewed warming at 8 Ma. The terrestrial record also supports the suggestion that the latest Miocene was a generally cool interval.

The oxygen isotopic record shows trends that are in general agreement with the nannofloral record (Figure 13.4). The warming trend between 22-20 Ma can be observed in the planktonic and benthic isotopic record from the South Pacific, high-latitude DSDP site 179 (Shackleton and Kennett, 1975), from the benthic and planktonic record at North Atlantic DSDP sites 398 (Vergnaud-Grazzini *et al.*, 1978) and 116 (C. Vergnaud-Grazzini, University of Pierre and Marie Curie, Paris, personal communication), and in the record at South Atlantic DSDP site 357 (Boersma and Shackleton, 1977). The most obvious and the warmest Miocene event is a marked decrease in $\delta^{18}O$ values at the early-middle Miocene boundary (shown by a line through the peaks in various oxygen isotope curves reproduced in Figure 13.4). This event corresponds also to the first incursion of mid-latitude assemblages into high latitudes between 17-16 Ma (see Figure 13.3), which

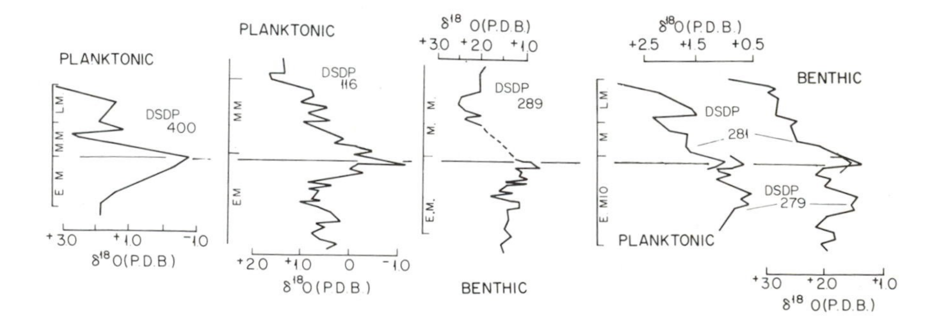

FIGURE 13.4 Miocene oxygen isotope stratigraphy of DSDP sites from various sources; planktonic foraminiferal curve for DSDP sites 400 (Vergnaud-Grazzini *et al.*, 1978) and 116 [Rabussier-Lointier and others (C. Vergnaud-Grazzini, University of Pierre and Marie Curie, Paris, personal communication, 1980)]; and planktonic and benthic foraminiferal curves from DSDP sites 279 and 281 (Shackleton and Kennett, 1975). Line connects the major warm acme between 17-16 Ma (see text).

is also observed in the southern Atlantic Ocean (Haq, 1980). It seems to have been the salient climatic event of the Miocene epoch. The relatively sharp drop in temperatures that followed this warm episode is also clearly indicated by isotopic curves (Figure 13.4) and seems to have been related to the onset of extensive glaciation on Antarctica. A late Miocene warming between 9-7.5 Ma and the cooling event that followed are also evident as well from the $\delta^{18}O$ curves from DSDP sites 400 and 281 (Figure 13.4) and from DSDP sites 357 (Boersma and Shackleton, 1977) and 398 (Vergnaud-Grazzini *et al.*, 1978).

These examples from the Cenozoic acmes and corroborative evidence from oxygen isotopic data indicate that most of these events were widespread and at least some of them were of global extent.

STRATIGRAPHIC IMPLICATIONS

Stratigraphy is primarily an exercise of arranging the succession of "events" aimed at achieving better correlation capability for the documentation of historical geology. Biostratigraphy, one aspect of stratigraphy, utilizes the sequences of biostratigraphic "events" (first and last occurrences of taxa and sometimes their dominances) and their inferred (or interpolated) ages to arrive at a utilitarian biochronology. The extent to which biostratigraphic zones or datum events can be applied for correlations is limited by the geographic distribution of the defining taxa. This is the basic cause of the difficulties in biostratigraphically correlating high- and low-latitude sequences.

The main reason for the latitudinal differences in faunal and floral assemblages is the ecologic exclusion of most low-latitude taxa from the higher latitudes; the high-stress environments of the high latitudes inhibit the development of diversified assemblages, and thus much of the time the high latitudes are populated with few, cosmopolitan, robust, and well-adapted eurythermal taxa, with long-stratigraphic ranges that are only marginally, if at all, useful in biostratigraphy. It is only during periods of marked climatic ameliorations (warm acme events) that tropical-temperate assemblages expand into higher latitudes, making lower latitude zonal schemes applicable in higher latitudes for the duration of the event. A good example of this is the late Paleocene-early Eocene and the early middle Miocene intervals when global warming episodes make tropical-subtropical biostratigraphies more readily applicable in higher latitudes. During other times in the Cenozoic the low- and high-latitude correlations are difficult at best.

The delineation of migrationary patterns of assemblages in response to climatic acmes can provide a solution to the high- to low-latitude correlation dilemma. As mentioned earlier, if episodes of extreme environmental changes are geologically contemporaneous, then acme events will approximate synchronous horizons. Acme horizons (Figure 13.5), or maxima in migrationary cycles, will define approximate time lines, and both the warm and cold acme events can be conveniently used for correlation purposes. The acme horizons are similar to, but more narrowly defined than, acme or peak zones, which are characterized by the "exceptional" abundance of a

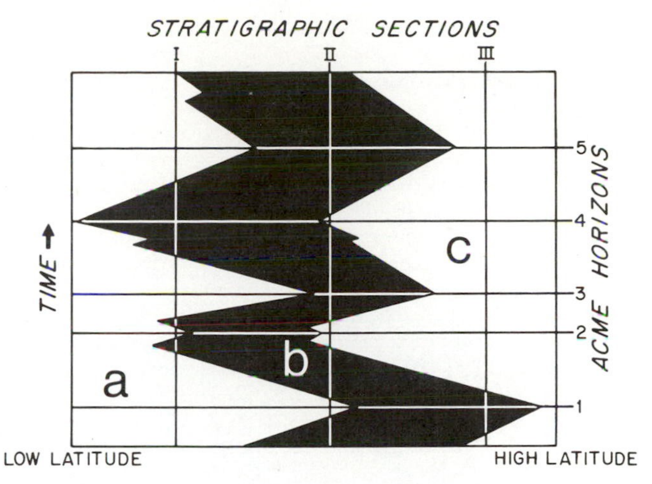

FIGURE 13.5 Acme horizons defined by maximum latitudinal migrations of assemblages in response to climatic acme events. Areas labeled a, b, and c delineate the latitudinal distribution of hypothetical assemblages and record latitudinal migrations of these assemblages through time. Acme horizons 1-5 approximate isochrons and help in the stratigraphic correlation of sections I, II, and III. A given acme horizon may be defined by abundances of different assemblages in different areas, allowing the horizon to be extended across all latitudes. At a given latitude it may not be possible to identify uniquely a given acme horizon from the many existing within a stratigraphic sequence. Thus stratigraphic sections must first be approximately correlated by the usual biostratigraphic methods, and then the acme horizons are used to refine the time stratigraphy. (After Haq and Lohmann, 1976.)

taxon rather than by the maximum abundance of an environmentally sensitive assemblage (Haq and Lohmann, 1976).

The use of acme horizons as a correlation tool has been illustrated in Figure 13.5. Essentially, the procedure involves three steps: (1) delineation of cycles of latitudinal migration of assemblages recorded in stratigraphic sequences, (2) approximate correlations of these cycles at different latitudes by the usual biostratigraphic methods, and (3) refinement of this approximate time framework using migrationary maxima as time lines. This scheme is analogous to the method of local time correlation that uses position within bathymetric cycles recorded in transgressive-regressive stratigraphic sequences.

ACKNOWLEDGMENTS

The author is indebted to C. Vergnaud-Grazzini and F. Woodruff for providing copies of the unpublished isotopic curves from DSDP sites 116 and 289, respectively. This paper was reviewed by W. A. Berggren, B. H. Corliss, and F. Thayer. The author's research is supported by a grant from the National Science Foundation, Division of Submarine Geology and Geophysics, No. OCE78-19769. This is Woods Hole Oceanographic Institution Contribution Number 4617.

REFERENCES

Berggren, W. A. (1969). Rates of evolution in some Cenozoic planktonic foraminifera, *Micropaleontology* 15, 351-365.

Boersma, A., and N. J. Shackleton (1977). Tertiary oxygen and carbon isotopic stratigraphy, Site 357 (mid-latitude South Atlantic), in *Initial Reports Deep Sea Drilling Project 39*, U.S. Government Printing Office, Washington, D.C., pp. 911-924.

Boersma, A., and N. J. Shackleton (1978). Oxygen and carbon isotope record through the Oligocene, DSDP Site 366, Equatorial Atlantic, in *Initial Reports Deep Sea Drilling Project 41*, U.S. Government Printing Office, Washington, D.C., pp. 957-962.

Boersma, A., N. J. Shackleton, M. Hall, and Q. Given (1979). Carbon and oxygen isotope records at DSDP site 384 (N. Atlantic) and some Paleocene paleotemperatures and carbon isotope variations in the Atlantic Ocean, in *Initial Reports Deep Sea Drilling Project 43*, U.S. Government Printing Office, Washington, D.C., pp. 698-717.

Buchardt, B. (1978). Oxygen isotope palaeotemperatures from the Tertiary period in the North Sea area, *Nature 275*, 121-123.

Fischer, A. G., and M. L. Arthur (1977). Secular variations in the pelagic realm, in *Deep Water Carbonate Environments*, H. E. Cook and P. Enos, eds., Soc. Econ. Paleontol. Mineral. Publ. No. 25, pp. 119-150.

Fredericksen, N. O. (in press). Mid-Tertiary climate of southeastern United States: The sporomorph evidence, *J. Paleontol.*

Haq, B. U. (1973). Transgressions, climatic change and diversity of calcareous nannoplankton, *Mar. Geol. 15*, 25-30.

Haq, B. U. (1980). Biogeographic history of Miocene calcareous nannoplankton and paleoceanography of the Atlantic Ocean, *Micropaleontology 26*, 414-443.

Haq, B. U., and G. P. Lohmann (1976). Early Cenozoic calcareous nannoplankton biogeography of the Atlantic Ocean, *Mar. Micropaleontol. 1*, 119-194.

Haq, B. U., I. Premoli-Silva, and G. P. Lohmann (1977). Calcareous plankton paleobiogeographic evidence for major climatic fluctuations in the Early Cenozoic Atlantic Ocean, *J. Geophys. Res. 82*, 3861-3876.

Shackleton, N. J., and J. P. Kennett (1975). Paleotemperature history of the Cenozoic and initiation of Antarctic glaciation: Oxygen and carbon isotope analyses in DSDP Sites 277, 279, and 281, in *Initial Reports Deep Sea Drilling Project 29*, U.S. Government Printing Office, Washington, D.C., pp. 743-755.

Vergnaud-Grazzini, C., C. Pierre, and R. Le'Tolle (1978). Paleoenvironment of the N.E. Atlantic during Cenozoic: Oxygen and carbon isotope analyses at DSDP sites 398, 400A and 401, *Oceanol. Acta 1*, 381-390.

Wolfe, J. A. (1978). A paleobotanical interpretation of Tertiary climates in the Northern Hemisphere, *Am. Sci. 66*, 694-703.

The Arctic Ocean and Post-Jurassic Paleoclimatology

14

DAVID L. CLARK
University of Wisconsin

INTRODUCTION

To what extent has the Arctic Ocean been a factor for pre-Pleistocene world climates? This paper examines the idea by asking three questions: (1) What is the role of the existing Arctic ice cover in modern climate? (2) What models are available for an ice-free Arctic Ocean, and what do these models suggest about the effect of an ice-free ocean on world climates? (3) What does the sediment record say about conditions of the Arctic Ocean in the post-Jurassic world?

Answers to these questions suggest a significant role for the Arctic Ocean in post-Jurassic paleoclimatology and, equally important, point to possible future dramatic climate changes if the Arctic Ocean environment is altered.

MODERN ARCTIC OCEAN AND EFFECT ON WORLD CLIMATE

The central Arctic Ocean (approximately 66° N latitude) differs most strikingly from other oceans because most of it is ice covered throughout the year. This includes the area above 70° N latitude, from 105° E to 135° W longitude. The area of the Greenland-Norwegian-Barents-Kara Seas has considerably more open water during summer months but only in the area south of 80° N (Figure 14.1).

Because atmospheric circulation largely is generated by the gradients resulting from heat gain at low latitudes and heat loss at high latitudes, the approximate 6×10^6 km^2 perennial ice cover of the Arctic Ocean has a profound effect on modern world climate. Flohn (1978) described details of this pattern for polar regions. Because it is more than twice as large, the Antarctic is a more significant climate modifier. In both areas, the ice cover restricts heat exchange between the atmosphere and the ocean. Heat loss by the ocean is suppressed in winter as is heat gain in summer. This results in at least three times as much atmospheric cooling with the ice as there would be without it (Fletcher and Kelley, 1978).

In addition to affecting general atmospheric circulation, short-term weather modifications caused by changes in extent of the ice cover illustrate direct effects. A decrease in area of Arctic ice correlates with a northward shift of storm tracks and a shift of mid-latitude rainfall patterns eastward (Fletcher, 1972; Fletcher and Kelley, 1978). A dramatic example of

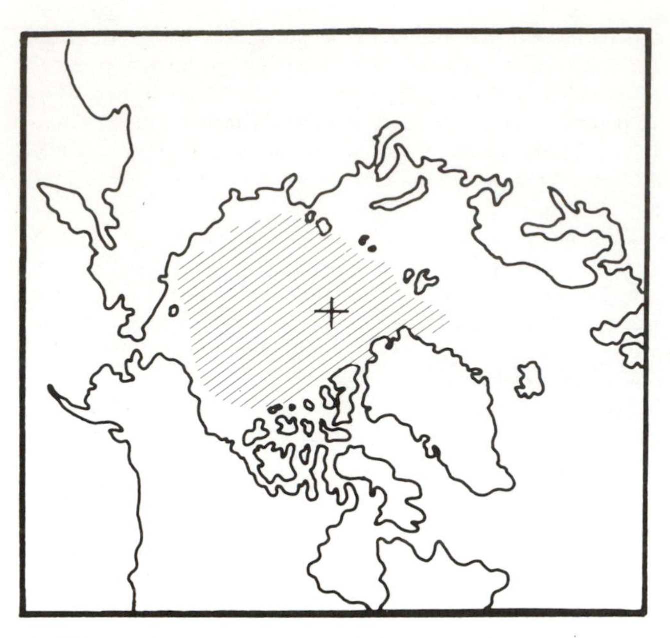

FIGURE 14.1 Approximate limits of year-round Arctic Ocean ice cover.

weather modification is based on satellite data that showed that the ice cover was 12 percent greater in 1971 than in 1970. This produced a surface-heat-exchange deficit for the Arctic that was correlated with anomalous weather patterns in lower latitudes during 1972 and 1973 (Kukla and Kukla, 1974).

Some of the effects are more complex. Other data show that the Arctic region (Greenland, in this example) and northern Europe are out of phase in terms of severity of winter temperatures. This has been described as the "seesaw effect." Colder Arctic (Greenland) winters commonly correlate with mild northern European winters and vice versa (van Loon and Rogers, 1978). The temperature anomaly is correlated with anomalous weather effects over wide areas including the southern Mediterranean, the Middle East, Central America, western North America, and Alaska. Explanations for this temperature anomaly are related to pressure anomalies as well as to general atmospheric circulation.

An interesting model of world climate and its relationship to astronomical theories of ice ages includes cooling and heating parameters; heat transport; seasonal variation in Arctic ice cover; and a variety of Earth obliquity, eccentricity, and precession factors (Pollard, 1978). A number of conclusions derived from the still imperfect model include the observations that variation in extent of ice cover would have a profound effect on world climate. Nonetheless, the totality of cryospheric processes including feedback mechanisms for the Arctic Ocean and the relationship to world climate is only partially understood (Polar Group, 1980).

However, both observations and models furnish suggestive data regarding the effect of the Arctic ice cover on world climates.

WORLD CLIMATE AND AN ICE-FREE ARCTIC OCEAN

Polar ice covers are basic to the generation of atmospheric circulation. Removal of the ice covers would reduce the thermal gradient and change circulation. The extent of the change that would be produced has not been quantitatively described, in fact ". . . a realistic model of the entire planetary circulation under the assumption of an ice-free Arctic Ocean is not yet available . . ." (Fletcher and Kelley, 1978, p. 103). Nonetheless, there are data suggesting how the absence of an Arctic ice cover would greatly modify world climates. For example, during the summer, an ice-free Arctic Ocean would absorb 90 percent of solar radiation reaching its surface in contrast to the present figure of 30 to 40 percent (Fletcher and Kelley, 1978). Such a change would affect the Earth's heat budget.

There is disagreement on the actual heat balance of an ice-free Arctic. One argument is that under ice-free conditions the Arctic Ocean would gain approximately 40 $kcal/cm^2$ in summer but lose a similar amount in the winter for a balance (Fletcher and Kelley, 1978). Others argue that there would be a smaller heat loss in the winter, resulting in a yearly net increase in ocean temperatures (Donn and Shaw, 1966).

Year-round open water in the Arctic would be a source for increased atmospheric moisture. The surface temperature of an ice-free Arctic would be critical in determining how much. At 0-5°C surface temperature, the Arctic would give only a fifth to a sixth as much moisture to the atmosphere as water at 25-30°C, under similar wind conditions (Lamb, 1974). Nonetheless, an ice-free Arctic Ocean would contribute to warmer atmospheric conditions that could produce an increase in precipitation for Canada, India, the Middle East, and China and a sharp decrease in precipitation for most of the United States, Eurasia, and much of northern Africa (Wigley *et al.*, 1980).

Such conclusions would be more impressive if framed in quantitative terms, but in the absence of complete modeling of these conditions, it may be safe to conclude that change in precipitation patterns in addition to other changes in atmospheric gradients produced by an ice-free Arctic would be a factor in amelioration of northern hemisphere climate.

DEVELOPMENT OF THE ARCTIC OCEAN

Having examined questions related to the effect of an ice-covered and ice-free Arctic Ocean on world climate, it is necessary to discuss the geologic development of the Arctic. What is known about the absence or presence of ice cover during the development of the Arctic Ocean? The size and position of the Arctic Ocean have changed as crustal plates have adjusted. By the Cretaceous, the present position of the Arctic Ocean was approached (Firstbrook *et al.*, 1972; Smith and Briden, 1977), but the ocean was only one half as large as it is at present. As the North Atlantic opened, spreading along the Nansen Ridge doubled the size of the Cretaceous Arctic Ocean (Figure 14.2) to its present dimensions (Clark, 1977a, 1977b, 1981); in this setting, sea ice was to form, but not quickly.

Pre-Cretaceous Conditions

Details of the pre-Cretaceous Arctic Ocean are unknown. Paleozoic and pre-Cretaceous Mesozoic rocks from the continents adjacent to the Arctic Ocean have yielded data that have been combined with geophysical surveys of the Arctic Ocean basin to give hints. The interpretation of all this information suggests "that the Arctic has been a center of marine sedimentation and an avenue for polar migration of faunas since at least the early Paleozoic" (Churkin, 1973, p. 497). Faunal elements from folded sedimentary rocks that apparently pass into the modern Arctic Ocean are similar to faunas from Eurasia and support the idea of a Paleozoic trans-Arctic connection. No profound ecologic differences have been suggested for these faunas. The pre-Cretaceous Arctic Ocean apparently may not have been a critical element in world climates. Paleogeographic reconstructions for the pre-Cretaceous Arctic are not known in sufficient detail to be of much help.

Late Cretaceous–Early Cenozoic Arctic Ocean

Maestrichtian and Paleocene deep-sea cores from the Alpha Cordillera of the central Arctic Ocean consist of opaline sediments representing biogenic silica accumulations and document periods of high primary productivity and intensified upwelling (Kitchell and Clark, in press). The sediment, originally defined as tuffaceous (Clark, 1974), has been more accurately calculated on a bulk-sediment basis to include from 43 to 78 percent biogenic silica (Kitchell and Clark, in press). There are no significant terrigenous nor carbonate components. The sediment is laminated, lacks burrowing, and most likely represents deposition in a deep basin.

Currently, the deep Pacific is the reservoir supplying silica through upwelling to the Bering Sea and the Seas of Okhotsk and Japan. The Late Cretaceous-Paleocene Arctic may have had a deep connection with the north Pacific, although the geography of the paleocirculation is unknown (Clark, 1977a, 1977b; Kitchell and Clark, in press).

Fossils in the biogenic sediment include diatoms, silicoflagellates, ebridians, and archaeomonads. They average 100×10^6 to 400×10^6 specimens/gram of bulk sediment. Modern density analogs include accumulations in the Gulf of California, on the southwest coast of Africa, and in the Antarctic Ocean, associated with strong upwelling conditions. The Cretaceous Arctic phytoplankton have affinities with known normal marine species in California, Russia, and the South Pacific and Indian Oceans (Clark and Kitchell, 1979a).

The oxygen isotope data suggest worldwide surface-water temperatures averaging around 20°C for at least part of the Cretaceous-Paleocene (Shackleton and Kennett, 1975). The

FIGURE 14.2 Tectonic setting for the Central Arctic Ocean. Nansen Ridge has been site of Cenozoic spreading in the Arctic, separating Lomonosov Ridge from the Barents Continental Shelf. Origin of Alpha Ridge is uncertain. Modified from Ostenso and Wold (1973).

similarity of the Arctic floral elements to those in the Pacific and Indian Oceans suggests a similar temperature for all of these areas.

These considerations are strong evidence for an open water, ice-free Arctic Ocean with ecologic conditions similar to those in lower latitudes. The uniformity of world climates during this time interval has been widely reported (e.g., Wolfe, 1975; Berggren and Hollister, 1977; Frakes, 1979). The question becomes one of whether the Cretaceous-Paleocene Arctic Ocean was a passive participant or a significant contributor to the worldwide temperature climate of the time. Theoretical considerations of an ice-free Arctic Ocean, discussed earlier, strongly suggest the latter, that is, the worldwide temperate climates of the Cretaceous-Paleocene may have been dependent, at least in part, on ice-free conditions of the Arctic Ocean. Atmospheric gradients generated by heat exchange from high latitudes to lower latitudes would be significantly less during ice-free Arctic conditions. A lower atmospheric gradient along with the general "commotion in the ocean" of the time (Berggren and Hollister, 1977) may be sufficient to explain the temperate Cretaceous-Early Cenozoic world climates.

Recently, considerable attention has been given to "injection events" in the world's oceans (e.g., Gartner and Keaney, 1978; Thierstein and Berger, 1978; Clark and Kitchell, 1979a, 1979b; Gartner and McGuirk, 1979; Clark and Kitchell, 1981). One model, the Arctic Spillover Model, suggests that restriction of Arctic circulation with lower-latitude oceans may have produced a "freshwater" mass eventually draining into the lower-latitude seas, forming a low-salinity surface layer that may have been a factor in extinction of certain pelagic organisms and even affecting terrestrial life. Although imaginative in concept, lack of data for a freshwater Cretaceous Arctic make the model highly speculative. In fact, the *only* data available for the Late Cretaceous Arctic Ocean indicate that water of normal salinity and high productivity was present close to the time a volume of freshwater is needed for the spillover.

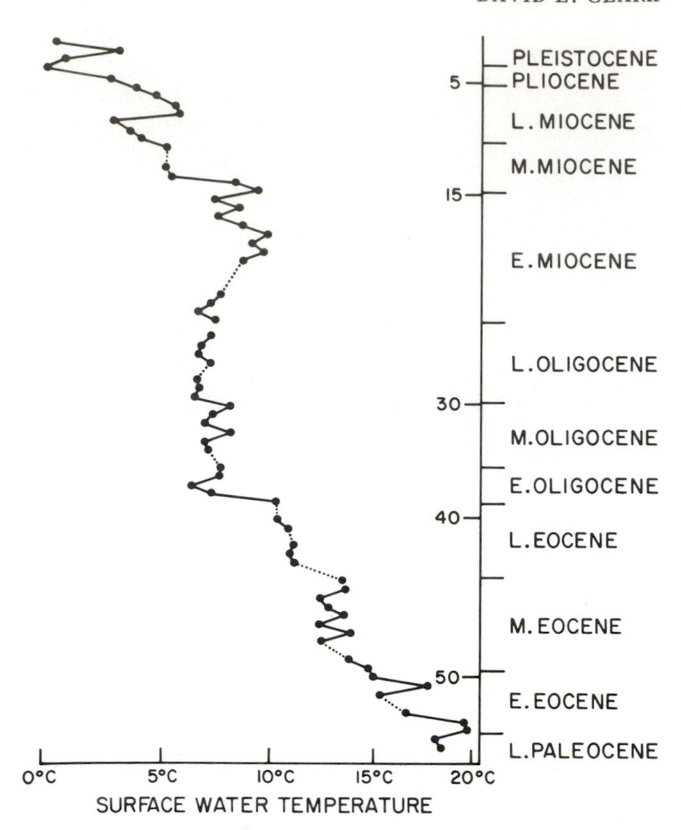

FIGURE 14.3 Surface-water temperatures for the Cenozoic determined from oxygen isotope data from planktonic foraminifera. Modified from Shackleton and Kennett (1975). Vertical numbers are in million years.

Middle Cenozoic Arctic Ocean

No Middle Cenozoic sediment has been identified from the Arctic Ocean. This includes the interval extending from the late Paleocene-Eocene through most of the Miocene. This was a critical time for world climates. Glaciation apparently was initiated in Antarctica. The polar cooling profoundly altered atmospheric circulation, and a strong thermal gradient, similar to that of the present, was established. Evidence for the mid-Cenozoic climate deterioration is based on development from polar glaciation as well as oxygen-carbon isotopes studies (Figure 14.3).

The Arctic Ocean ice cover possibly many have developed during this time. However, the evidence is circumstantial and includes recognition of general lowering of worldwide ocean-water temperature as well as knowledge that in the central Arctic the middle Cenozoic interval is bounded by older Paleocene, ice-free conditions and younger Miocene conditions that included glacial ice rafting as the dominant sedimentation process.

The central Arctic during the middle Cenozoic was approaching a land-locked condition with restrictive Pacific connections. There may have been a short period during this critical interval that Pacific circulation was sufficiently restricted to prevent effective exchange of water. Also, perhaps the growing Eurasian Basin connection with the Atlantic was not yet sufficient to allow more than surficial exchange. If this preceded Pliocene uplift of the straits of Panama, no warm Gulf Stream boundary current was available to the North Atlantic (Berggren and Hollister, 1975). This coupled with Eurasian freshwater river flow into the Arctic may have led to development of a salinity stratification as discussed by Aagaard and Coachman (1975), which permitted the first winter ice to form. Nonetheless, any surface freshening and freezing during the Middle Cenozoic probably did not affect deeper water of the Arctic as indicated by the presence of benthic foraminifera in sediment of probable late Miocene age (O'Neill, 1981).

ARCTIC OCEAN AND LATER CENOZOIC CLIMATES

The role of the Arctic Ocean for the immediate pre-Pleistocene world climate is fairly well understood. Probably, the Arctic Ocean became ice covered in the middle Cenozoic and together with the Antarctic icecap provided significant areas of net heat loss to propel the atmospheric circulation that has led to modern climates.

The Arctic Ocean sediment record for the past approximate 5 million years has been studied in detail. Some 13 lithostratigraphic units have been defined for the late Miocene to Holocene (Clark *et al.*, 1980) (Figure 14.4). These sedimentary units, correlatable over 400,000 km^2, show variations on a single theme, ice rafting. The 13 lithostratigraphic units indicate alternating times of more and less ice rafting but no significant deviation from this mode of sedimentation. Whether times of greater ice rafting correlate with thinner or thicker ice cover has not been answered definitely (Clark, 1971).

Recently, a few coccoliths have been reported from late Pliocene and Pleistocene central Arctic sediment (Worsley and Herman, 1980). Although this is interpreted to indicate episodic ice-free conditions for the central Arctic, the occurrence of ice-rafted debris with the sparse coccoliths is more easily interpreted to represent transportation of coccoliths from ice-free continental seas marginal to the central Arctic. The sediment record as well as theoretical considerations make strong argument against alternating ice-covered and ice-free conditions (Donn and Shaw, 1966; Clark *et al.*, 1980). Although the time of development of the pack ice for the central Arctic Ocean is unknown, to date there is no evidence that precludes a Miocene origin.

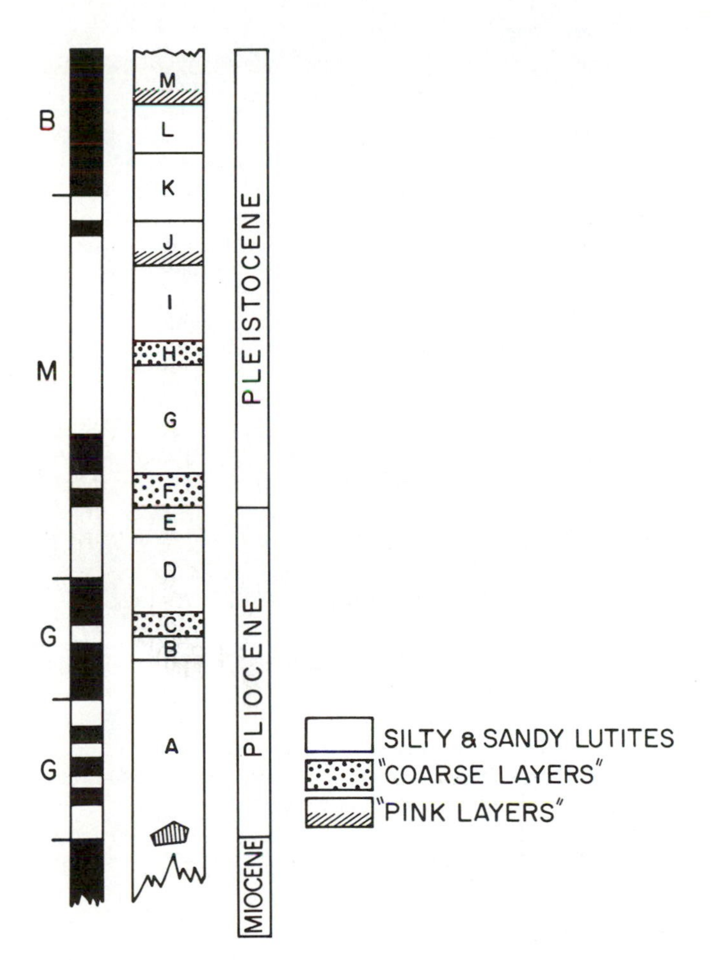

FIGURE 14.4 Lithostratigraphy of the central Arctic Ocean. Units A-M consist of glacial-marine sediment, suggesting alternating times of relatively rapid and slow rates of deposition. Magnetic stratigraphy correlation indicated. Modified from Clark (1977a).

Geologically significant invertebrate faunas developed in this glacial-marine environment. This includes a predominantly arenaceous (textularid) biofacies in the late Miocene to Pliocene and a calcareous biofacies (milioids and rotalids) in the Pliocene and Pleistocene (O'Neill, 1981). There is no evidence for other than normal salinity Arctic bottom water since the late Miocene.

Productivity of Arctic water, as measured by foraminifera abundance, has increased to the present condition, and at no time during the Late Cenozoic was productivity any greater than that of the present (Clark, 1971). Reported high-bottom-water productivity during the late Miocene is grossly overstated (Margolis and Herman, 1980). Modern productivity is not measured by the standard of other oceans. Nonetheless, it is high compared with productivity for any other part of the Late Cenozoic for which there is a record. Modern productivity has been achieved under year-round ice-covered conditions.

Correlation of pre-Pleistocene climate events of lower latitudes with that of the central Arctic is difficult. For example, waxing and waning of continental glaciers occurred while the Arctic ice-cover remained relatively constant. Perhaps, thickening and thinning of ice may have been the only central Arctic response to glacial and interglacial climates in lower latitudes. The Ewing-Donn (1956) theory suggested that the Arctic Ocean played a central role in northern hemisphere glaciation, with an ice-free Arctic producing glaciation and an ice-covered Arctic (such as that of the present) producing interglacial times. The evidence available no longer supports this model. Rather, Late Cenozoic glacial and interglacial stages developed with a constant, central Arctic ice cover. It appears that the Arctic ice cover, while responsible in part for Pleistocene atmospheric circulation patterns, was only a participant in the drama of the recent ice ages; certainly it did not play the leading role.

SUMMARY

The conclusion that the Arctic Ocean had an important role in pre-Pleistocene climates appears justified. The open-water Late Cretaceous-Paleocene Arctic was one of the factors contributing to worldwide climate uniformity. The probable Middle Cenozoic development of an ice cover, accompanied by Antarctic ice development and a late shift of the Gulf Stream to its present position, were important events that led to the development of modern climates.

The record suggests that altering the present ice cover would have profound effects on future climates. The technology is available to melt the Arctic Ocean ice cover (e.g., Arnold, 1961). In addition, there have been proposals by the Soviets to divert major rivers from their normal courses toward the Arctic Ocean, which would significantly alter the Arctic Ocean water-mass density structure. A change in the Arctic water budget could affect the ice cover. Yearly ice-free conditions could develop because of more vertical convection and release of increased amount of heat to the atmosphere (Aagaard and Coachman, 1975). Is the world ready for new temperature and precipitation patterns? Execution of any of these proposals would seriously affect the Earth's climate, a fact made clear by the record of the Arctic Ocean's ice cover.

REFERENCES

Aagaard, K. and L. K. Coachman, (1975). Towards an ice-free Arctic Ocean, *EOS Trans. Am. Geophys. Union 56*, 484-486.

Arnold, K. C. (1961). An investigation into methods of accelerating the melting of ice and snow by artificial dusting, in *Geology of the Arctic*, G. O. Raasch, ed., U. of Toronto Press, Toronto, pp. 989-1013.

Berggren, W. A., and W. A. Hollister (1977). Plate tectonics and paleo-circulation—commotion in the ocean, *Tectonophysics 38*, 11-48.

Churkin, M. (1973). Geologic concepts of Arctic Ocean Basin, in *Arctic Geology*, M. G. Pitcher, ed., Am. Assoc. Petrol. Geol. Mem. 19, pp. 485-499.

Clark, D. L. (1971). Arctic Ocean ice cover and its Late Cenozoic history, *Geol. Soc. Am. Bull. 82*, 3313-3324.

Clark, D. L. (1974). Late Mesozoic and early Cenozoic sediment cores from the Arctic Ocean, *Geology 2*, 41-44.

Clark, D. L. (1977a). Paleontologic response to post-Jurassic crustal plate movements in the Arctic Ocean, in *Paleontology and Plate Tectonics*, R. M. West, ed., Milwaukee Pub. Museum Spec. Pap. Biol. Geol., no. 2, pp. 55-76.

Clark, D. L. (1977b). Climatic factors of the Late Mesozoic and Cenozoic Arctic Ocean, in *Polar Oceans*, M. J. Dunbar, ed., Arctic Inst. of North America, Calgary, pp. 603-615.

Clark, D. L., (1981). Geology and geophysics of the American Basin, in *The Ocean Basins and Margins 5*, A. E. M. Nairn, M. Churkin, Jr., and F. G. Stehli, eds., Plenum, New York, pp. 599-634.

Clark, D. L., and J. A. Kitchell (1979a). Injection events in ocean history, *Nature 278*, 669.

Clark, D. L., and J. A. Kitchell (1979b). The terminal Cretaceous events: A geologic problem with an oceanographic solution (comment), *Geology 7*, 228.

Clark, D. L., and J. A. Kitchell (1981). Terminal Cretaceous extinctions and the Arctic spillover model, *Science 212*, 577.

Clark, D. L., R. R. Whitmann, K. A. Morgan, and S. D. Mackey (1980). Stratigraphy and glacial-marine sediments of the Amerasian Basin, Central Arctic Ocean, *Geol. Soc. Am. Spec. Pap. 181*, 57 pp.

Donn, W. L., and D. M. Shaw (1966). The stability of an ice free Arctic Ocean, *J. Geophy. Res. 71*, 1086-1096.

Ewing, M., and W. L. Donn (1956). A theory of ice-ages, *Science 128*, 1061-1066.

Firstbrook, P. L., B. M. Funnell, A. M. Harley, and A. G. Smith (1972). Paleo-oceanic Reconstructions, 160-0 Ma, National Science Foundation, National Ocean Sediment Coring Program, Scripps Inst. Oceanogr.

Fletcher, J. O. (1972). Ice on the ocean and world climate, in *Beneficial Modifications of the Marine Environment*, National Academy of Sciences, Washington, D.C., pp. 4-49.

Fletcher, J. O., and J. J. Kelley (1978). The role of the polar regions in global climate change, in *Polar Research*, M. A. McWhinnie, ed., AAAS Selected Symposium Series, Westview Press, Boulder, Colo., pp. 97-116.

Flohn, H. (1978). Comparison of Antarctic and Arctic climate and its relevance to climate evolution, in *Antarctic Glacial History and World Palaeoenvironments*, E. M. van Zinderen Bakker, ed., A. A. Balkeman, Rotterdam, pp. 3-13.

Frakes, L. A. (1979). *Climates Throughout Geologic Time*, Elsevier, Amsterdam, 310 pp.

Gartner, S., and J. Keaney (1978). The terminal Cretaceous event: A geologic problem with an oceanographic solution, *Geologie 81*, 708-812.

Gartner, S., and J. P. McGuirk (1979). Terminal Cretaceous extinction, scenario for a catastrophe, *Science 206*, 1272-1276.

Kitchell, J. A., and D. L. Clark (in press). Late Cretaceous-Paleocene paleogeography and paleocirculation: Evidence of North Polar upwelling, *Paleogeogr. Paleoclimatol. Palaeoecol.*

Kukla, G. L., and H. J. Kukla (1974). Increased surface albedo in the northern hemisphere, *Science 183*, 709-714.

Lamb, H. H. (1974). Atmospheric circulation during the onset and maximum development of the Wisconsin/Wurn Ice Age, in *Marine Geology and Oceanography of the Arctic Seas*, Heiman, ed., Springer-Verlag, New York, pp. 349-358.

Margolis, S. V., and Y. Herman (1980). Northern hemisphere sea-ice and glacial development in the late Cenozoic, *Nature 286*, 145-149.

O'Neill, B. J. (1981). Pliocene and Pleistocene benthic foraminifera from the central Arctic Ocean, *J. Paleontol. 55*.

Ostenso, N. A., and R. J. Wold (1973). Aeromagnetic evidence for origin of the Arctic Ocean Basin, in *Arctic Geology*, M. G. Pitcher, ed., Am. Assoc. Petrol. Geol. Mem. 19, pp. 506-516.

Polar Group (1980). Polar atmosphere-ice-ocean processes: A review of polar problems in climate research, *Rev. Geophys. Space Phys. 18*, 525-543.

Pollard, D. (1978). An investigation of the astronomical theory of the ice ages using a simple climate-ice sheet model, *Nature 272*, 233-235.

Shackleton, N. J., and J. P. Kennett (1975). Paleotemperature history of the Cenozoic and the initiation of Antarctic glaciation: Oxygen and carbon isotope analysis in DSDP sites 277, 279 and 281, in *Initial Report of the Deep Sea Drilling Project 29*, U.S. Government Printing Office, Washington, D.C., pp. 743-755.

Smith, A. G., and J. C. Briden (1977). *Mesozoic and Cenozoic Paleocontinental Maps*, Cambridge U. Press, Cambridge.

Thierstein, H. R., and W. H. Berger (1978). Injection events in ocean history, *Nature 276*, 461-466.

van Loon, H., and J. C. Rogers (1978). The seesaw in winter temperatures between Greenland and Northern Europe, Part I: General description, *Mon. Weather Rev. 106*, 296-310.

Wigley, T. M. L., P. D. Jones, and P. M. Kelly (1980). Scenario for a warm, high CO_2 world, *Nature 283*, 17-21.

Wolfe, J. A. (1975). An interpretation of Tertiary climates in the northern hemisphere, *Geoscience and Man 11*, 160-161.

Worsley, T. R., and Y. Herman (1980). Episodic ice-free Arctic Ocean in Pliocene and Pleistocene time: Calcareous nannofossil evidence, *Science 210*, 323-325.

Interpreting Paleoenvironments, Subsidence History, and Sea-Level Changes of Passive Margins from Seismic and Biostratigraphy*

15

JON HARDENBOL *and* PETER R. VAIL
Exxon Production Research Company

J. FERRER
Exxon Production Research-European

INTRODUCTION

Along passive margins, successions of marine deposits separated by unconformities are commonly found at elevations above present sea level. These sediments were deposited on a slowly subsiding margin while sea level was at a much higher level than at present. In addition to long-term falls or rises of sea level, rapid short-term changes occur that are responsible for the unconformities bounding the marine depositional sequences.

The magnitude of changes in eustatic sea level can be determined from the sedimentary section if the chronostratigraphic and paleobathymetric history can be restored in sufficient detail. Long-term changes in sea level can be estimated from the present elevation of ancient marine deposits after the subsidence history for those deposits is reconstructed. Short-term, sea-level changes are in general more elusive, since continuous marine sedimentation during the sea-level change is the exception rather than the rule. Rarely is the entire short-term, sea-level cycle recorded in marine sediments laid down in water depths favorable for a high stratigraphic and paleobathymetric resolution.

Eustatic sea-level changes, basement subsidence, and sediment supply are the principal interacting variables in a marine basin (Figure 15.1). The interaction is recorded in the sedimentary section deposited in the basin. The resulting depositional sequences, bounded by unconformities or their correlative conformities (Figure 15.2), can be recognized in outcrops and wells and on seismic reflection profiles because of important changes in areal distribution, lithology, and depositional environment. Retracing the interaction of eustatic changes and basement subsidence from the sedimentary record is efficiently handled by the geohistory analysis technique. This technique integrates the stratigraphic and paleobathymetric data in a time-depth framework and reproduces the subsidence of basement as a result of basement faulting, crustal and mantle cooling, and sediment loading.

*This paper was originally published in *Oceanologica Acta*, Supplement to Volume 3 (1981). Proceedings 26th International Geological Congress, Geology of Continental Margins Symposium, Paris, July 7-17, 1980, pp. 33-44, and is reprinted here with permission and with minor editorial changes.

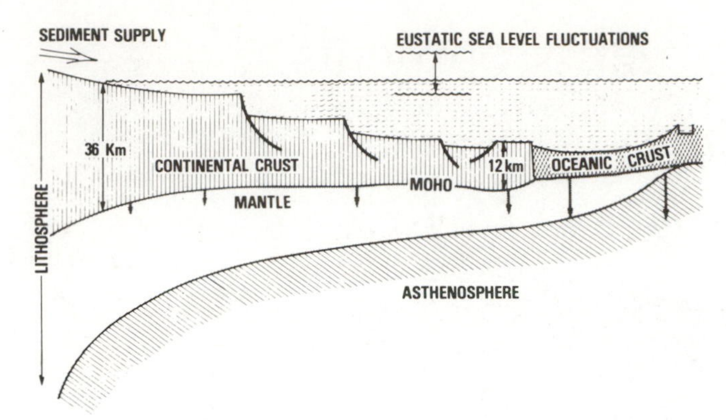

FIGURE 15.1 Schematic cross section of a passive margin illustrating the variables controlling the distribution of paleoenvironments and the type of depositional sequence boundaries.

This total subsidence also includes the apparent vertical movement effects of sediment compaction and eustatic sea-level changes. By quantification of the effects of sediment compaction, sediment loading, and thermal subsidence, the eustatic changes can be determined. An example of this analysis shows that in the Early Cenomanian from offshore northwestern Africa sea level could have been nearly 300 m higher than the present sea level.

The selection of northwest Africa to illustrate the procedure was motivated by the amount and quality of seismic and paleontological data available. The absence of the Late Cretaceous and Tertiary at the selected well site was offset by the unusual quality of data for the Early Cretaceous and Jurassic section. The older section is much more significant in determining the subsidence history of a margin than is the younger section, although a young tectonic event may escape attention.

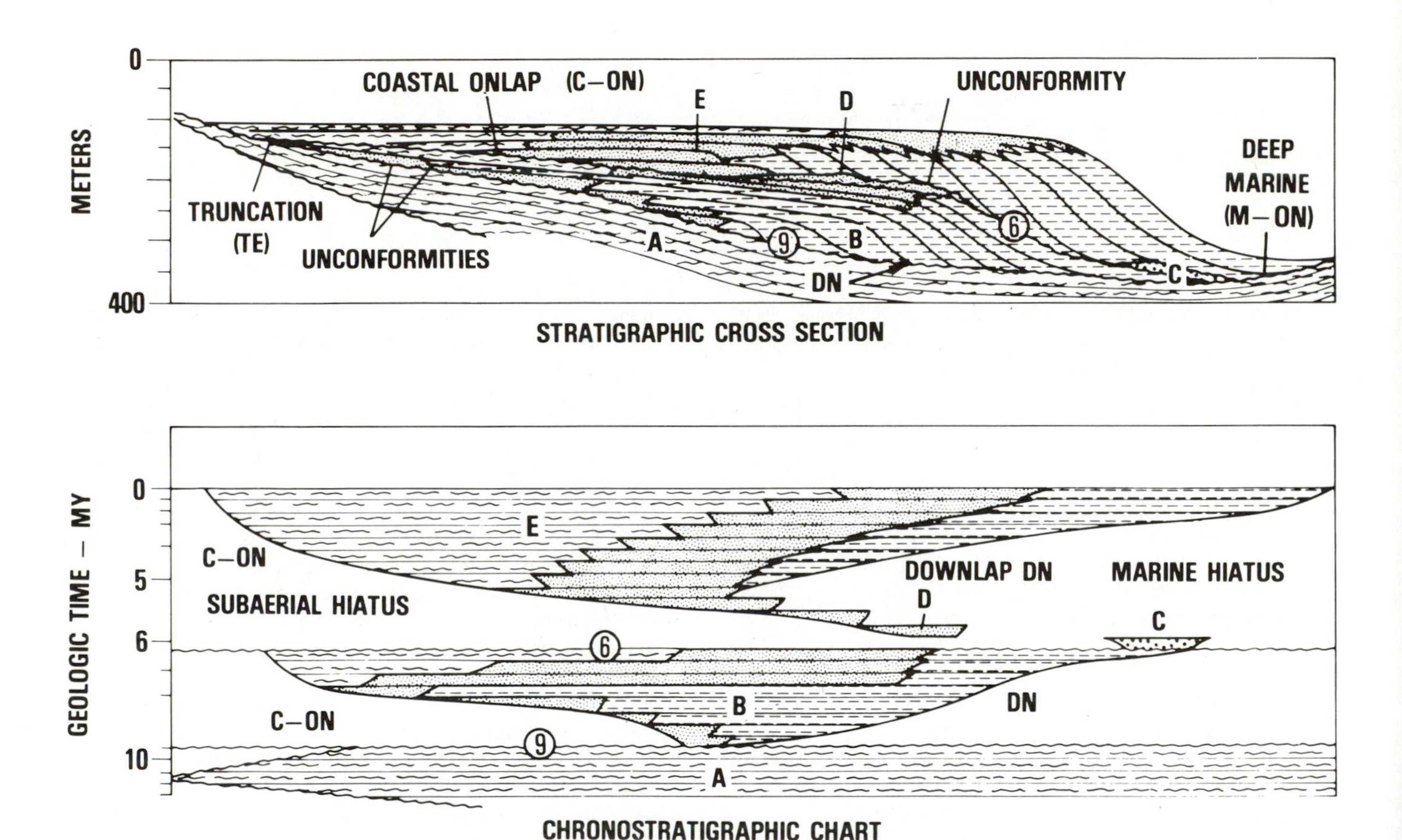

FIGURE 15.2 Interaction between basement subsidence and eustatic sea-level changes is recorded in the sediments filling a basin. The upper illustration shows a schematic depth-distance stratigraphic cross section of three sedimentary sequences along a basin margin. The lower figure shows a time-distance or chronostratigraphic reconstruction of the same sedimentary sequences. After Vail and Hardenbol (1979).

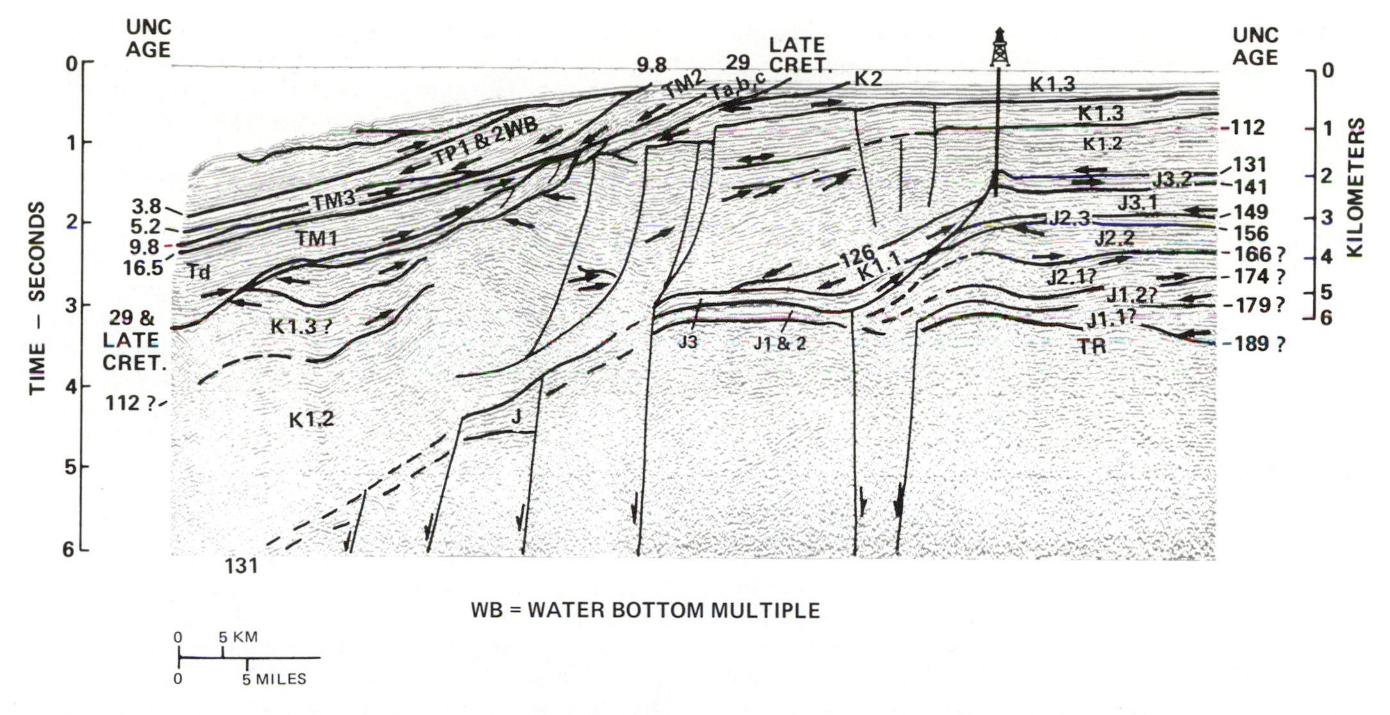

FIGURE 15.3 Seismic section from offshore northwest Africa showing interpreted seismic stratigraphic sequences. The sequence boundaries are determined by cycle terminations indicated by arrows. The ages are indicated in millions of years. For sequence designations, see Figure 15.12. After Vail and Mitchum (1980).

EUSTATIC SEA-LEVEL CHANGES

Changes in coastal onlap patterns along basin margins are mainly the result of relative changes of sea level (Vail *et al.*, 1977). These relative changes of sea level are in effect produced by the interaction between eustatic changes, basement subsidence, and sediment supply. Without eustatic fluctuations, coastal onlap would be continuous and regular without the downward shifts commonly observed on seismic reflection profiles. Sedimentation would be controlled only by available space, created by basement subsidence, and sediment supply. Quantifying eustatic sea-level changes from measured changes in coastal onlap does not provide an accurate measure, because of variations in subsidence in different basins (Vail *et al.*, 1977). The eustatic sea-level curves from the Tertiary by Vail and Hardenbol (1979) and for the Jurassic through Early Cretaceous by Vail and Todd (in press) are based on estimates from changes in coastal onlap and from paleontological studies. Quantified basement subsidence was not considered for those estimates.

Sedimentary sequences can be viewed in a depth-distance or in a time-distance framework (Figure 15.2). Well control, outcrop data, and seismic reflection profiles view the sequences and their unconformable boundaries in a depth-distance framework that does not do justice to the time that is not represented by sediments. If sufficient age data can be obtained, a time-distance record or chronostratigraphic chart can be produced that shows the distribution of the depositional sequences and the extent of the unconformities in space and time.

STRATIGRAPHIC RESOLUTION

A stratigraphic framework that integrates seismic stratigraphic and biostratigraphic information produces a stratigraphic resolution that could not be obtained by either method alone. Seismic stratigraphy delineates depositional sequences by identifying their boundaries from reflection termination patterns such as downlap, onlap, truncation, and toplap of strata (Mitchum *et al.*, 1977), (Figures 15.2 and 15.3). Repeating this procedure for an extensive network of reflection seismic lines defines the sequences and their boundaries over a wide area of varying environments of deposition. This detailed seismic stratigraphic network provides a comprehensive record of relative sea-level changes. The relative magnitudes of a succession of relative sea-level changes can be used to assign tentative ages to the sequences by comparing the succession with one obtained from an area where the stratigraphy is well known. Where possible, however, the predicted seismic stratigraphic ages need to be confirmed by biostratigraphic control.

Well control available in the study area needs to be correlated in detail with the seismic stratigraphic network before the biostratigraphic information can be integrated with the seismic stratigraphic framework. If a number of wells are available, the higher-resolution biostratigraphy from the deeper water portion of the basins, where more age-diagnostic taxa are found, can be correlated with the basin margin with the help of the seismic stratigraphic framework. The area chosen to demonstrate the procedure for determining the magnitude of eustatic sea-level changes is the same offshore northwest Africa area used to document Mesozoic sequences by Todd and Mitchum (1977) and Vail and Mitchum (1980).

PALEOBATHYMETRIC RESOLUTION

Water depth represents the distance between seafloor and sea level and is the portion of the column that is not filled with sediment. Therefore, at any given time, no record of that column is preserved in sediment thickness measurements. The position of the seafloor relative to sea level can, however, be restored from paleontological and facies records for any significant time horizon in the basin history. Paleobathymetric reconstructions need to be sufficiently detailed (Figure 15.4) to give an acceptable accuracy to calculations of the magnitude of eustatic sea-level changes.

Paleobathymetric estimates are based on studies by Bandy (1953), Tipsword *et al.* (1966), and Pflum and Frerichs (1976). These studies use the modern depth distribution of moslty foraminiferal taxa as a potential water-depth key to fossil occurrences of the same or related taxa. The accuracy of this method seems acceptable for shallow-water deposits of up to 200 m. In deeper environments, the resolution decreases rapidly. For a successful study of eustatic changes in sea level, a well has to be available in a portion of the basin that was filled within 200 m of sea level during eustatic highstands. In such shallow-water conditions, the chronostratigraphic resolution is often rather poor and has to be complemented with ages obtained with seismic stratigraphy.

Facies and paleobathymetric information can also be obtained from the reflection seismic record by careful analysis of the cycle termination at the sequence boundaries and the internal configuration (Figure 15.5) of the sequence (Bubb and Hatlelid, 1977; Sangree and Widmier, 1977, 1979).

GEOHISTORY ANALYSIS

Geohistory analysis is a quantitative stratigraphic technique that combines the stratigraphic and paleobathymetric information in a time-depth framework. The technique in its modern quantified form was first described by van Hinte (1978), although relative age-depth diagrams were published much earlier (Lemoine, 1911; Bandy, 1953). Further improvements, such as corrections for eustatic sea-level changes, sediment compaction, and sediment loading, were added more recently.

Quantitative stratigraphy became feasible with the introduction of reliable time-scale models based on a careful integration of biostratigraphy, magnetostratigraphy, seismic stratigraphy, and radiometric dating (Jurrasic and Cretaceous: van Hinte, 1976a and 1976b; Tertiary: Berggren, 1972; Hardenbol and Berggren, 1978). The resulting linear time scale from the base of the Jurassic to the recent is incorporated in the geohistory analysis base form.

Geohistory diagrams show effectively the interaction between sediment supply, eustatic sea-level changes, and basement subsidence through time. Corrections for sediment compaction (Horowitz, 1976) are necessary to obtain a correct total subsidence history. The total subsidence is the sum of all vertical movement and represents the real movement of base-

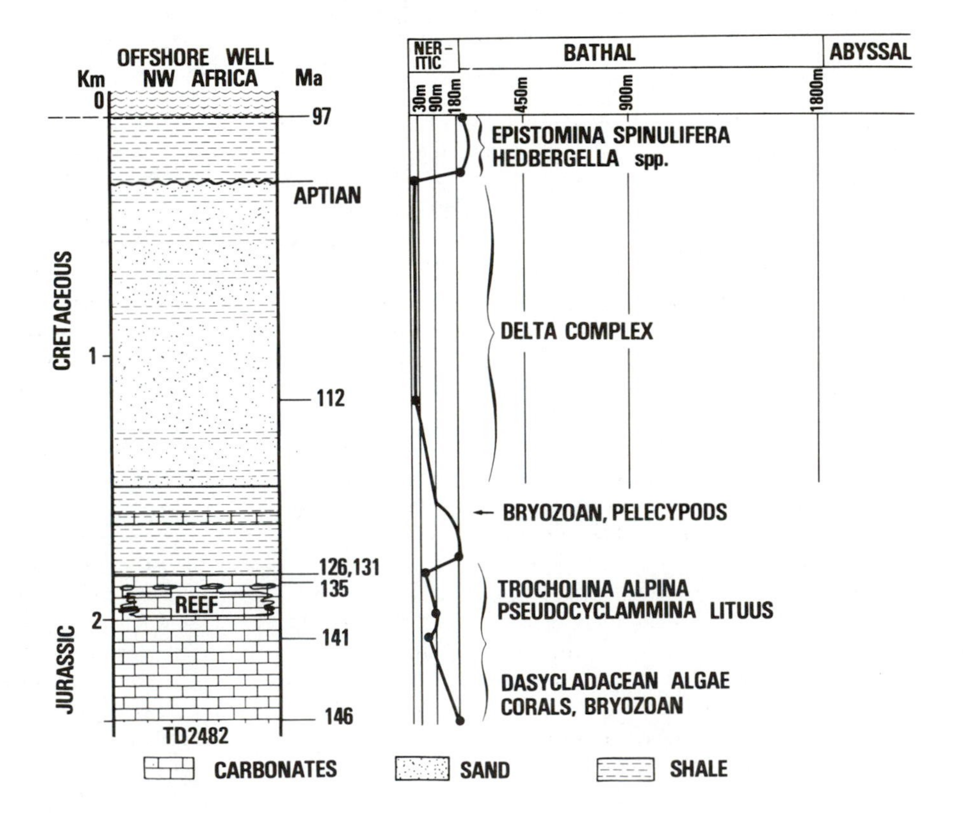

FIGURE 15.4 Well section with lithologic, stratigraphic, and paleobathymetric interpretations. The linear paleobathymetric scale indicates decreasing resolution with increasing water depth. The well section shows high paleobathymetric resolution throughout.

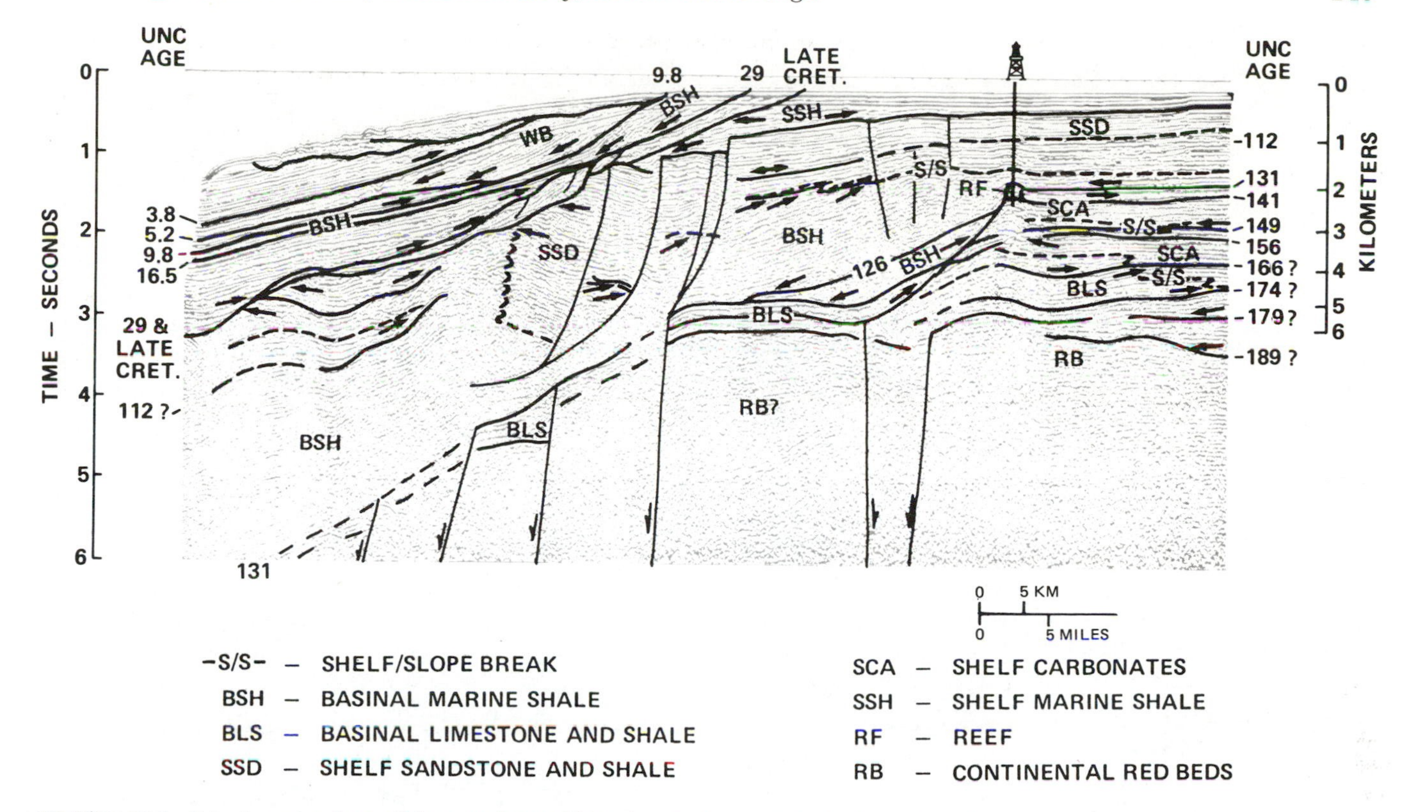

FIGURE 15.5 Seismic section from offshore northwest Africa showing interpreted seismic stratigraphic sequences with seismic facies interpreted from the nature of the sequence boundaries and the internal reflection characteristics within the sequence. For sequence designations, see Figure 15.12. After Vail and Mitchum (1980).

ment through time. After the effects of sediment loading are subtracted (Horowitz, 1976), a thermotectonic subsidence is obtained, which would be the subsidence of basement if no sediment had accumulated in the basin. Crustal cooling and basement-involved faulting are the main components of thermotectonic subsidence. Growth faults and salt and shale movements cause anomalies in the thermotectonic subsidence curve but do not affect its real magnitude. After allowances are made for anomalies caused by faulting and/or mobile substrata, the curve can be compared with one of a number of crustal subsidence curves for different amounts of lithospheric injection and modified after the dike injection model (Figure 15.6) of Royden *et al.* (1980). The stretching model of the same authors did not match the Early and Middle Jurassic subsidence observed at the well site. Matching the thermotectonic data points with one of the dike injection model curves allows quantification of the thermal component of the observed subsidence history, independent of anomalies caused by eustatic sea-level changes and inaccuracies in chronostratigraphic and paleobathymetric interpretations.

CONSTRUCTION OF GEOHISTORY DIAGRAM

To construct a geohistory diagram (Figure 15.7), first a paleobathymetric interpretation for each stratigraphic datum is plotted below the present sea level. The line connecting these points represents the paleobathymetric history through time. The stratigraphic information for a given location, gathered from all available sources, is entered in a stratigraphic column, which shows the subdivisions and thicknesses as they are encountered at present in a well or on a reflection seismic line. This stratigraphic information is also entered in the diagram and plotted against the linear time scale beginning with the base of the Sinemurian [189 million years ago (Ma)], which is the oldest horizon that can be correlated within the area. The underlying Hettangian was at the surface just before the first marine sediment was laid down. If the first sediment is nonmarine, an elevation relative to sea level at that time remains unknown. At each subsequent datum, the base of the Sinemurian is plotted at the depth it reached below the seafloor at that time as a result of basement subsidence amplified by sediment loading. The sediment columns above the Hettangian are restored to their original depositional thickness. The line connecting the basal Sinemurian depth plots depicts the total subsidence of basement that occurred at this location from the earliest Jurassic to the present.

The correction for sediment compaction with depth of burial is lithology dependent. Lithologies for the section for which the geohistory diagram is made are determined from well samples and well logs. Lithologies for the section below the total well depth are determined from seismic data. The Ceno-

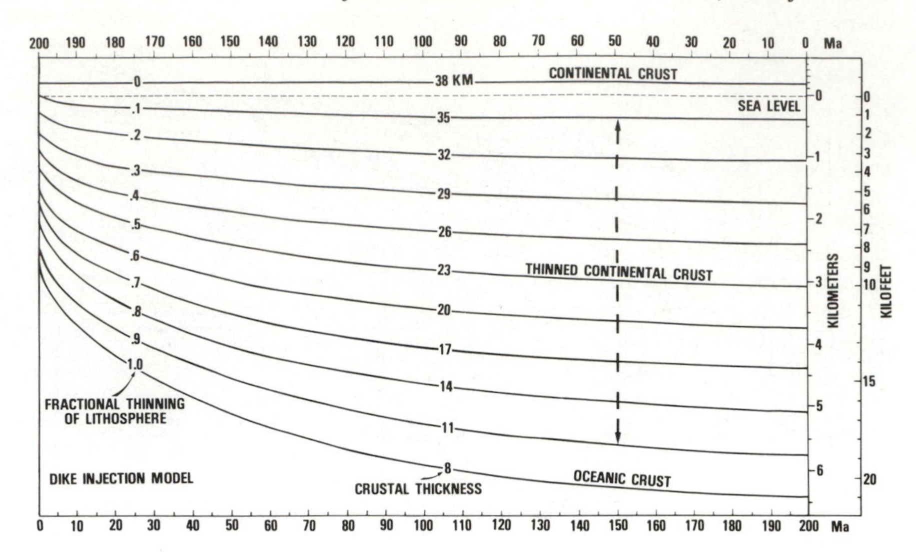

FIGURE 15.6 Subsidence curves showing the relationship between subsidence and injection of the lithosphere for the dike injection model. Modified from Royden *et al.* (1980).

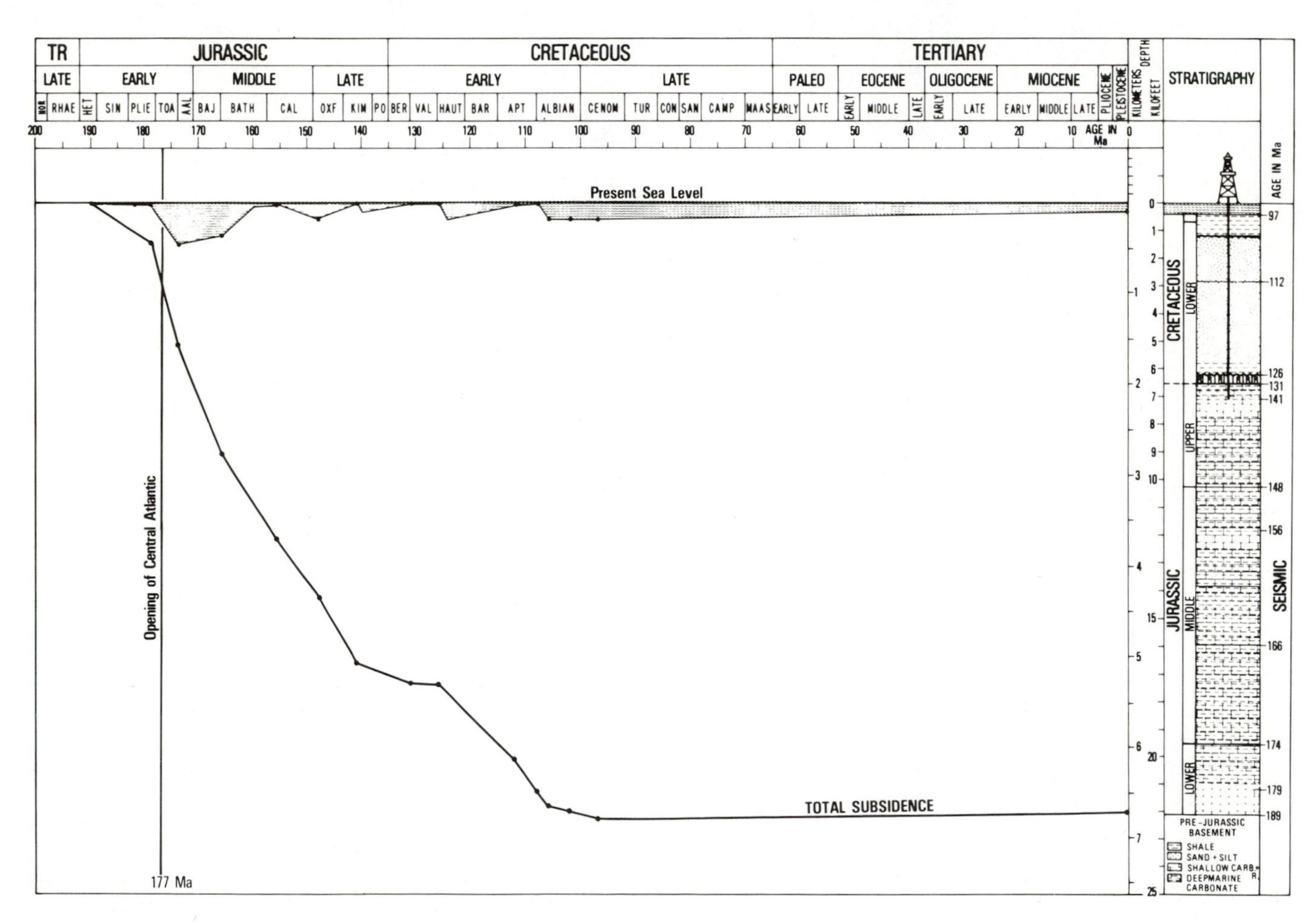

FIGURE 15.7 Geohistory analysis diagram with stratigraphic and paleobathymetric histories for a combined well and seismic location. The total subsidence curve depicts the subsidence of the Early Jurassic base through time.

manian through Hauterivian section is predominantly clastic. The compaction for shales is determined with $\phi Z = 0.7/(1 + 0.0017Z)$ (Horowitz, 1976) and for sand with $\phi Z = 0.38 \exp(-5 \times 10^{-5})$ (D. H. Horowitz, Exxon Production Research Company, personal communication, 1980), where ϕ is porosity and Z is depth in meters.

The Berriasian through Sinemurian section consists of different types of carbonate rocks. Reef carbonates are assumed to undergo minimal compaction comparable to the porosity reduction of sand. Grain carbonates are assumed to compact like sand containing 30 percent shale, and micrites like shale containing 35 percent sand.

SEDIMENT LOADING CORRECTIONS

Subsidence resulting from sediment loading represents an important portion of the total subsidence observed anywhere in a sedimentary basin. Since sedimentation is a function of sediment supply and available space, sediment fill histories can be highly variable depending on the position in the basin. For meaningful comparisons of subsidence histories between different locations within a basin, the effect of differences in sediment fill histories has to be eliminated.

An isotatic loading model assuming an Airy-type crust (Horowitz, 1976) seems to be adequate for a loading correction in most basins as long as sediments are more or less uniformly distributed (Figure 15.8). In areas adjacent to localized depocenters, such as major deltas building out in deep water or along compressional plate boundaries, an isostatic correction is not sufficient and a flexural loading model should be used (Watts and Ryan, 1976). With a sediment density of 2.7 g/cm^3 (at zero percent porosity), the mantle displacement as a result of sediment loading is 0.726 times the thickness of the solid-sediment column.

The subsidence resulting from sediment loading is subtracted from the total subsidence in the geohistory diagram. The points thus obtained show the subsidence of the basal Sinemurian if no sediment had been deposited and only water

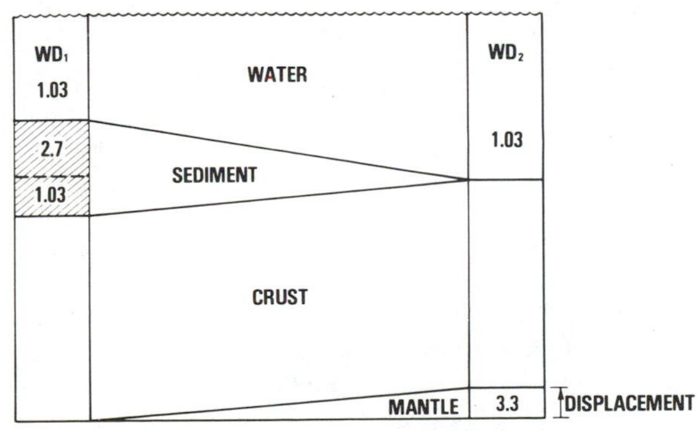

FIGURE 15.8 Isostatic backstripping model for Airy-type crust.

had filled the basin. This resultant curve is the thermotectonic subsidence curve (Figure 15.9), which shows mostly the effect of cooling of the crust and mantle, but effects of eustatics, faulting, and salt or shale movement should be identified if present.

Thermotectonic subsidence combines the effects of cooling of the crust and mantle and of tectonics such as basement-involved faulting and mobile salt and shale. The tectonic effects will, however, vary much more from place to place within a basin than the effects of crustal cooling. Constructing a number of thermotectonic curves for different locations in a basin will facilitate the distinction between thermal and tectonic subsidence. A detailed familiarity with the geologic history of the basin will further help in the distinction. The type of basin is another important factor in the interpretation. Passive margins with a single crustal-thinning event, such as the Atlantic margin off northwest Africa, are dominated by tectonic subsidence during the initial rifting, but soon after the formation of the first oceanic crust, the subsidence is entirely due to thermal contraction unless interrupted by a tectonic event.

ESTIMATE OF LONG-TERM EUSTATIC SEA-LEVEL CHANGES

The thermotectonic subsidence curve for the offshore northwest Africa example shows a high-subsidence rate during the Early Jurassic that rapidly decays to a much slower rate in the Cretaceous and Tertiary (Figure 15.9). The general shape of the curve resembles the exponential subsidence curves resulting from crustal and mantle cooling (Royden *et al.*, 1980), even though the positions of the data points in the example are affected by eustatic changes of sea level.

The initial rifting phase along the northwest African margin probably started in the Triassic, whereas the age of the earliest oceanic crust is generally quoted as 180 Ma (Pitman and Talwani, 1972; Hayes and Rabinowitz, 1975). The earliest magnetic anomaly is M 26 at 153 Ma, but the presence of a magnetic quiet zone does not preclude an earlier onset of spreading. Seismic stratigraphic and seismic facies studies support an Early Jurassic opening because of indications of Late Pliensbachian to Toarcian deeper water deposits in the area. Although the rapid subsidence in the Early Jurassic could have a minor tectonic component, most of the constructed curve is the result of crustal and mantle cooling, albeit with a clear influence from the long-term eustatic change in sea level.

The history of relative changes of coastal onlap suggests (Vail *et al.*, 1977) that eustatic sea level underwent considerable changes in the Mesozoic and Cenozoic. Their study shows that in the Early and early Middle Jurassic, sea level was close to the present sea level. A major rise began in the Callovian (156 Ma) and, with short interruptions in the Valanginian, Aptian, and Cenomanian, continued into the Late Cretaceous. This general rise of sea level from late Middle Jurassic to Late Cretaceous has a steepening effect on thermotectonic subsidence curves that are not corrected for eustatic changes. This eustatic effect complicates the comparison between

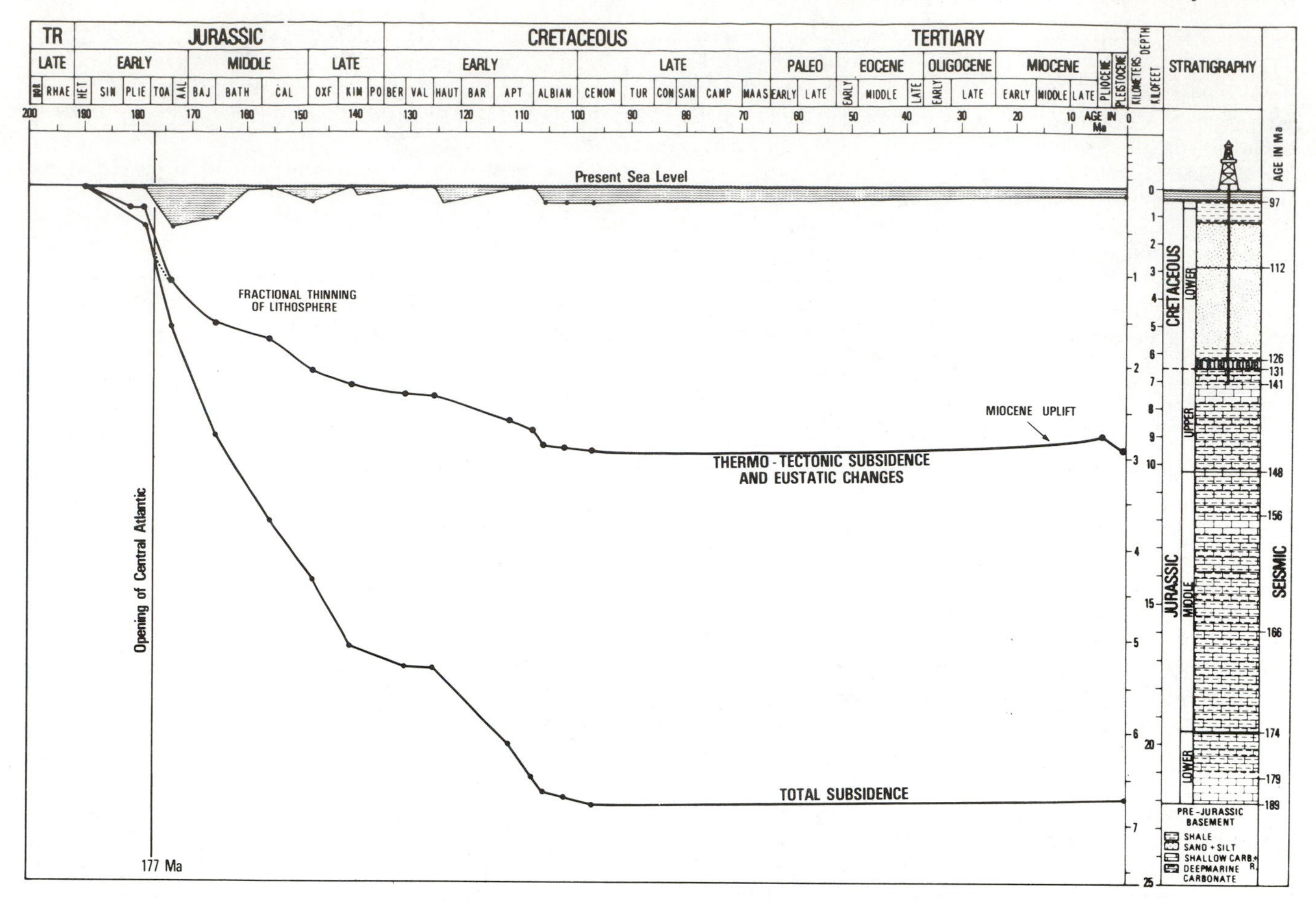

FIGURE 15.9 Geohistory diagram showing the thermotectonic subsidence curve of the base independent of sediment loading and depicting the subsidence of the Early Jurassic base if the basin was not filled with sediment. A small compressional uplift is assumed for the Late Tertiary.

reconstructed thermotectonic curves and model curves for crustal and mantle thermal subsidence (Royden *et al.*, 1980).

If sea level in the Early and Middle Jurassic was as close to the present sea level as is suggested by the magnitude of relative changes of coastal onlap, the data points for that portion of the section should match one of the model curves for crustal and mantle cooling. This assumes that the observed thermotectonic subsidence curve represents only thermal subsidence with long-term eustatic departures that started immediately after the formation of the first oceanic crust. The geohistory diagram (Figure 15.10) suggests that the first oceanic crust was formed at 177 Ma between the 179 Ma and 174 Ma sequence boundaries. The model curve for a 60 percent injection of the lithosphere matches the Early and Middle Jurassic data points (Figure 15.10). All other data points beginning in the Late Jurassic fall significantly below the 60 percent model curve, which is consistent with a general rise in eustatic sea level in the Late Jurassic and Early Cretaceous.

Other model curves, such as those for 50 percent and 40 percent injection of the lithosphere, do not match the steep initial portion of the constructed curve. These two model curves indicate an initial thermal subsidence for an opening event at 177 Ma that is less than the observed total subsidence.

The present-day data point in the geohistory diagram (Figure 15.10) falls about 100 m above the 60 percent model curve. This is interpreted as the results of Late Tertiary uplift, which can be substantiated by several lines of evidence. Compressional tectonics associated with the formation of the High Atlas Mountains may have begun in the Late Oligocene and continued through the Late Miocene, resulting in significant erosion of older deposits. Only the clinoform toes of the Middle Miocene sea-level highstand sequences are preserved (Figure 15.3). Toplap associated with the Pliocene highstand sequence seaward of the well location suggests that the Cenomanian at the well site was at or slightly above sea level at that time. The subsidence of the Cenomanian surface since the Middle Pliocene can be estimated at 177 m if we add Middle Pliocene eustatic sea level [+ 80 m (Vail and Hardenbol, 1979)] and present-day water depth (97 m). This subsidence is faster than the thermal contraction from an Early Jurassic opening event as is suggested by the 60 percent model curve. By the end of the Miocene, the total compressional uplift may have exceeded 200 m (Figure 15.10).

Independent qualitative evidence for an uplift is provided by seismic interval velocities that suggest that the Early Cretaceous sediments are overcompacted for their present

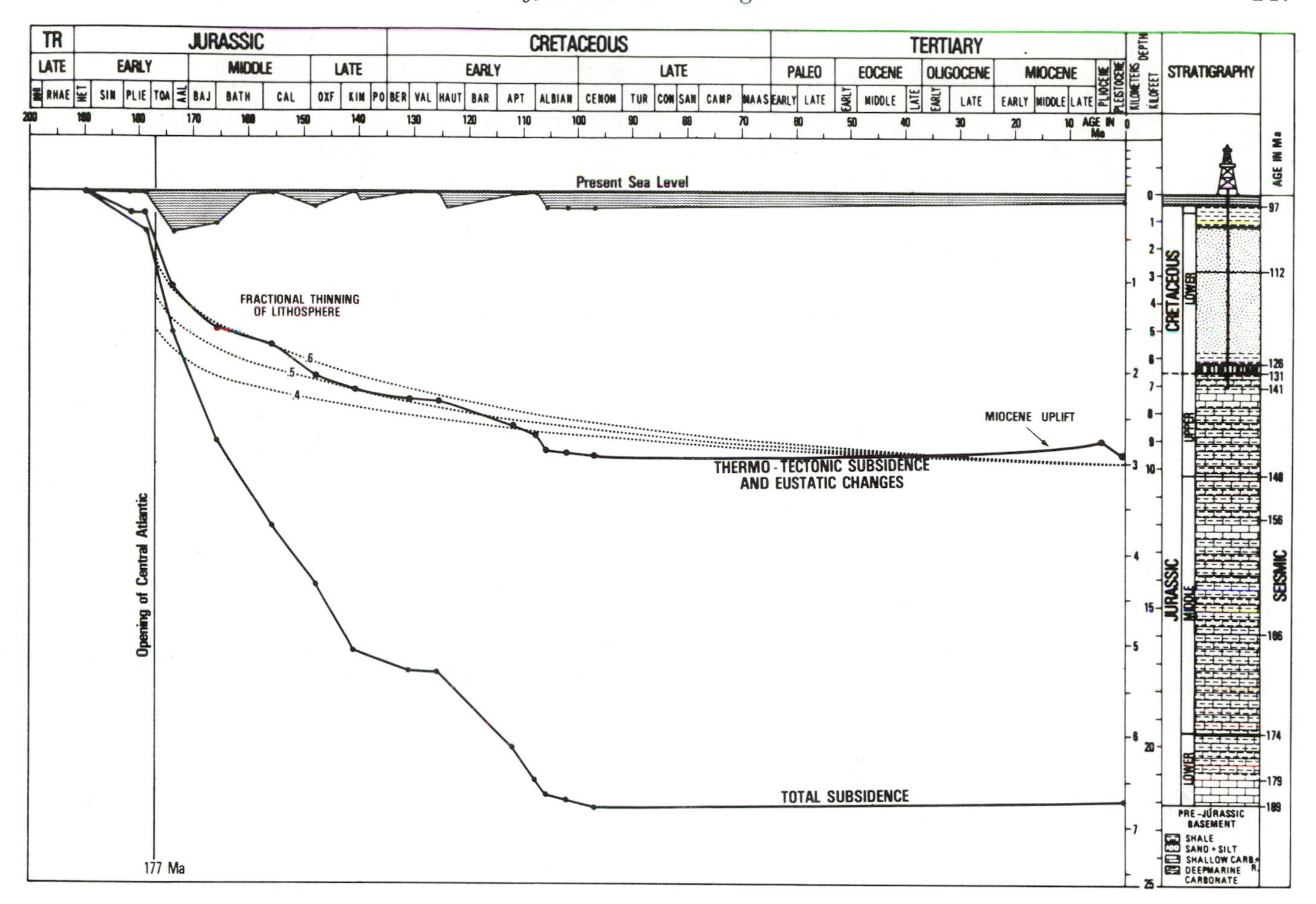

FIGURE 15.10 Geohistory diagram comparing thermotectonic subsidence with model curves for three different amounts of injection of the lithosphere. A small compressional uplift is assumed for the Late Tertiary.

burial depth. The observed interval velocities require a burial depth between 450 and 750 m some time between the Cenomanian and the present. Without an uplift, this burial depth could have been accomplished only if the basin at the well site was filled to sea level in the Late Cretaceous and subsequently eroded during lowstands in the Tertiary, especially in the Late Oligocene. All Upper Cretaceous and Lower Tertiary deposits encountered in the area are deposited in bathyal water depths, and the configuration of the seismic sequences suggests that the basin at the well site was not filled to sea level until the Oligocene. Several lines of evidence seem to agree that at least some uplift occurred in the Late Tertiary. The amount of uplift cannot be determined with accuracy because the Tertiary stratigraphic information is very incomplete as a result of the uplift. In the calculations for eustatic sea-level changes, a 100-m uplift is used. This value is estimated from the position of the present-day data point relative to the 60 percent model subsidence curve in Figure 15.10. The resulting values are compared with those obtained if no uplift is taken into account.

The distance between the thermotectonic subsidence curve and the 60 percent model subsidence curve in Figure 15.10 is a measure of the long-term eustatic sea-level change in the Jurassic and Early Cretaceous. Sea level is rising in the late Middle Jurassic and begins falling in the latest Jurassic and Berriasian. A new sea-level rise begins in the Hauterivian or mid-Valanginian and continues into the Early Cenomanian. Measuring eustatic changes directly from Figure 15.10 produces values that require correction for the loading effect of eustatic sea-level changes as follows:

$$MD = \Delta \text{sea level} + (\rho w/\rho m) \times \Delta \text{sea level} = 1.309\, \Delta \text{sea level}$$

or

$$\Delta \text{sea level} = MD/1.309,$$

where MD is the measured distance of the data point below the model subsidence curve in Figure 15.10.

The calculation of long-term eustatic sea-level changes is based on an isostatic comparison on an Airy-type crust (Horowitz, 1976; Watts and Ryan, 1976). At 97 Ma (Early Cenomanian) depositional water depth as determined from benthonic microfossils is 180 m. Early Cenomanian beds are still (or again) at the seafloor covered by 97 m of water.

The isostatic comparison is given by the equation:

$$WD_1 + Ts = WD_2 + EF + MC + Ru, \qquad (15.1)$$

where WD_1 is the original water depth, Ts is the thermal subsidence from the 60 percent model curve in Figure 15.10, WD_2 is the present water depth, EF is the eustatic difference, MC is the mantle compensation after an eustatic change, and Ru is the present-day expression of the regional uplift in the Late Tertiary (100 m). The effects of mantle compensation can be determined by the equation

$$MC = EF\,(\rho w/\rho m - \rho w) = 0.45\,EF, \qquad (15.2)$$

where ρw is the density of seawater (1.03 g/cm^3) and ρm is the average density of the mantle (3.33 g/cm^3). Equation (15.1) can be written as

$$EF = [(WD_1 = WD_2) + Ts \pm Ru]/1.45. \qquad (15.3)$$

Substituting values for the Early Cenomanian and the present in Eq. (15.3) results in an apparent eustatic sea-level fall of 281 m (Figure 15.11). Equation (15.3) calculates eustatic sea-level change for water-filled basins only. For other data points in the Early Cretaceous and Jurassic, additional corrections are required for sediments deposited during the respective time intervals. Isostatic comparisons can be made by removing the sediment and restoring the water depth as follows:

$$EF = [WD_1 - (WD_2 + WR) + Ts - Ru]/1.45, \qquad (15.4)$$

where WR is the restored water column after backstripping and taking into account unloading adjustments and porosity restoration of the underlying section. The apparent sea-level changes for data points in the Early Cretaceous, Late Jurassic, and Middle Jurassic determined in several different ways are listed in Table 15.1. The eustatic changes were initially calculated for a 100-m uplift and for the case that no uplift occurred. The values measured in Figure 15.10 and corrected for eustatic water loading should agree with the calculated values if no uplift occurred. Table 15.1 shows that the measured values fall between those calculated for a 100-m uplift and for no uplift, thus confirming some uplift but not as high as 100 m. If the present-day data point on the thermotectonic curve were 70 m above the model curve, there would be close agreement between calculated and measured eustatic sea-level changes.

The eustatic sea-level values thus obtained represent the highest sea-level stand in each sequence. The changes observed are therefore predominantly changes in the long-term eustatic sea level. There is close agreement with previous results obtained by different methods by Hays and Pitman (1973), Vail *et al.* (1977), Pitman (1978), and Vail and Todd (in press). A significant difference exists, however, with the values obtained by Watts and Steckler (1979) using a similar method based on the subsidence history of the Atlantic Ocean margin of the North American east coast. A possible explanation for their much lower results is that their assumption of the thermotectonic subsidence of the Atlantic Ocean margin being a single thermal contraction event is not valid. Thermotectonic subsidence histories for Georges Bank and Baltimore Canyon COST wells suggest considerable tectonic activity in the Early Cretaceous and Middle Tertiary.

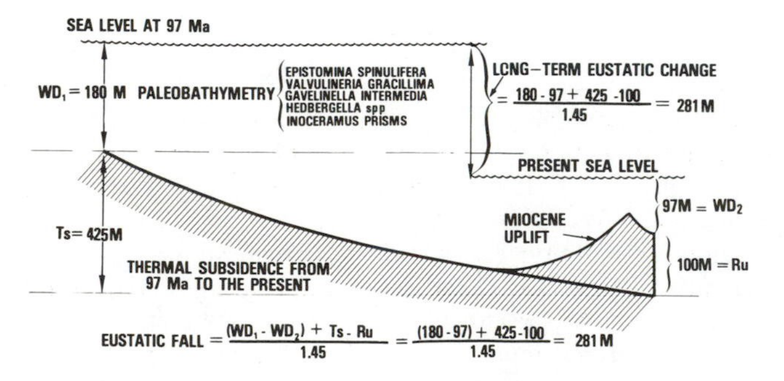

FIGURE 15.11 Calculation of eustatic sea-level change from the Early Cenomanian to the recent.

TABLE 15.1 Eustatic Sea Level in Meters above the Present Sea Level Calculated with Equation (15.4) for Three Different Amounts of Late Tertiary Compressional Uplift Compared with the Eustatic Sea Level Measured from the Geohistory Diagram in Figure 15.10 (Corrected for Eustatic Highstand Loading Effects)[a]

Age	No Uplift	100-m Uplift	70-m Uplift	From Figure 15.10
Early Cenomanian	350	281	302	302
Late Albian	359	290	311	306
Top Aptian	259	190	210	206
Middle Aptian	195	126	147	138
Top Valanginian	129	60	81	69
Top Berriasian	179	110	130	107
Mid-Kimmeridgian	247	178	199	149
Top Callovian	210	141	162	107
Top Bathonian	77	8	29	0

[a]Figures for the 70-m uplift are the preferred values.

IDENTIFICATION OF SHORT-TERM EUSTATIC CHANGES

Previous sections of this paper discussed the paleoenvironments, subsidence history, and long-term sea-level changes determined from a well located on seismic line C, offshore northwestern Africa. The well log and seismic section show that there are many abrupt vertical changes in depositional environments and lithofacies. In order for these abrupt changes to occur, a rapid change is necessary in one or more of the three variables: rate of subsidence, rate and type of deposition, or rate of eustatic sea-level changes. Figure 15.10 shows that subsidence in the area of the well follows a normal thermal contraction curve from the time of initial formation of oceanic crust (± 177 Ma) to the Late Oligocene, when uplift associated with the High Atlas Mountains commended. All changes in subsidence are gradual except for the initial subsidence following the formation of oceanic crust in the Atlantic Ocean (± 177 Ma) and the subsidence following the uplift associated with the High Atlas Mountains (± 3.8 Ma). The type and rate of deposition of the sediments in the northwest African margin also change through time. These changes are, however, gradual when the total basin is studied. In any one location, abrupt changes in sediment type and sedimentation rate are common occurrences. Global studies of the Jurassic and Tertiary (Vail *et al.*, 1977, Part 4, Figure 6; Vail and Hardenbol, 1979; Vail and Todd, in press) show that rapid eustatic sea-level changes occur periodically (Figure 15.12). The timing of these global changes coincides with the changes observed in northwest Africa. We conclude that most of the abrupt changes observed in the well and on seismic line C are caused by these rapid changes in eustatic sea level.

Three types of eustatic changes that cause unconformities can be distinguished: (1) rapid falls of eustatic sea level usually greater than the rate of subsidence, (2) slower falls of eustatic sea level commonly less than the rate of subsidence in basins with significant subsidence, and (3) rapid rises of eustatic sea level.

Rapid rates of fall of eustatic sea level are characterized by unconformities associated with subaerial exposure on the shelf, canyon cutting, submarine erosion, lowstand deltas and fans, and shifts in submarine depositional patterns where fans and lowstand deltas are absent (Figure 15.13). Slow falls less than the rate of subsidence are similar except that the outer portion of the shelf may not be subaerially exposed and lowstand deltas, fans, and canyon cutting are rare. Marine regressions commonly underlie these types of unconformities. Rapid rises are commonly associated with marine transgressions. Basal transgressive deposits overlain by local submarine unconformities may be present (Vail and Todd, in press).

The following paragraphs will describe the stratigraphy and facies of the well and seismic line C (Figure 15.5) and discuss what may have caused the abrupt changes in depositional environments and lithofacies. Seismic line C is used to illustrate these changes for the Mesozoic section. Two seismic lines shown on Figure 15.14 are used for the Tertiary and Cretaceous.

The first major abrupt vertical change occurs at a unconformity labeled 189(?) Ma on Figure 15.5. On other seismic sections, this unconformity can be traced across the shelf close to a well drilled on land near the coastline (Vail and Mitchum, 1980) where Middle to Late Jurassic clastics overlie the unconformity and Hettangian-Triassic continental red beds underlie it. These beds are locally truncated. To the south, the uncon-

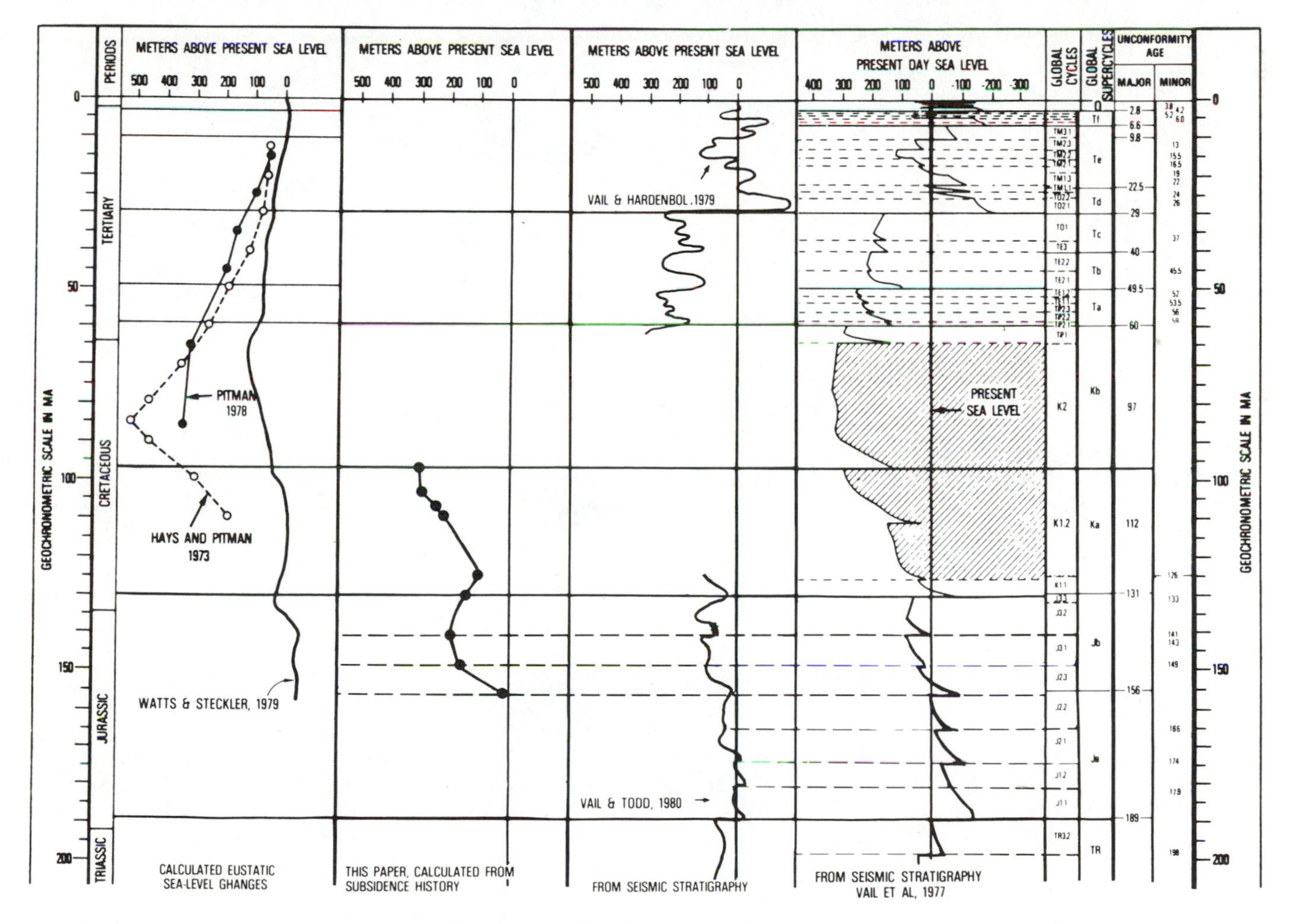

FIGURE 15.12 Comparison of eustatic sea-level changes determined from the subsidence history of the northwest African margin with previously published estimates for global changes of sea level.

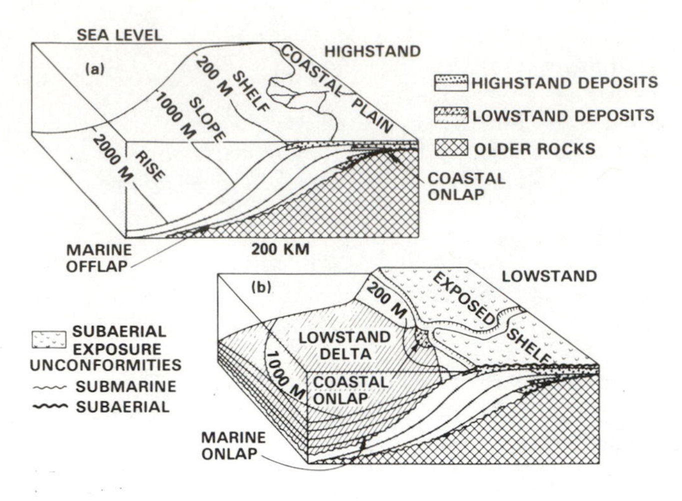

FIGURE 15.13 Depositional patterns during highstand (a) and lowstand (b) of sea level.

formity can be traced below a thick salt section. Well data to the south show that an interval of shallow-water carbonates, anhydrites, and variegated shales, dated by biostratigraphy as probably Early Jurassic, overlie the salt. Seismic correlations indicate that the onlapping reflections between sequence boundaries 177 Ma and 189 Ma (on line C, Figure 15.5) correspond to this interval. The unconformity is believed to correlate with the basal Sinemurian (189 Ma) unconformity. In the area of study, it marks an abrupt change from continental red beds to a marine section containing salts, anhydrites, and carbonates. It is interpreted on the basis of global studies to be caused by a rapid fall and rise of eustatic sea level that occurred in the Early Sinemurian [between the *Arietites bucklandi* and *Arnioceras semicostatum* ammonite zones (A. Hallam, University of Birmingham, personal communication, 1981)].

The next abrupt vertical change is best observed on seismic data, such as line C, Figures 15.3 and 15.5. It occurs at the sequence boundary labeled 179(?) Ma and is marked by downlap over parallel reflections. The downlap is interpreted as the

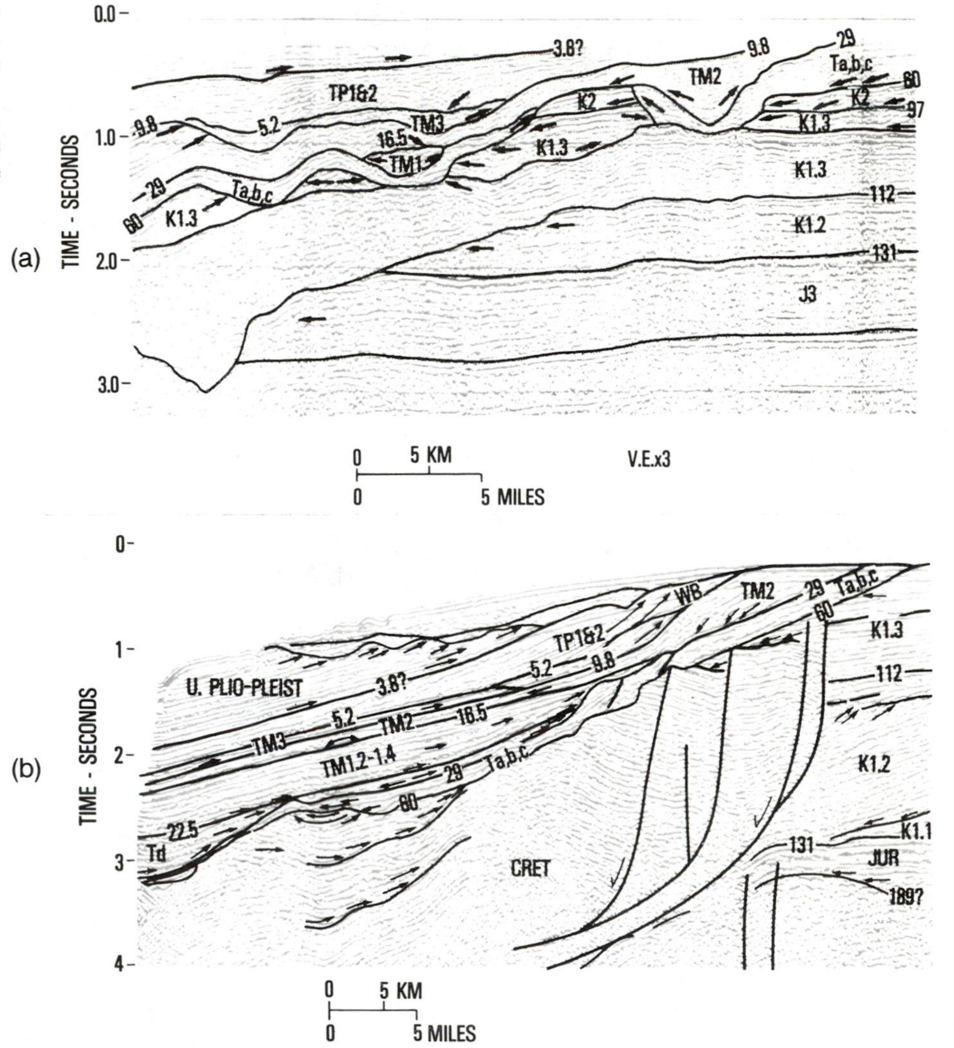

FIGURE 15.14 (a) Seismic line parallel to shelf edge, offshore northwest Africa, showing major erosional patterns in the Early Cretaceous and Tertiary. (b) Seismic line perpendicular to the shelf edge, offshore northwest Africa, showing major erosional patterns in the Tertiary. For sequence designations see Figure 15.12.

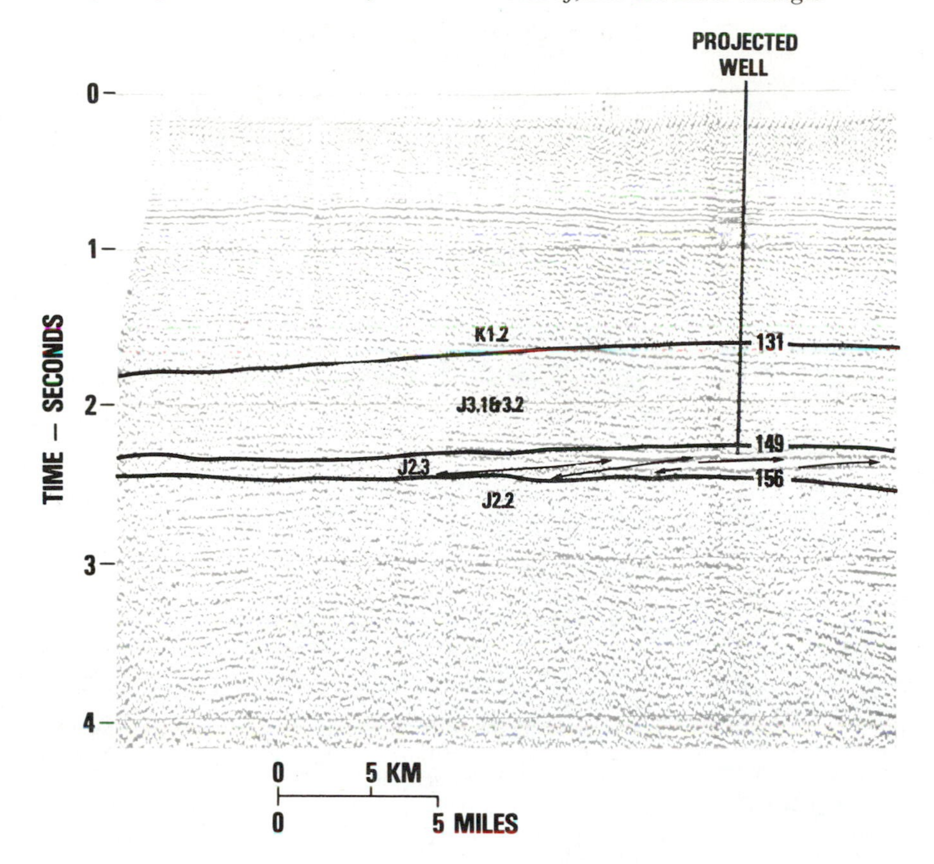

FIGURE 15.15 Offshore northwest Africa seismic line showing prograding Callovian shelf margin. For sequence designations see Figure 15.12.

toes of prograding clinoforms. The relief of these clinoforms indicates upper bathyal paleowater depths. This boundary is interpreted as an abrupt change from shallow-water carbonates, variegated shales, and evaporites to deep-water carbonate muds. Well data to the south document a change from the shallow-water carbonates and evaporites to massive micritic limestones, but no fossils were identified to verify the deep-water interpretation of this interval. The sequence boundary is correlated with the global fall and rise of eustatic sea level that occurred within the early part of the Late Pliensbachian at 179 Ma. Paleontological data from wells within the area were undiagnostic and could only be dated as probably Early Jurassic. The rapid deepening and landward shift of the shelf edge is believed to be related to the increased rate of subsidence following the formation of oceanic crust to the west.

The next major abrupt vertical change is well documented by wells as a change from Bathonian shelf carbonates to upper bathyal Callovian shales. Seismic data show a continuous high-amplitude reflection on the shelf with evidence of truncation at the shelf edge. The sequence boundary labeled 156 Ma on line C (Figure 15.5) indicates the abrupt change from carbonate to shale. Landward, this sequence boundary can be traced on seismic data below a prograding sequence that is verified in a well as Callovian marine shale (Figure 15.15). The relief on the clinoforms, shown on this section, indicates water depths of approximately 300 m. The rapid change from shallow shelf to upper bathyal water depths (±50 to 300 m) from the Bathonian to the Callovian must have taken place early in the Callovian, since the seismic patterns indicate mainly progradation during the Callovian. This water-depth change of 250 m probably occurred in not more than 2-4 million years (m.y.), during which time the thermotectonic subsidence of 23 m/m.y. (Figures 15.6 and 15.10) would amount to 45 to 90 m. Between 160 and 200 m of the water-depth change must have been caused by a rise of eustatic sea level in 2 m.y. to 4 m.y., which requires rates of eustatic rise from 4 to 10 cm per 1000 yr and could be much more rapid.

The next major abrupt vertical change is also well documented by well control and occurs between Berriasian shallow-water shelf carbonates and upper bathyal (200-250 m) Hauterivian shale on the shelf, and between Berriasian deep marine micritic carbonates and Valanginian reddish-brown, deep-marine mudstones in the basin. The well on the shelf edge encountered a 6.3-m cavern, the base of which was 70 m below the top of the Berriasian carbonate. The sequence boundary labeled 131 Ma on line C (Figures 15.3 and 15.5) marks this change. The Valanginian sequence is shown as a slopefront-fill seismic facies pattern seaward of the shelf-edge reef encountered in the well. The Valanginian unit (131-126 Ma) is interpreted as the edge of a lowstand delta diagrammatically illustrated in Figure 15.13. It indicates that sea level fell below the shelf edge at the beginning of Valanginian time. The karst cavern drilled in the well is also evidence for a sea-level fall at this time. Since the thermotectonic subsidence in the Valanginian was 0.99 cm per 1000 years (Figures 15.6 and 15.10), sea level must have dropped at a greater rate to fall below the shelf edge. The lowest marine shales overlying the Berriasian carbonates are dated as Hauterivian. The above observations are interpreted as indicating a rapid rise of sea level during the Valanginian of at least a few centimeters per 1000 yr following a fall that exceeded 1 cm per 1000 yr. The magnitude of the fall cannot be estimated from these data, but studies from other areas indicate it may approximate 100 m. The water-depth increase over the shelf is approximately 200 m. The subsidence curves (Figures 15.6 and 15.10) indicate that 47 m were due to thermotectonic subsidence and thus 153 m must be due to eustatics rising at a rate of 3-8 cm/1000 yr. The amount of rise over the upper slope must be added to this to obtain the total rise, but it cannot be determined from the available data.

The Hauterivian-Aptian interval consists of a large upward-coarsening, prograding delta. A rapid fall and rise of sea level within this delta is indicated by a large canyon cut, shown on Figure 15.14(a), which is dated as occurring within the Mid-Aptian (112 Ma). Upper bathyal Middle Albian shales overlying the delta indicate a rise in sea level at this time.

Evidence of many other rapid changes of sea level is shown in Figure 15.14. Slope-front-fill facies of Late Oligocene-Early Miocene (29-16.5 Ma), Late Miocene (9.8 Ma), and Late Pliocene (3.8 Ma), together with their underlying unconformities showing erosional truncation, indicate lowstands. Prograding units and their equivalent deep marine draping shale, such as the Middle Miocene (16.5-9.8 Ma) and Early-Middle Pliocene (6.6-3.8 Ma), indicate highstands.

The present flooded shelf is believed to be due to subsidence following the Miocene uplift associated with the High Atlas Mountains. A widespread truncated surface marks the erosions of tilted beds following that uplift.

DISCUSSION

The magnitude of long-term, sea-level changes between the Early Cretaceous and the present, as determined from the subsidence history of a basin offshore northwest Africa, is in line with previously published estimates. Values obtained by Hays and Pitman (1973) and Pitman (1978) for the Mid-Cretaceous from rates of ocean spreading and by Vail and Todd (in press) for the Jurassic and earliest Cretaceous are very close to the estimates in this paper. They exceed, however, the values obtained by Watts and Steckler (1979), using a similar method based on margin subsidence.

The models and assumptions used in this study are tentative and require further evaluation and improvements. However, repeating this method for a number of passive margins with adequate stratigraphic and paleobathymetric control should demonstrate trends in the magnitude of long-term eustatic sea-level changes. Short-term changes in sea level can be determined from detailed studies within a quantified framework of subsidence and long-term eustatic sea-level changes. Rates of change of the short-term, sea-level fluctuations may be readily determined in areas with detailed age and paleobathymetric control. The magnitude of short-term, sea-level changes is much more difficult to estimate.

ACKNOWLEDGMENTS

We are grateful to R. M. Mitchum and C. R. Tapscott for reviewing the manuscript and suggesting improvements. D. H. Horowitz and C. R. Tapscott provided valuable assistance in solving problems with the compaction, loading, and subsidence models. We thank our colleagues in Houston and Bordeaux, France, for providing stratigraphic or paleobathymetric information. We also thank N. J. Douglas and B. K. Trujillo for their care in preparing the manuscript and illustrations.

REFERENCES

Bandy, O. L. (1953). Ecology and paleoecology of some California foraminifera, Part II, Foraminiferal evidence of subsidence rates in the Ventura basin, *J. Paleontol. 27*, 200-203.

Berggren, W. A. (1972). A Cenozoic time-scale: Some applications for regional geology and paleobiogeography, *Lethaia 5*, 195-215.

Bubb, J. N., and W. G. Hatlelid (1977). Seismic stratigraphy and global changes of sea level, Part 10: Seismic recognition of carbonate buildups, in *Seismic Stratigraphy—Application to Hydrocarbon Exploration*, C. E. Payton, ed., Am. Assoc. Petrol. Geol. Mem. 26, pp. 185-204.

Hardenbol, J., and W. A. Berggren (1978). A new Paleogene numerical time scale, *Am. Assoc. Petrol. Geol. Stud. Geol. 6*, 213-234.

Hayes, D. E., and P. D. Rabinowitz (1975). Mesozoic magnetic lineations and the magnetic quiet zone off northwest Africa, *Earth Planet. Sci. Lett. 28*, 105-115.

Hays, J. D., and W. C. Pitman III (1973). Lithospheric plate motion, sea-level changes and climatic and ecological consequences, *Nature 246*, 18-22.

Horowitz, D. H. (1976). Mathematical modeling of sediment accumulation in prograding deltaic systems, in *Quantitative Techniques for the Analysis of Sediments*, D. F. Merriam, ed., Proc. IX Int. Sedimentol. Congr., Nice, France, 1975, pp. 105-119.

Lemoine P. (1911). *Geologie du Bassin de Paris*, Librarie Scientifique, Herman, Paris.

Mitchum, R. M., Jr., P. R. Vail, and S. Thompson III (1977). Seismic stratigraphy and global changes of sea-level, Part 2: The depositional sequence as a basic unit for stratigraphic analysis, in *Seismic Stratigraphy—Application to Hydrocarbon Exploration*, C. E. Payton, ed., Am. Assoc. Petrol. Geol. Mem. 26, pp. 53-62.

Pflum, C. E., and W. E. Frerichs (1976). Gulf of Mexico deep-water foraminifers, W. V. Sliter, ed., Cushman Found., *Foraminiferal Res., Special Publ. 14*, pp. 1-125.

Pitman, W. C., III (1978). Relationship between eustacy and stratigraphic sequences of passive margins, *Geol. Soc. Am. Bull. 89*, 1389-1403.

Pitman, W. C., III, and M. Talwani (1972). Sea-floor spreading in the North Atlantic, *Geol. Soc. Am. Bull. 83*, 619-646.

Royden, L., J. G. Sclater, and R. P. von Herzen (1980). Continental margin subsidence and heat flow: Important parameters in formation of petroleum hydrocarbons, *Am. Assoc. Petrol. Geol. Bull. 64*, 173-187.

Sangree, J. B., and J. M. Widmier (1977). Seismic stratigraphy and global changes of sea level, Part 9: Seismic interpretation of clastic depositional facies, in *Seismic Stratigraphy—Application to Hydrocarbon Exploration*, C. E. Payton, ed., Am. Assoc. Petrol. Geol. Mem. 26, pp. 165-184.

Sangree, J. B., and J. M. Widmier (1979). Interpretation of depositional facies from seismic data, *Geophysics 44*, 131-160.

Tipsword, H. L., F. M. Setzer, and F. L. Smith, Jr. (1966). Interpretation of depositional environment in Gulf Coast petroleum exploration from paleoecology and related stratigraphy, *Trans. Gulf Coast Assoc. Geol. Soc. 16*, 119-130.

Todd, R. G., and R. M. Mitchum, Jr. (1977). Seismic stratigraphy and global changes of sea level, Part 8: Identification of Upper Triassic, Jurassic and Lower Cretaceous seismic sequences in Gulf of Mexico and offshore west Africa, in *Seismic Stratigraphy—Application to Hydrocarbon Exploration*, C. E. Payton, ed., Am. Assoc. Petrol Geol. Mem. 26, pp. 145-163.

Vail, P. R., and J. Hardenbol (1979), Sea-level changes during the Tertiary, *Oceanus* 22, 71-79.

Vail, P. R., and R. M. Mitchum (1980). Global cycles of sea-level change and their role in exploration, *Proc. Tenth World Petrol. Congr., 2, Expl. Supply and Demand*, 95-104.

Vail, P. R., and R. G. Todd (in press). Northern North Sea Jurassic unconformities, chronostratigraphy and sea-level changes from seismic stratigraphy, *Proc. Petrol. Geol. Cont. Shelf, N.W. Europe Conf.*, Heydon, Philadelphia.

Vail, P. R., R. M. Mitchum, Jr., and S. Thompson III (1977). Seismic stratigraphy and global changes of sea level, Part 3: Relative changes of sea-level from coastal onlap, in *Seismic Stratigraphy—Application to Hydrocarbon Exploration*, C. E. Payton, ed., Am. Assoc. Petrol. Geol. Mem. 26, pp. 63-81.

van Hinte, J. E. (1976a). A Jurassic time scale, *Am. Assoc. Petrol. Geol. Bull. 60*, 489-497.

van Hinte, J. E. (1976b). A Cretaceous time scale, *Am. Assoc. Petrol. Geol. Bull. 60*, 498-516.

van Hinte, J. E. (1978). Geohistory analysis-application of micropaleontology in exploration geology, *Am. Assoc. Petrol. Geol. Bull. 62*, 201-222.

Watts, A. B., and W. B. F. Ryan (1976). Flexure of the lithosphere and continental margins, *Tectonophysics 36*, 25-44.

Watts, A. B., and M. S. Steckler (1979). Subsidence and eustasy at the continental margin of eastern North America, in *Deep Drilling Results in the Atlantic Ocean: Continental Margins and Paleoenvironment*, M. Talwani, W. F. Hay, and W. B. F. Ryan, eds., Maurice Ewing Ser. 3, American Geophysical Union, Washington, D.C., pp. 218-234.

Tertiary Marine and Nonmarine Climatic Trends

16

JACK A. WOLFE *and* RICHARD Z. POORE
U.S. Geological Survey

INTRODUCTION

Climates of the Tertiary period have been more thoroughly studied and discussed than those of any pre-Quaternary interval. Despite this seeming wealth of data (or perhaps because of such wealth), many researchers have arrived at differing conclusions about both the intensity of warm intervals and temperature fluctuations during the Tertiary. Some of the differences are related to sampling at widely spaced stratigraphic intervals, imprecise correlations of samples, and/or the use of different techniques to arrive at paleotemperature estimates. These problems are greatly intensified in attempting to relate marine and nonmarine trends; to overcome one or more of these problems, our discussion will concentrate on two geographic regions.

PALEOGENE OF THE ATLANTIC BASIN

The Paleogene of the Mississippi embayment of southeastern North America contains an extensive sequence of land floras. These flora were, for the most part, included by Berry (1916) in his "Wilcox flora" and were thus considered by him to be approximately synchronous. Work by many other stratigraphers in recent decades has demonstrated that Berry's "Wilcox flora" was based on material from rocks of early Paleocene through late Eocene Age.

These Mississippi embayment floras all represent lowland vegetation, and thus the altitudinal variations that can affect analysis of floras in continental interiors are negated. In turn, the data on land floras can be related to paleoclimatic models based on data from deep-sea sediments in the North Atlantic. Reliance on data from the North Atlantic and its borderlands also has the advantage that this region has had a simple and well understood plate-tectonic history during the Tertiary in comparison with that of many other regions, and thus paleolatitudes are well known.

The interpretation of climatic change during the Paleogene in the North Atlantic is based both on paleontological evidence and on isotopic data (Figure 16.1). The paleontological interpretations largely follow previously published data (Haq and Lohmann, 1976; Haq *et al.*, 1977), which involve recognizing distinctive latitudinally zoned nannofossil assemblages and tracing the latitudinal migrations of these assemblages through time. However, an exception is our interpretation of the mid-

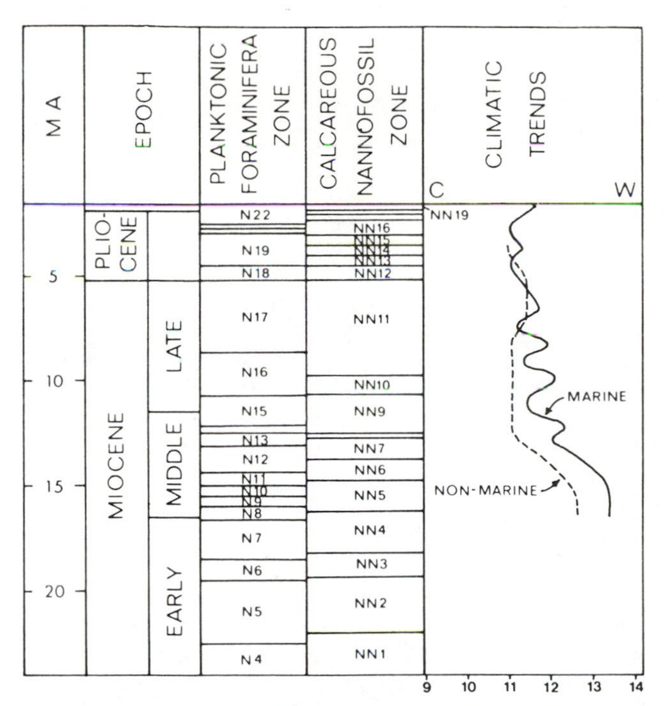

FIGURE 16.1 Comparison of temperature trends in the North Atlantic Paleogene with inferred temperatures of Paleocene and Eocene land floras from the Mississippi embayment. Extent of bars indicates possible age range of land floras in terms of planktonic time scale.

dle Eocene *Discoaster* assemblage (Haq *et al.*, 1977) as representing low- rather than middle-latitude climate. Our interpretation is in accord with the isotopic record from the Deep Sea Drilling Project (DSDP) site 398 (Vergnaud-Grazzini, 1979); data on the early Paleocene climate is also based largely on somewhat meager isotopic records.

Interpretations of the land floras are based primarily on physiognomic analyses of assemblages (Wolfe, 1971, 1978). Such interpretations rely on the present observed correlation between the physical features of leaf assemblages and particular climatic regimes rather than on the identification of particular taxa in fossil assemblages and the present distribution of related taxa.

The marine and nonmarine records are, overall, strikingly similar in the timing of major temperature fluctuations during the Paleocene and Eocene. Marked warm intervals occurred in the early Paleocene, latest Paleocene-early Eocene, late middle Eocene, and latest Eocene. Note that our biostratigraphic and radiometric calibrations indicate that a major decrease in mean annual temperature occurred at about 39 million years ago (Ma); we suggest that the "terminal Eocene" cooling of some workers (e.g., Thierstein and Berger, 1978) is in fact this major cooling within the late Eocene. The subsequent latest Eocene warm intervals were also followed by a major cooling event following the Eocene-Oligocene boundary (34-35 Ma) that led to temperatures cooler than any known in the Eocene.

The Mississippi embayment sequence is not known to contain land floras of early Oligocene age; however, sequences of radiometrically dated floras in western North America document this cooling event (Wolfe, 1971). The event at the Eocene-Oligocene boundary is particularly pronounced on land because the event was characterized by both a major decrease in mean annual temperature and a major increase in mean annual range of temperature (Wolfe, 1978).

The degree of warmth indicated for the Paleogene has been much discussed; although the consensus is that all the Paleocene and Eocene was warmer than present, some workers consider that the level of warmth was only moderately higher than now and that the highly equable climates were the primary factor in allowing the poleward excursions of tropical and subtropical organisms (e.g., Axelrod and Bailey, 1969). Our data indicate that, during warm intervals in the Eocene, mid-latitude temperatures were significantly higher than today. Tropical-subtropical nannoplankton assemblages in the North Atlantic that are the analogs of the Paleogene tropical-subtropical assemblages are now confined to latitudes 25-28° N. During times of the Paleocene and Eocene, tropical-subtropical species form a significant part of nannoplankton assemblages far poleward of 30° N.

The early Eocene appears to represent the warmest part of the Tertiary. The marine record in the North Atlantic, as well as elsewhere, indicates that the most poleward (50-55° N) excursion of low-latitude assemblages occurred at this time; the tropical-subtropical belt may have been double its present latitudinal extent. Data from land plants are also indicative of a major expansion of tropical climates (i.e., mean annual temperature ~25°C)—climates that are currently found within 20-25° of latitude from the equator. The Mississippi embayment floras that occur as far north as 36° latitude are tropical during the warm intervals of the early and middle Eocene. Floras from the west coast of North America were, during the late middle Eocene, tropical to north of latitude 40-45° (Wolfe, 1978); latest Eocene floras indicate paratropical climate at these latitudes. Note that extratropical latest Eocene planktonic foraminiferal assemblages are associated with paratropical land floras in the Pacific Northwest. Thus the juxtaposition of relatively warm land climates with relatively cool marine conditions characteristic of the U.S. Pacific margin today was established in the late Eocene.

On the Gulf of Alaska borderlands (60° N), land floras indicating mean annual temperatures as high as 22°C were thought to be of middle Eocene age based on provincial correlations by marine mollusks (Wolfe, 1977). In the planktonic chronology, however, these Alaskan paratropical floras and the associated mollusks are early Eocene, and other Alaskan floras referable to the late middle Eocene, although indicative of considerable warmth, represent cooler conditions than the early Eocene floras. Similarly, planktonic foraminiferal and nannoplankton assemblages from early Eocene rocks of the eastern Gulf of Alaska contain a major low-latitude component (Poore and Bukry, 1979). Early Eocene floras from Ellesmere Island certainly represent mean annual temperature in excess of 3°C, i.e., at least 22°C higher than today (L. J. Hickey, Smithsonian Institution, personal communication).

In combination with significantly higher mean annual temperatures during parts of the Eocene, mean annual range of temperature and latitudinal temperature gradients were less. Western Washington, which today has mean annual ranges of temperature of 11-17°C, had a mean annual range of temperature as low as 4-5°C during the Eocene (Wolfe, 1978). The latitudinal temperature gradient from middle to high latitudes along western North America was about half the present gradient.

Not only were Eocene temperatures at middle to high latitudes significantly different than present temperatures, precipitation patterns were different. Areas of the western United States that now experience summer drought have Eocene floras that indicate abundant precipitation throughout the year. In contrast, the Paleogene floras from the Mississippi embayment—a region that now has abundant precipitation distributed throughout the year—exhibit a trend from tropical moist to tropical dry climate.

The temperature fluctuations indicated by our analysis of the North Atlantic marine record are difficult to quantify, but the latitudinal differences between the occurrences of tropical-subtropical assemblages indicate fluctuations in mean annual temperature of at least 10-15°C at high middle latitudes; these fluctuations were, thus, at least as great as modern seasonal fluctuations. No fluctuations are observable at low latitudes, i.e., the present tropical belt was always tropical during the Paleogene. Wolfe (1978) has suggested that the land floras from middle latitudes indicate fluctuations in mean annual temperatures of about 7°C.

The elevated temperatures suggested for Paleogene warm intervals concommitant with the major temperature fluctuations are difficult to explain by purely geographic or meteorological mechanisms endemic to the Earth. That is, low latitudes were consistently tropical, and high latitudes were at the same times significantly warmer; redistribution of heat to warm high latitudes should produce cooling at low latitudes, no matter what mechanisms resulted in redistribution. Geographic factors (e.g., the circumglobal circulation at low latitudes, generally lower altitudes of the continents, and greater extent than now of epicontinental seas) could have produced the somewhat elevated global temperature of the cool intervals during the Paleocene and Eocene compared with the global temperature today, but the high levels of warmth during, for example, the early Eocene cannot be so readily explained.

Re-expansion of low-latitude planktonic assemblages during the latest Eocene indicates a renewed warning, but these assemblages did not reach poleward as far as during warm intervals earlier in the Paleogene. Land-plant data from western North America also indicate that the latest Eocene warm interval (Kummerian Stage) was not so warm as previous warm intervals (Wolfe, 1978). This renewed warming was followed by a cooling that, for much of the Oligocene, resulted in low-latitude planktonic assemblages being restricted to about their present latitudes. The land floras that followed this terminal Eocene cooling indicate that not only was mean annual temperature on land significantly lowered, mean annual range of temperature dramatically increased to values even higher than today at middle and high latitudes of western North America (Wolfe, 1978). By the end of the Oligocene, some warming had occurred, as indicated both by planktonic microfossil and isotopic data from the North Atlantic and land floras in western North America.

NEOGENE RECORD OF THE NORTHEASTERN PACIFIC

Interpretations of temperature changes in the northeastern Pacific (Figure 16.2) during the Miocene and Pliocene are based on a variety of data. These include the varying latitudinal distributions of assemblages of calcareous nannoplankton, diatoms, silicoflagellates, and planktonic foraminifers from DSDP Legs 18 and 63 (Ingle, 1973; Barron, 1981; Bukry, 1981; Poore, 1981). We have also used isotopic data on benthic foraminifers from DSDP Site 289 [Cenozoic Paleo-oceanography Research Project (CENOP), unpublished data] to aid and supplement interpretations based on marine floras and faunas. The isotopic data presumably monitor intermediate waters at this site (Ontong Java Plateau) and thus, in part, mid- to high-latitude surface-water temperatures. Our interpretations of the marine record can be compared with analyses, again largely physiognomic, of land floras from the Pacific Northwest (Wolfe, in press) and Alaska (Wolfe and Tanai, 1980).

Temperature trends for the oceans during the early Miocene are problematic, primarily because of the lack of adequate time control. Meager isotopic data from Site 289 and elsewhere indicate that the early Miocene may have been somewhat cooler than the early part of the middle Miocene. What is not known is the warmth of the early Miocene relative to the late Oligocene. On land, the early Miocene (early Seldovian Age) floras are of cooler aspect than those of the middle Miocene as well as those of the late Oligocene. Age control on these floras, however, is uncertain relative to the planktonic time scale.

The late part of the early Miocene and the early part of the middle Miocene has generally been regarded as the warmest interval of the Neogene, although not so warm as any part of the Paleocene or Eocene. This warm interval is clearly recorded in deep-sea records in the northeastern Pacific. Following planktonic foraminiferal zone *N*11, a cooling trend set in that continued through the end of the Neogene, although some minor warmings of short duration occurred during the late Miocene. The youngest marked late Miocene warming at 8 Ma was followed by a gradual cooling; minor warm intervals occurred in the early Pliocene and in the early Pleistocene.

Whether the late Miocene through early Pleistocene temperature fluctuations seen in the northeastern Pacific represent worldwide climatic events is difficult to ascertain. At least some of these fluctuations may relate to changing current patterns in the northeastern Pacific, including changes in the intensity of upwelling—an indication that at least some of the late Miocene fluctuations may not represent global events in Haq's (Chapter 13) data on Miocene nannoplankton assemblages from the Atlantic. The latitudinal distribution of these assemblages indicates that apparently only one significant warming occurred during the late Miocene rather than the three that occurred in

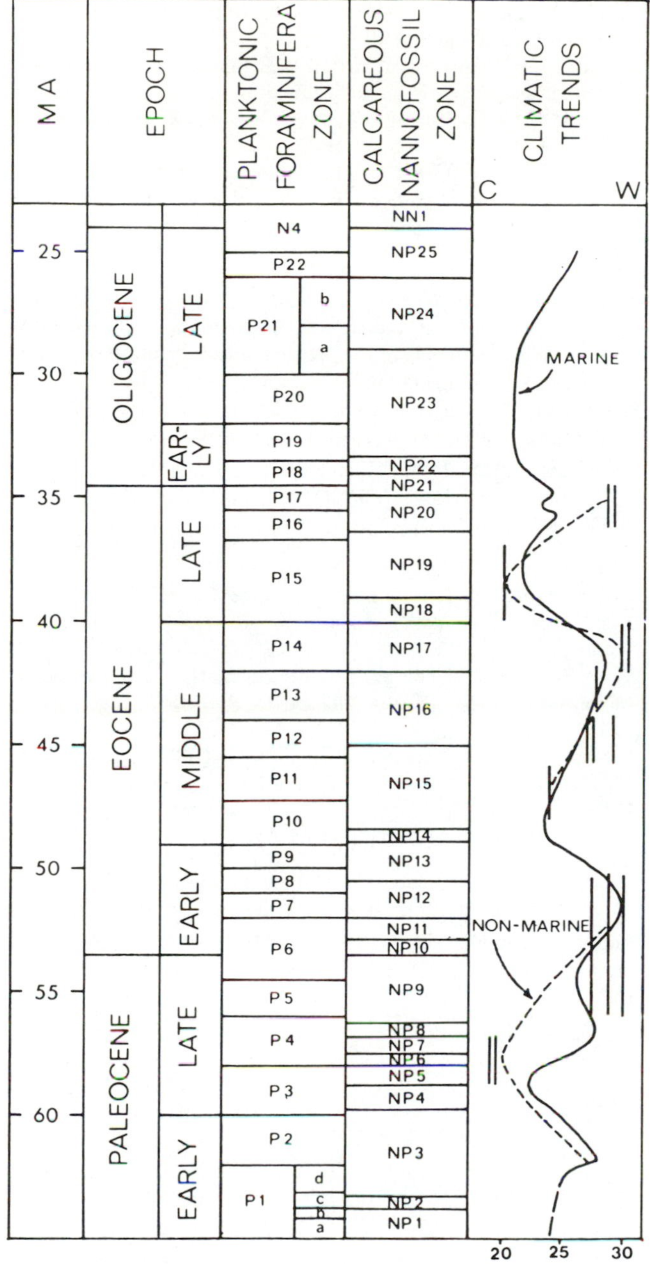

FIGURE 16.2 Comparison of temperature trends in the Northeastern Pacific middle Miocene through Pliocene with inferred temperatures of land floras.

the northeastern Pacific. Haq's data, however, match the Pacific record in showing a latest Miocene cool event followed by a renewed warming into the early Pliocene.

In general, the land-plant record from the Neogene of the northeastern Pacific margin compares well with the deep-sea record: a temperature high in the middle Miocene (floras radiometrically dated at 17-15 Ma) followed by cooling, a slight renewed warming at about 8 Ma, and followed by a cooling at about 5 Ma. The land floras, however, also furnish important data on two other climatic factors.

First, in this region equability has clearly increased during the Oligocene and Neogene from the highly inequable temperature regime of the early Oligocene. At latitude 60° N, the increase in equability has largely resulted from a decline in summer temperatures (a decrease of about 9°C in the mean of the warm month), and at latitude 45° N summer temperature has decreased concommitant with an increase in winter temperature. The decline of summer temperature at higher latitudes must have been a major factor in the initiation of widespread northern hemisphere glaciation during the late Cenozoic.

The trends of temperature changes at various latitudes have also resulted in an increase in latitudinal temperature gradients. In turn, the effect of this increase may have been to intensify the subtropical high-pressure systems that are largely responsible for summer drought along the west coasts of the continents. The land floras from western United States definitely record a shift from a regime in which much of the precipitation was received during the summer to a regime of summer drought even in coastal areas.

The temperature fluctuations in the marine record during the late Miocene and Pliocene do not appear to be so significant as those in the Paleocene and Eocene. The Neogene fluctuations may represent changes in mean annual temperature of as much as 5°C for marine water at middle latitudes. This condition probably results from intensification of a well-defined and relatively stable eastern boundary current system—the California current—sometime in the late middle Miocene. This area of the northeastern Pacific exhibited great stability during the Quaternary, as evidenced by comparison of sea-surface temperatures at the last glacial maximum (18,000 yr ago) and today (CLIMAP Project Members, 1976, Figure 2). The land floras, however, do not indicate as much fluctuation of mean annual temperature; changes in this parameter at middle latitudes appear to be no more than 1-2°C.

SUMMARY

Paleotemperature trends interpreted from deep-sea marine organisms and from land floras are generally similar to one another, both in the timing and direction of changes. Both groups of organisms indicate that a thermal maximum occurred in the early Eocene and that major fluctuations in mean annual temperature occurred during the Paleogene. The land floras also indicate that mean annual range of temperature was low during the Paleocene and Eocene (i.e., a low seasonal contrast of temperature), although mean annual range increased to values greater than present by the Oligocene. Following a middle-Miocene warm interval, mean annual temperatures overall have decreased in extratropical areas, although this trend was interrupted by some minor reversals. The land floras indicate that, during the Neogene, the seasonal contrast of temperatures has decreased. Latitudinal temperature gradients were low during the Paleocene and Eocene and increased to their present values during the Neogene.

Precipitation patterns derived from studies of land floras indicate that an increasing drying trend occurred in southeastern North America during the Paleocene and Eocene; at the same time, western North America appears to have received abundant precipitation throughout the year. In the Oligocene and earlier half of the Miocene, western North America appears to have had abundant summer precipitation, but during the later Neogene summer precipitation decreased.

ACKNOWLEDGMENTS

John A. Barron released unpublished data on Neogene diatoms of the northeastern Pacific and extensively discussed with us their paleoclimatic significance. David Bukry furnished unpublished data on Neogene silicoflagellates. Information on Atlantic Neogene nannoplankton was furnished by Bilal Haq.

REFERENCES

Axelrod, D. I., and H. P. Bailey (1969). Paleotemperature analysis of Tertiary floras, *Palaeogr. Palaeoclim. Palaeoecol. 6*, 163-195.

Barron, J. A. (1981). Late Cenozoic diatom biostratigraphy and paleoceanography of the middle-latitude eastern North Pacific, DSDP Leg 63, in *Initial Reports of the Deep Sea Drilling Project 63*, U.S. Government Printing Office, Washington, D.C., pp. 507-538.

Berry, E. W. (1916). The lower Eocene floras of southeastern North America, *U.S. Geol. Surv. Prof. Pap. 91*, 481 pp.

Bukry, D. (1981). Silicoflagellate stratigraphy of offshore California and Baja California, Deep Sea Drilling Project Leg 63, in *Initial Reports of the Deep Sea Drilling Project 63*, U.S. Government Printing Office, Washington, D.C., pp. 539-557.

CLIMAP Project Members (1976). The surface of the ice-age Earth, *Science 191*, 1131-1137.

Haq, B. U., and G. P. Lohmann (1976). Early Cenozoic calcareous nannoplankton biogeography of the Atlantic Ocean, *Mar. Micropaleontol. 1*, 119-194.

Haq, B. U., I. Premoli-Silva, and G. P. Lohmann (1977). Calcareous plankton paleobiogeographic evidence for major climatic fluctuations in the early Cenozoic Atlantic Ocean, *J. Geophys. Res. 82*, 3961-3876.

Ingle, J. C., Jr. (1973). Summary comments on Neogene biostratigraphy, physical stratigraphy, and paleo-oceanography in the marginal northeastern Pacific Ocean, in *Initial Reports of the Deep Sea Drilling Project 19*, U.S. Government Printing Office, Washington, D.C., pp. 949-960.

Poore, R. Z. (1981). Miocene through Quaternary planktonic foraminifera from offshore southern California and Baja California, in *Initial Reports of the Deep Sea Drilling Project 63*, U.S. Government Printing Office, Washington, D.C., pp. 415-436.

Poore, R. Z., and D. Bukry (1979). Preliminary report on Eocene calcareous plankton from the eastern Gulf of Alaska continental slope, *U.S. Geol. Surv. Circ. 804-B*, pp. B141-B143.

Thierstein, H. R., and W. H. Berger (1978). Injection events in ocean history, *Nature 276*, 461-466.

Vergnaud-Grazzini, C. (1979). Cenozoic paleotemperatures at site 398, eastern North Atlantic: Diagenetic effects on carbon and oxygen isotopic signal, in *Initial Reports of the Deep Sea Drilling Project 47*, U.S. Government Printing Office, Washington, D.C., pp. 507-511.

Wolfe, J. A. (1971). Tertiary climatic fluctuations and methods of analysis of Tertiary floras, *Palaeogr. Palaeoclim. Palaeoecol. 9*, 27-57.

Wolfe, J. A. (1977). Paleogene floras from the Gulf of Alaska region, *U.S. Geol. Surv. Prof. Pap. 997*, 108 pp.

Wolfe, J. A. (1978). A paleobotanical interpretation of Tertiary climates in the northern hemisphere, *Am. Sci. 66*, 694-703.

Wolfe, J. A. (in press). Climatic significance of the Oligocene and Neogene floras of northwestern United States, in *Evolution, Paleoecology, and the Fossil Record*, K. J. Niklas, ed., Praeger, New York.

Wolfe, J. A., and T. Tanai (1980). The Miocene Seldovia Point flora from the Kenai Group, Alaska, *U.S. Geol. Surv. Prof. Pap. 1105*, 52 pp.

The Jurassic Climate

17

ANTHONY HALLAM
University of Birmingham, England

INTRODUCTION

All the reliable evidence that we can muster points strongly to the conclusion that the Jurassic climate was appreciably more equable than that of the present day, with tropical-subtropical conditions extending far into the present temperate belts and temperate conditions occurring in polar regions. There is no evidence of polar ice caps, and, at least partly for this reason, the ocean surface stood at a higher level with respect to the continents. This in itself must have contributed to the higher equability of continental climates. In addition, there appears to have been in general more extensive aridity on the continents whose distribution was, of course, appreciably different from that of today.

Unfortunately, with the present state of knowledge it is difficult to go beyond broad qualitative statements. For the Jurassic, we are denied the excellent information of oceanic surface- and bottom-water temperatures obtained from oxygen isotope analysis of microfossils in deep-sea drilling cores of Cretaceous and Cenozoic strata. Furthermore, Jurassic fossils have been extinct too long to have close modern relatives whose climatic tolerances are precisely known. In the following sections of this paper the principal climatic criteria are briefly outlined and the evidence for climatic changes through space and time discussed. A fuller account of some topics, with additional references, is given by Hallam (1975).

CLIMATIC CRITERIA

By far the best climatic indicators among sedimentary rocks found in the Jurassic are evaporites and coals (Frakes, 1979). Substantial deposits of evaporites (notably gypsum, anhydrite, and halite) indicate conditions of both warmth and aridity, whereas coals indicate swampy conditions in generally humid regimes, though there is no particular temperature connotation. On the other hand, the abundance and extent of deposition of limestones is not particularly reliable. In particular, the greater spread of limestone facies in the late Jurassic, far from signifying increased temperature, is probably no more than a consequence of the greater extent of epicontinental seas at that time (Hallam, 1975).

Laterites and bauxites have been widely considered to be good indicators of humidity, because of the intensity of chemical weathering required for their formation. However, in southern Israel there is a horizon of reworked laterites, including pisolitic conglomerates, sandwiched between evaporite-bearing Upper Triassic and Lower Jurassic deposits and attributed to the basal Jurassic. Following the development of a karstic surface after an episode of regression, leading to the subaerial exposure of hypersaline mud flats, so-called flint clays were generated by chemical weathering in the vadose zone (Goldbery, 1979). Whereas swampy conditions might have occurred locally, no drastic regional increase of humidity after the late Triassic, followed by a return to aridity, apparently needs to be invoked.

Aeolian sandstones with large-scale dune cross bedding should be good indicators of desert conditions, but doubt has been thrown on the aeolian origin of at least part of the well-known Lower Jurassic Navajo Sandstone of the Rocky Mountain states. Stanley *et al.* (1971) recorded ripple-marked and wavy-bedded horizons, shale seams associated with widespread truncation planes, and dolomitic carbonate lenses, suggesting a subaqueous origin for the deposits containing them. Some fossil organisms are good temperature indicators, notably hermatypic corals in marine and ferns in terrestrial deposits. Reef-building corals suggest a minimum water temperature of about 20°C, and there are abundant Jurassic ferns whose living relatives cannot tolerate frost. The occurrence of genera or, better, species of a wide variety of organisms over a broad range of latitude is in itself a strong argument for climatic equability.

Considerable attention has been paid to oxygen isotope paleotemperature determinations on Jurassic belemnites obtained from rocks currently exposed on the continents. There is such a wide disparity in the results of various workers, however, presumably as a consequence of significant postdepositional alteration, that I have argued at some length that they contribute little to a further understanding of Jurassic climates (Hallam, 1975). Unfortunately, there is only a negligible record of Jurassic microfossils in deep-sea drilling cores, from which one might expect to obtain more reliable results.

CLIMATIC CHANGES IN SPACE

The fact that rich Jurassic terrestrial fern and gymnosperm floras are known from both polar regions is a strong argument in favor of general warmth and equability, and this is strongly supported by the wide distribution of fern genera whose modern relatives are intolerant of cold (Barnard, 1973). Thus the basal Jurassic *Dictyophyllum* ranges from 50° N to 60° S, and a number of Middle Jurassic genera are almost as widespread, from 40° N to 50° S. These distributions imply a tropical-subtropical climate extending far beyond the present limits. According to Vakhrameev (1964), the plant record indicates that winter temperatures in Siberia probably never fell below 0°C. Equability is also indicated by the wide latitudinal range of large reptiles (Colbert, 1964) and ceratodontid lungfishes. The latter, whose living relatives are confined to the tropical-subtropical zone, are more or less worldwide in distribution (Schaeffer, 1971).

If the continents enjoyed a warm, equable climate, the same should be true of the marine realm; and indeed the majority of invertebrate genera are cosmopolitan in distribution. While substantial carbonate buildups partly composed of corals are confined to deposits in what are inferred on palaeomagnetic grounds to have been low latitudes, such as the Pliensbachian of Morocco, the Oxfordian of the Swiss Jura and southern Poland, and the Bajocian and Oxfordian of the Paris Basin (Figures 17.1 and 17.2), reef-building corals are also found as far as 60° N paleolatitude, in Sakhalin, some 30° beyond the present limits (Beauvais, 1973). The absence of such corals from comparably high latitudes in the North Atlantic region and southern hemisphere is quite probably due to factors other than low temperature.

The bivalves are instructive to study, both because they are the most abundant and diverse macroinvertebrate group in the Jurassic and because they include many extant families and even genera whose climatic tolerance is well known. In marked contrast to the present-day situation, there is no sharp reduction in diversity with increasing latitude, and many genera and even some species have a wide latitudinal range. A particularly good example is provided by the pectinid genus *Weyla*, largely confined to the eastern Pacific margins, with the same species extending all the way from Chile to southern Alaska. The best candidates for a stenothermal tropical group, restricted to a belt within 30° of the Jurassic equator, are a minority of thick-shelled genera including rudists (Hallam, 1977). There is also a group of distinctive foraminifera more or less confined to the zone of the Tethys and thought to be a stenothermal tropical group (Gordon, 1970).

There has been considerable controversy about the environmental cause of the Tethyan and Boreal provinciality exhibited by ammonites and belemnites, with the Boreal Realm (or superprovince) being confined to the northern part of the northern hemisphere. The majority opinion is that ambient

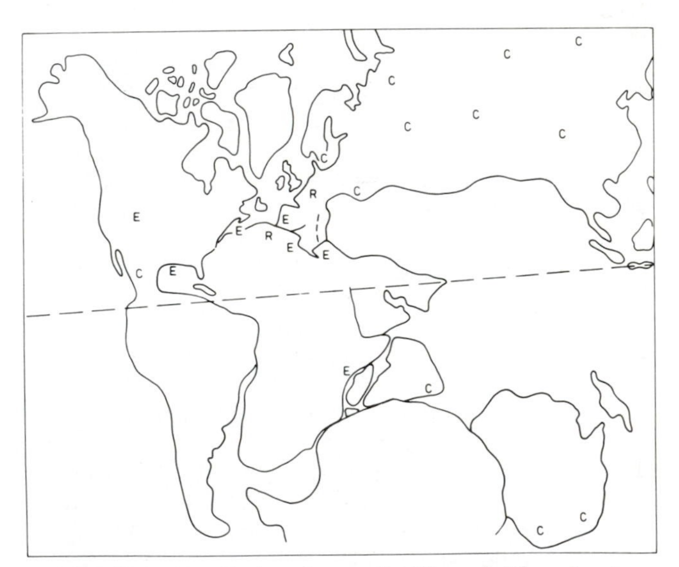

FIGURE 17.1 Distribution of evaporites (E), coals (C), and major coral reefs (R) for the Lower and Middle Jurassic. Broken line signifies approximate position of equator.

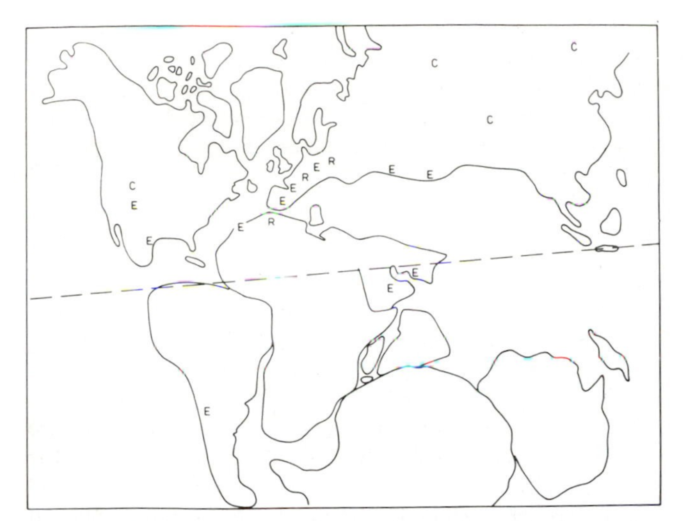

FIGURE 17.2 Distribution of evaporite, coals, and major coral reefs for the Upper Jurassic. Symbols as in Figure 17.1.

temperature was the primary control, but there are difficulties with such a simple interpretation. One of the most important objections is that a drastic change from one faunal realm to the other may take place within a mere degree of latitude, yet the evidence of the climatically more sensitive plants indicates strongly that latitudinal temperature gradients were much more modest than today. It seems necessary to invoke a complex of factors, including paleogeographic configurations, environmental stability, and perhaps such latitudinally related factors as changing patterns of diurnal illumination and constancy of food resources throughout the year (Hallam, 1975). Climate is unlikely to have been the dominant control in all events.

With regard to the distribution of arid and humid belts, the best evidence comes from evaporite and coal deposits (Figures 17.1 and 17.2). According to Gordon (1975), evaporites range between 45° N and 45° S, but the deposits are concentrated in a narrower zone 10-20° from the palaeoequator. Nearly all of the evaporites are confined to the western parts of Laurasia and Gondwana. Among the more substantial deposits in North America are the Lower Lias Argo Salt of the Scotia Shelf and the probably Middle Jurassic Louann Salt and equivalents around the margins of the Gulf of Mexico, while important deposits of gypsum and anhydrite occur in the Bajocian, Callovian, and Oxfordian of the United States Western Interior. There are also evaporites in the Andean Jurassic, in the Oxfordian and Kimmeridgian of northern Chile. Turning to the Old World, there are thick sequences of pre-Bathonian evaporites around the northwestern, northern, and eastern margins of Africa and in the Jurassic of the southern U.S.S.R., southern Iran, and Arabia (Hallam, 1975; Leeder and Zeidan, 1977). Thinner deposits occur also in the basal and late Jurassic of western, southern, and central Europe.

By far the most abundant coal measures occur over a wide area of the Soviet Union, especially in the Lower and Middle Jurassic, and there are also important Lower Jurassic coals in eastern Australia. In Europe thin Lower Jurassic coal beds are known in the so-called Gresten facies of northern Austria and the Mecsek Mountains of Hungary, also in the basal Liassic of southern Scandinavia and the Middle Jurassic of the northern North Sea. In the New World, coals are much rarer, but thin coal seams occur in the Lower and Middle Jurassic of southern Mexico as well as in the Upper Jurassic of Montana, the Dakotas, Alberta, and British Columbia (Jansa, 1972).

The overall geographic distribution of evaporites and coals bears quite a close resemblance to that of the Triassic, and so Robinson's (1971) inference of a western arid belt and two eastern humid belts seems to apply also to the Jurassic. Hence Robinson's climatic model is relevant. She suggests that winds reaching the eastern parts of Laurasia and Gondwana, on either side of the Tethyan Ocean, might have brought monsoon-type summer rains to areas of middle and low latitude, while a dry, hot season would occur in winter as winds blew offshore. The central and western parts of the two supercontinents would have tended to have a much less humid climate because the dominant easterly winds would have traveled over land for a considerable distance or, blowing toward the equator without the intervention of mountains, could not readily have jettisoned their moisture. Coal occurrence is largely restricted to the eastern, peninsular parts of the landmasses in middle to high latitudes, where the temperature was more and the rainfall less strictly seasonal. It is worth adding that coals form less readily in the tropical than in the temperate zone (Frakes, 1979).

CLIMATIC CHANGES THROUGH TIME

There is no convincing evidence of any notable global temperature change through the course of the Jurassic. The best evidence available concerns the areal distribution of terrestrial plant provinces in Eurasia. Vakhrameev (1964) drew a boundary between a northern, possibly temperate, Siberian Province and a southern, presumed subtropical, Indo-European Province. He detected a slight northward shift of this boundary from the early to the mid Jurassic and an appreciably greater northward shift from the mid to the late Jurassic (Figure 17.3). The implied slight warming trend through the period continues into the Cretaceous.

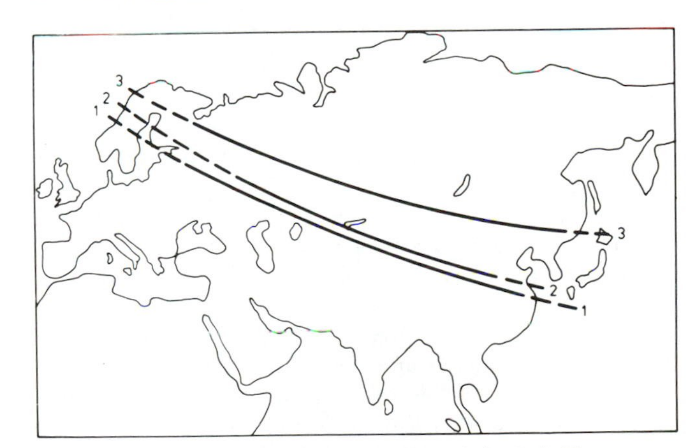

FIGURE 17.3 The shift in the boundary of the Indo-European and Siberian floral provinces in Eurasia from the early (1) and mid (2) to late (3) Jurassic. Adapted from Vakhrameev (1964).

More can be said about humidity-aridity distribution. From the greater spread of evaporite facies in the late Jurassic, for instance into Chile and the southern parts of the Soviet Union, Frakes (1979) inferred an overall trend toward a drier climate from early in the period. This is confirmed by the occurrence of xerophytic plants in the late Jurassic of the latter area (Vakhrameev, 1964). However, the regional picture may be more complicated. Thus, in Israel the Lower Jurassic contains evaporites, while the Middle and Upper Jurassic contains coals, so that the climatic trend through time is inferred by Goldberg and Friedman (1974) to have been the reverse of what is assumed to be the general picture. Yet there are abundant late Jurassic evaporites in southern Europe and the Middle East. Goldberg and Friedman stressed the importance of regional climatic change and draw an analogy with the Gulf of Mexico margins. In southern Texas, for instance, a dry climate is recorded by gypsum deposits in the Laguna Madre, whereas the moister climate of Louisiana is reflected by salt-marsh deposits. Perhaps local swampy conditions in an area of moderately dry climate can promote the formation of thin coals, in which case the validity of coal distribution as a climatic indicator needs to be more closely investigated.

Frakes (1979) argued for a continuation of the global trend toward greater aridity into the Cretaceous. Yet evaporite-bearing deposits in the Jurassic of the western interior of the United States are succeeded by coal-bearing deposits in the Lower Cretaceous. On the other hand, the facies change from the Upper Triassic to the Lower Jurassic in western Europe supports Frakes's postulation of a global change toward increased humidity. Thus the Keuper red beds contain evaporites and a suite of clay minerals, in which kaolinite is absent, suggestive of postdepositional magnesium enrichment in hypersaline water (Jeans, 1978). Substantial quantities of kaolinite, suggesting intensive leaching on a land experiencing a warm, humid climate, first appear in the topmost Triassic (Rhaetian) marginal marine deposits and continue into the Lias (Will, 1969). A humid climate is confirmed by the occurrence in northern Europe of Rhaeto-Liassic plant beds including coals and perhaps also by the more widespread occurrence of Liassic ironstones (Hallam, 1975).

With regard to the oceans, much interest has been provoked by Fischer and Arthur's (1977) model of cyclic alternations, lasting about 32 million years and ranging back to the Triassic, between what they term *polytaxic* and *oligotaxic* episodes. Polytaxic episodes are characterized by high organic diversity, higher and more uniform oceanic temperatures, with continuous pelagic deposition, widespread marine anoxicity, and eustatic sea-level rises. In contrast, oligotaxic episodes, such as at present, are characterized by lower marine temperatures with more pronounced latitudinal sedimentation, marine regression, and a lack of anoxicity. During polytaxic episodes, warm, globally equable climates result in reduced oceanic convection, causing expansion and intensification of the oxygen minimum layer, while colder climatic intervals give rise to increased circulation rates and better oxygenation of ocean waters.

Whereas there may well be some merit in the Fischer and Arthur model for the Cretaceous and Cenozoic, for which we have an ample record from deep-ocean cores, the evidence they cite for the Jurassic, such as oxygen isotope data from belemnites, is dubious, and I see no grounds for their invocation of an oligotaxic episode in Bathonian-Callovian times. I am rather inclined to believe that the whole of the Jurassic was a polytaxic episode, at least with regard to climate and oceanic circulation.

CONCLUDING REMARKS

Perhaps the greatest advance in the future will come from paleoclimatic modeling of the type outlined by Gates (Chapter 2). The geographic location of the continents and oceans is accurately known, and reasonably accurate estimates can be made of the spread of epicontinental seas, which toward the end of the period was much greater than today. A fair approximation to mean annual temperature distributions in different zones of latitude can be achieved by utilizing data on fossil distributions, though it may prove more difficult to quantify temperature, seasonality, and rainfall. Reasonable estimates can also be made about the location of mountain belts.

One of the questions of most obvious interest is the extent to which the climatically equable world of the Jurassic, with its eastern humid and western arid belts, is a function primarily of the different geography of the time, compared with today. In addition, it would be instructive to enquire into the climatic effects of a more or less progressive rise of sea level through most of the period, with a concomitant flooding of continental lowlands and the creation of a continuous, low-latitude oceanic girdle in the latter part of the period following opening of the oldest, central sector of the Atlantic.

REFERENCES

Barnard, P. D. W. (1973). Mesozoic floras, in *Organisms and Continents Through Time*, N. F. Hughes, ed., Palaeontol. Spec. Pap. No. 12, Palaeontol. Soc., London, pp. 175-188.

Beauvais, L. (1973). Upper Jurassic hermatypic corals, in *Atlas of Palaeobiogeography*, A. Hallam, ed., Elsevier, Amsterdam, pp. 317–328.

Colbert, E. H. (1964). Climatic zonation and terrestrial faunas, in *Problems of Palaeoclimatology*, A. E. M. Nairn, ed., Wiley, New York, pp. 617-637.

Fischer, A. G., and M. A. Arthur (1977). Secular variations in the pelagic realm, in *Deep-Water Carbonate Environments*, H. E. Cook and P. Enos, eds., Soc. Econ. Paleontol. Mineral. Spec. Publ. 25, pp. 19-50.

Frakes, L. A. (1979). *Climates Throughout Geologic Time*, Elsevier, Amsterdam, 310 pp.

Goldberg, M., and G. M. Friedman (1974). Paleoenvironments and paleogeographic evolution of the Jurassic System in southern Israel, *Geol. Surv. Israel Bull. 61*, 44 pp.

Goldbery, R. (1979). Sedimentology of the Lower Jurassic flint clay-bearing Mish hor Formation, Makhtesh Ramon, Israel, *Sedimentology 26*, 229-251.

Gordon, W. A. (1970). Biogeography of Jurassic foraminifera, *Geol. Soc. Am. Bull. 81*, 1689-1704.

Gordon, W. A. (1975). Distribution by latitude of Phanerozoic evaporite deposits, *J. Geol. 53*, 671-684.

Hallam, A. (1975). *Jurassic Environments*, Cambridge U. Press, London, 269 pp.

Hallam, A. (1977). Jurassic bivalve biogeography, *Paleobiol. 3*, 58-73.

Jansa, L. F. (1972). Depositional history of the coal-bearing Upper Jurassic-Lower Cretaceous Kootenary Formation, Southern Rocky Mountains, Canada, *Geol. Soc. Am. Bull. 83*, 3199-3222.

Jeans, C. V. (1978). The origin of the Triassic clay assemblages of Europe with special reference to the Keuper Marl and Rhaetic of parts of England, *Phil. Trans. R. Soc. Lond. A 289*, 549-639.

Leeder, M. R., and R. Zeidan (1977). Giant late Jurassic sabkhas of Arabian Tethys, *Nature 368*, 42-44.

Robinson, P. L. (1971). A problem of faunal replacement on Permo-Triassic continents, *Palaeontology 14*, 131-153.

Schaeffer, B. (1971). Mesozoic fishes and climate, *Proc. N. Am. Paleontol. Conv. Sept. 1969, Chicago, Part D*, 376-388.

Stanley, K. O., W. M. Jordan, and R. H. Dott (1971). New hypothesis of early Jurassic paleogeography and sediment dispersal for the western United States, *Bull. Am. Assoc. Petrol. Geol. 55*, 10-19.

Vakhrameev, V. A. (1964). Jurassic and early Cretaceous floras of Eurasia and the paleofloristic provinces of this period, *Tr. Geol. Inst. Moscow 102*, 1-263 (in Russian).

Will, H. J. (1969). Untersuchungen zur Stratigraphie und Genese des Oberkeupers in Nordwestdeutschland, *Beih. Geol. Jahrb. 54*, 245 pp.

Stable Isotopes in Climatic Reconstructions

18

SAMUEL M. SAVIN
Case Western Reserve University

INTRODUCTION

The calcium carbonate-water oxygen isotope geothermometer has become the most widely applied quantitative tool for estimating ancient ocean temperatures and has been applied increasingly often to studies of paleoclimate and paleo-oceanography. For many years the greatest impact of isotope paleoclimatology on geologic thinking was in studies of the Quaternary period. More recently, analyses of marine carbonates of Tertiary and late Mesozoic age have permitted refinement of our knowledge of marine temperatures during the past 100 million years (m.y.). This quantification of pre-Pleistocene marine climates has been especially timely, as it evolved when our growing understanding of plate motions and resultant changing oceanic geometry, and of sea levels, has encouraged the development of theories to explain the causes of climatic change. Sufficient data soon will be available to provide boundary conditions for mathematical models of atmospheric circulation, at least during Neogene time.

With progressively older sediments, the occurrence of carbonate material suitably preserved for oxygen isotope paleotemperature studies becomes increasingly scarce. The detailed and quantitative paleotemperature records that have been reconstructed for Tertiary and late Cretaceous time cannot be envisaged for pre-Cretaceous time with samples now available or likely to become available. In addition, paleoclimatic information from nonisotopic sources becomes more difficult to obtain as we proceed backward through the geologic record and knowledge of climatic history becomes correspondingly poorer. Hence, while the kinds of paleoclimatic information obtainable using isotopic techniques becomes increasingly imprecise as we proceed back through time, the important questions about the climate of those earlier times can be meaningfully answered with less precisely interpretable data. For these earlier times, other isotope paleoclimatic techniques, in addition to the calcium carbonate-water paleothermometer, become useful. Most notable of these so far has been the paleoclimatic interpretation of the oxygen (and hydrogen) isotopic compositions of cherts. Some isotopic methods can provide information about terrestrial climates. In this paper the calcium carbonate-water paleothermometer and the climatic record it has yielded are reviewed. Other isotopic techniques that have provided, or that have the potential to provide, useful paleoclimatic data are also discussed.

THE CALCIUM CARBONATE-WATER PALEOTHERMOMETER AND MARINE PALEOCLIMATES

The use of the calcium carbonate-water isotope paleothermometer has been reviewed many times since the technique was proposed by Urey (1947) and developed and applied by Epstein *et al.* (1951) and Urey *et al.* (1951). Critical reviews and discussions of various aspects of the method include those by Craig (1965), Bowen (1966), Teis and Naidin (1973), Savin and Stehli (1974), Hecht (1976), Hudson (1977), Savin (1977), and Berger (1979). The basic principles are straightforward. If calcium carbonate is deposited in isotopic equilibrium with seawater the difference between the $^{18}O/^{16}O$ ratio of the carbonate and that of the seawater is strictly a function of temperature. If the temperature dependence of that difference has been calibrated, if the $^{18}O/^{16}O$ ratio of the seawater can be estimated, and if the $^{18}O/^{16}O$ ratio of the carbonate has not been altered since formation, the temperature of carbonate deposition can be calculated. It is this calculated temperature that is referred to as an *isotopic temperature.*

In practice, uncertainties are encountered when the isotopic paleotemperature method is applied to the study of marine carbonates. These uncertainties, discussed in the reviews mentioned above, lead to ambiguities in relating isotopic temperatures to climatically meaningful temperatures at specific localities and depths within the water column. Somewhat less uncertainty is entailed in estimating the water-temperature *changes* than in estimating absolute values of water temperature.

The most successful applications of isotope paleoclimatology have been in the study of foraminifera from deep-sea sediments. An assumption in most of these studies has been that planktonic foraminifera deposit their tests in isotopic equilibrium with seawater. Although there is some indication that this is not always completely true (van Donk, 1970; Shackleton *et al.*, 1973; Grazzini, 1976; Williams *et al.*, 1977), it seems a sufficiently close approximation to reality that for most purposes it can be taken as if true. Most taxa of benthic foraminifera clearly show departures from isotopic equilibrium (Duplessy *et al.*, 1970; Woodruff *et al.*, 1980). Departures may be as great as 1 per mil or more. Fortunately, departures from equilibrium may be taken as approximately (but not exactly) constant for a species, and isotopic compositions may therefore be interpreted in terms of temperatures.

The $^{18}O/^{16}O$ ratio of water in which foraminifera grew must always be estimated. To a first approximation the open oceans can be taken to be well mixed and, hence, their isotopic compositions to be constant through at least Phanerozoic time and through space. Whereas this sort of approximation may be sufficient (and unavoidable) for early Cretaceous and older paleotemperature studies, it is woefully inadequate in the investigation of Tertiary and Quaternary climates where the important problems require more accurate paleoclimatic knowledge. The oxygen isotopic composition of the hydrosphere has probably remained constant through much of, at least Phanerozoic, time. However, that of the oceans has varied in response to the formation and disappearance of ^{16}O-rich continental icecaps. The extent to which this has affected seawater $^{18}O/^{16}O$ ratios during Pleistocene ice advances and retreats has been a matter of controversy for many years (Savin and Yeh, 1981). As this paper does not deal with Pleistocene climates, this controversy can be largely ignored. Uncertainty in the magnitude and isotopic composition of the Antarctic icecap in middle Miocene and later times does create ambiguities when interpreting the middle and late Miocene and Pliocene isotopic record in terms of temperature changes. (As discussed below, most, but not all, investigators involved with the isotopic record would argue that, prior to middle Miocene time, late Mesozoic and Cenozoic continental ice was never so extensive as to introduce ambiguities into the interpretation of the isotopic data.) Isotopic paleotemperature data for pre-Tertiary glaciations are so sparse that discussion of uncertainties resulting from glacially induced variations in the isotopic composition of the oceans is unwarranted.

Locally, the $^{18}O/^{16}O$ ratio of surface seawater varies in response to evaporation (increased $^{18}O/^{16}O$), precipitation (decreased $^{18}O/^{16}O$), and freshwater runoff (decreased $^{18}O/^{16}O$). These variations in seawater $^{18}O/^{16}O$ can cause errors in estimated water temperatures of a few degrees if they are not properly taken into account. Until now, this problem has frequently been largely ignored or dealt with in rudimentary fashion, by assuming that local variations in the past have been analogous to those of today. The time appears to be approaching when local variations in Neogene seawater $^{18}O/^{16}O$ can be estimated from paleo-oceanographic (including isotopic) data, and estimates of water temperatures can be refined.

Well-preserved calcium carbonate suitable for isotopic analysis is common in Neogene deep-sea sediments, but alteration rendering samples unsuitable becomes progressively more common in older sediments. Suitably preserved samples of any age are rare (but *do* exist) in rocks exposed on the continents. Additional work is needed to develop techniques for the recognition of minor amounts of alteration that affect but do not obliterate the original isotopic record.

When all criteria have been satisfied, and an accurate isotopic temperature has been obtained, it is still not always a straightforward matter to report a climatologically meaningful ocean temperature. The living habits, and especially the site and water depth of carbonate secretion, must be known. For organisms for which modern counterparts are extant, this can be relatively straightforward, as, for example, for Tertiary and late Cretaceous planktonic foraminifera (Douglas and Savin, 1978). When modern counterparts are not extant, establishment of the environment of carbonate deposition can be more difficult and may in some cases depend on isotopic analyses of large numbers of taxa within fossil assemblages. As noted above, because questions about early Mesozoic and older climates do not usually require answers as precise as do questions about Tertiary climates, meaningful information may be obtained in many cases without solution of these ecological problems.

THE ISOTOPIC RECORD

Isotopic studies of Tertiary and late Cretaceous climates have been concentrated on deep-sea sediments. While in some instances useful data have been obtained from deep-sea piston cores and from rocks exposed on the continents, it is the Deep Sea Drilling Project (DSDP) that has provided the largest collection of samples for isotope paleoclimatic study. The availability of these DSDP samples, more than anything else, has been responsible for the enormous increase in the number of isotopic investigations of pre-Pleistocene climates during the past 10 yr. A compilation of much published (and some unpublished) isotopic data on Cretaceous and Tertiary foraminifera is shown in Figure 18.1. The relationship between the isotopic data and other aspects of paleo-oceanography such as oceanic anoxic events, the biotic crisis at the Cretaceous-Tertiary boundary, Eocene-Oligocene extinctions, the Messian "crisis," and sea-level changes has recently been reviewed by Arthur (1979).

The temperature trend during the past 130 m.y. has been generally downward, but it has been neither smoothly nor monotonically downward. Bottom-water temperatures have decreased to their modern low values from values of more than 15°C, which prevailed during much of the Cretaceous. Late Cretaceous time saw significant cooling of bottom waters to values of 10 to 12°C. This cooling was not especially abrupt or intense compared to subsequent Tertiary events. From late middle Eocene through early Miocene time there was a series of warmings and coolings of bottom waters. Waters seldom warmed as much as they had cooled, and the net drop in bottom temperature between the middle Eocene high and the late Oligocene low was 10 or 11°C. [An alternative interpretation of these data, offered by Matthews and Poore (1980) is that the temperature drop was not so great as this and that a portion of the oxygen isotopic change reflected significant growth of continental ice during Eocene and Oligocene time rather than a temperature drop.] Many of the decreases in Paleogene bottom-water temperatures appear quite abrupt. Best documented of these is that near the Eocene-Oligocene boundary, where Kennett and Shackleton (1976) proposed a cooling of 5°C in 100,000 yr. An aspect of interpretation of these data that remains incompletely resolved is the extent to which the bottom-water temperature fluctuations record the surface-temperature history of a single source region (perhaps at the coast of Antarctica) and the extent to which they record alternations in the source area for bottom-water production (e.g., from high northern to high southern latitudes).

The Miocene benthic isotopic record from DSDP Site 289 (Woodruff *et al.*, 1981) is an especially striking one and is shown in Figure 18.2. Between 15 and 13.5 million years ago (Ma) a large net increase in benthic foraminiferal $^{18}O/^{16}O$ occurred. This isotopic shift probably reflects both bottom-water cooling and rapid accumulation of ice on Antarctica. There is every reason to believe that a major Antarctic icecap has persisted from that time to the present. However, variations of its size and isotopic composition during Miocene and Pliocene times are not well known. This introduces a degree of uncertainty about the extent to which late Miocene and Pliocene benthic foraminifera isotopic variations should be taken to reflect bottom-water temperature variations as opposed to variations in continental ice volume and isotopic composition.

Through most of Cretaceous and Tertiary time, benthic and planktonic oxygen isotopic compositions fluctuated in roughly parallel manner. This pattern changes during middle Miocene time. Although high-latitude planktonic foraminiferal $^{18}O/^{16}O$ ratios increase, as do those of the benthics (Shackleton and Kennett, 1975), tropical values decrease (Savin *et al.*, 1975), indicating a warming of tropical surface waters. Hence, middle Miocene time is characterized not only by the rapid growth of ice on Antarctica but by a change in the way heat is distributed on the surface of the Earth and by a marked increase in the equator-to-pole temperature gradient. Meridional heat transfer must have been sharply reduced, causing high latitudes to cool while low latitudes warmed. An understanding of the cause of this major change in the Earth's thermal regime is fruitful ground for future research.

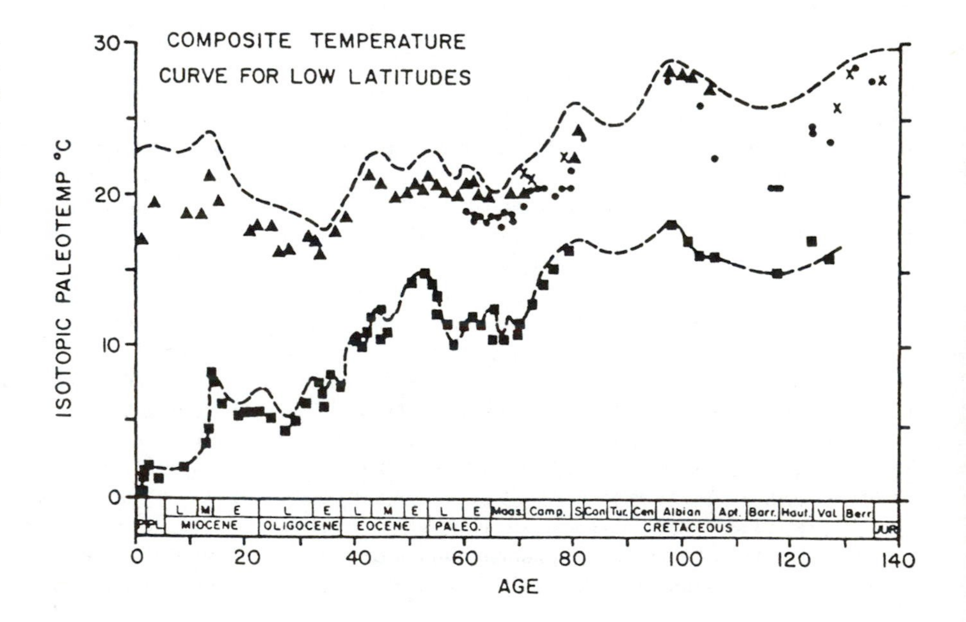

FIGURE 18.1 Compilation of oxygen isotope paleotemperature data obtained by analyses of benthic and planktonic foraminifera (and some nannofossils) from DSDP cores. Most data are for the Pacific Ocean. Bottom curve is drawn through bottom-water data; upper curve is estimate of tropical sea-surface temperatures. Figure from Douglas and Woodruff (1981).

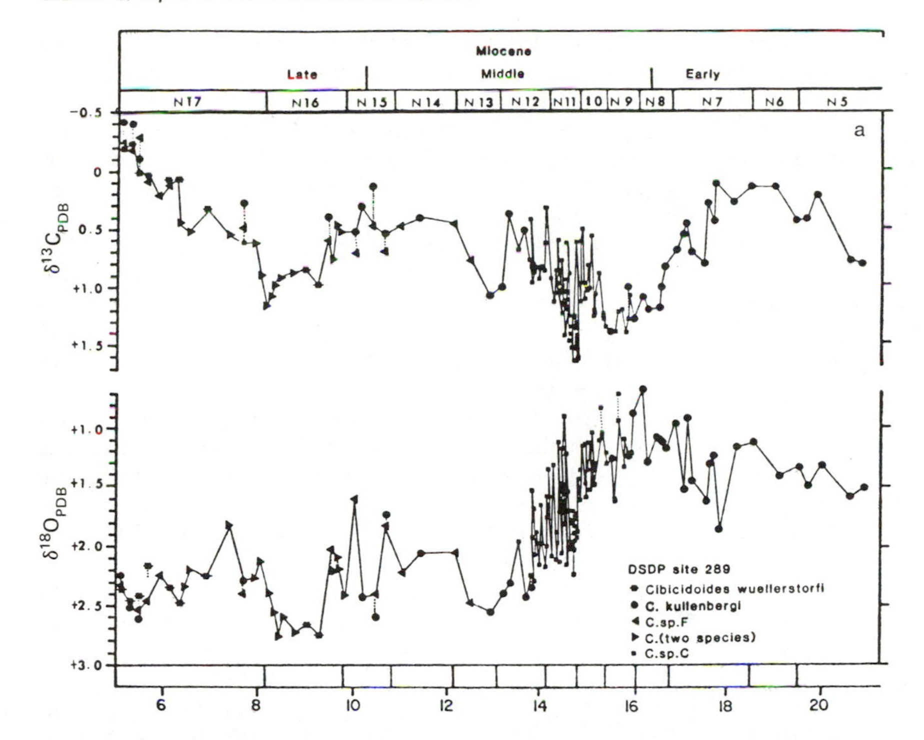

FIGURE 18.2 High-resolution oxygen isotopic study of Miocene benthic foraminifera from DSDP Site 289 (Ontong-Java Plateau). From Woodruff *et al.* (1981), copyright 1981, American Association for the Advancement of Science.

The earliest application of oxygen isotope measurements to paleoclimates was the study of Cretaceous climates by Urey *et al.* (1951) followed by that of Lowenstam and Epstein (1954). This study included numerous analyses of belemnites sampled from outcrops. A characteristic of this study, as well as other similar studies, is the large amount of scatter in the isotopic data, which makes interpretation difficult. Among possible reasons for isotopic variation are postdepositional alteration; variations in the temperature and isotopic composition of seawater in the relatively shallow, nearshore sedimentary environments in which most or all of the samples were deposited; and migration of belemnites of different ontological stages into different growth environments. A synthesis by Stevens and Clayton (1971) of paleotemperature trends derived from several studies of Jurassic and Cretaceous megafossils is shown in Figure 18.3. Some of the results that appear discrepant from study to study may reflect real climatic differences from place to place, but in many instances this is unlikely. Further work is needed on the isotope systematics of megafossils in outcrop and the factors (especially diagenetic alteration) that affect their isotopic compositions. Some work on these problems has been done in the past several years, using the scanning electron microscope and cathodoluminescence-equipped microscope, which have become readily available. With these new tools it may well be possible to develop criteria to identify and eliminate samples that have undergone postdepositional isotopic alteration.

There exists in the literature a small number of analyses of pre-Jurassic carbonate shells. All of these are, of necessity, from the continents, owing to the lack of oceanic crust this old. Hence, extreme caution is needed to avoid the effects of postdepositional alteration. Because of the small number of analyses of samples widely scattered in space and time there has been little impetus to synthesize such data. However, the search for suitable samples for study and the synthesis of the existing and new data should be encouraged. Both may yield valuable information about early Mesozoic and Paleozoic climates.

SILICA-WATER ISOTOPIC PALEOTEMPERATURES

Biogenic silica is a common constituent of marine sediments and is frequently found where biogenic carbonate is absent (e.g., below the calcium carbonate compensation depth). In recent years there has been a great deal of progress made on the development of a biogenic silica-water oxygen isotope geothermometer, chiefly by Labeyrie and coworkers (Labeyrie, 1974; Mikkelsen *et al.*, 1978; Labeyrie and Juillet, 1980). However, analytical methods for isotopic analysis of opaline silica are difficult and arduous. As yet, applications of this approach to paleoclimatological problems have been limited.

Opaline silica becomes diagenetically altered to opal-cristobalite (opal-CT) and then to microcrystalline quartz. Both opal-CT and quartz can be analyzed by methods that have become routine for isotopic analysis of silicates. Studies of the isotopic effects accompanying the conversion of opaline silica through opal-CT to microcrystalline quartz in DSDP sediments have been done by Knauth and Epstein (1975) and Kolodny and Epstein (1976). These studies have shown that isotopic exchange with pore waters accompanies mineralogical reactions at depths of tens to hundreds of meters below the sediment-water interface. Furthermore, the reactions can occur some tens of millions of years following deposition. Hence, isotopic temperatures obtained from opal-CT and microcrystal-

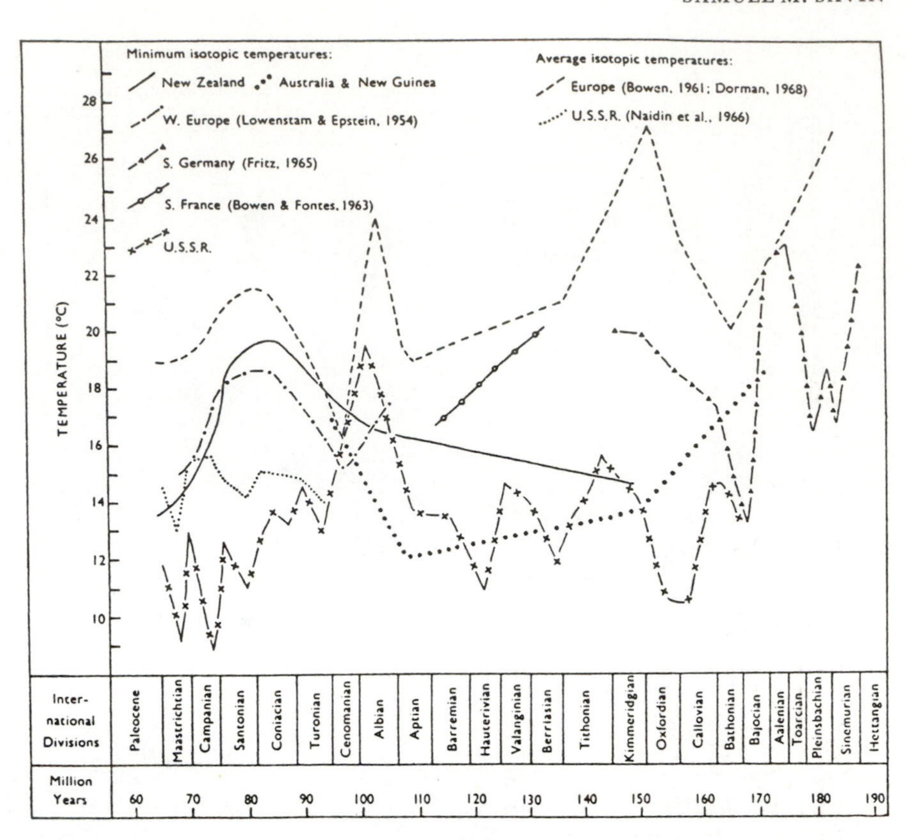

FIGURE 18.3 Compilation of isotope paleotemperatures estimated from analyses of belemnite guards by several investigators. Figure from Stevens and Clayton (1971), with permission.

line quartz should be related to, but somewhat higher than, bottom temperatures at a time a few millions to tens of millions of years more recent than the time of deposition of the silica. Kolodny and Epstein (1976) have presented a comparison between Tertiary and Cretaceous benthic foraminiferal isotopic temperatures and chert isotopic temperatures based on the study of DSDP materials, and the comparison is consistent with the behavior outlined above. The cherts do contain a climatic signal that is sufficiently imprecise to be of little use in the study of Tertiary or Cretaceous climates. However, for older times, for example the Precambrian, for which quantitative climatic data are indeed scanty, this approach can provide climatic information of considerable use. As a cautionary note, in some geologic settings the conversion of diatomaceous silica to chert may occur under conditions that at least partially mask the climatically relevant isotopic signal. Murata *et al.* (1977) analyzed two sections of opal, opal-CT, and chert from the Miocene Monterey Formation of California and found $^{18}O/^{16}O$ ratios in the quartz consistent with diagenetic formation at 80°C. Using the quartz-water isotopic fractionation curve given by Knauth and Epstein (1976) that temperature would be reduced to 66°C. Even that is, of course, substantially warmer than any reasonable estimate of Miocene surface or bottom temperatures.

Perry and coworkers (most recently, Perry *et al.*, 1978) and Knauth and coworkers (most recently, Knauth and Lowe, 1978) have published large numbers of isotopic analyses of Precambrian cherts. Both of these series of papers have served to document a striking tendency toward lower $^{18}O/^{16}O$ ratios with increasing age. A summary of these data is given in Figure 18.4. Although the data of these two research groups are consistent with one another, their interpretations differ. Knauth and Lowe have argued that the data are best explained if the oxygen isotopic composition of the ocean has remained essentially constant, with time, and that the low $^{18}O/^{16}O$ ratios of early Precambrian charts reflect warm temperatures, perhaps as warm as 80°C. Muehlenbachs and Clayton (1976) have proposed that the $^{18}O/^{16}O$ ratio of the modern hydrosphere is determined by two seawater-lithosphere reactions: low-temperature weathering reactions, which deplete the oceans in ^{18}O and high-temperature hydrothermal alteration of basalt, which enriches the oceans in ^{18}O. Knauth and Lowe (1978) have suggested that if similar processes occurred throughout Precambrian time the $^{18}O/^{16}O$ ratio of seawater would have remained constant. Gregory and Taylor (1981) have concluded that $^{18}O/^{16}O$ ratio of seawater is constrained to a value similar to today's by the interaction between water and rock associated with the seafloor spreading process. Perry *et al.* (1978) on the contrary, have argued that the $^{18}O/^{16}O$ ratio of seawater was substantially lower in early Precambrian time than today and that ocean temperatures need not have been markedly greater than Phanerozoic temperatures. Perry *et al.* have suggested that in the Archaean intense weathering and low-temperature alteration of volcanic rocks were the dominant processes controlling the $^{18}O/^{16}O$ ratio of the oceans. This, they argued, could entail a depletion of ^{18}O in seawater by perhaps as much as 12 to 24 per mil relative to today's values.

It is of course possible that higher temperatures *and* lower oceanic $^{18}O/^{16}O$ ratios have both contributed to the oxygen isotope record of Precambrian cherts. Resolution of the relative importance of these two variables at various times during Precambrian time is of the utmost importance to Precambrian paleoclimatology. The approach of Knauth and Epstein (1976) in which both $^{18}O/^{16}O$ and D/H ratios of cherts are used, may provide the best solution. However, even without resolution of this question, useful paleoclimatic conclusions can be drawn. As an example, Oskvarek and Perry (1976) concluded from the analyses of cherts from the 3800 Ma Isua Series (Greenland) that the highest possible surface temperature during Isua time was 150°C. This is based on chert precipitation from a hypothetical ocean of +6 per mil, the approximate value of water degassed from the mantle, and the highest $^{18}O/^{16}O$ ratio for seawater given by any reasonable model for ocean formation. Lower estimates of the $^{18}O/^{16}O$ ratio in seawater would give lower isotopic temperatures. A minimum temperature of 0°C can be estimated from the fact that the Isua series is made up of apparently water-laid sediments. For any time in the Phanerozoic, an ocean temperature estimate of 0-150°C would be trivial. For a 3800-m.y.-old ocean it is not. (Keep in mind that 150°C is the temperature of water in equilibrium with an atmospheric p_{H_2O} of 4.65 atm and that only approximately 46 m of seawater would have to be evaporated in order to achieve that p_{H_2O}.) Even crude estimates of early Archaean ocean temperatures can be useful in constraining models of the atmosphere and of the Sun during the Earth's earliest history.

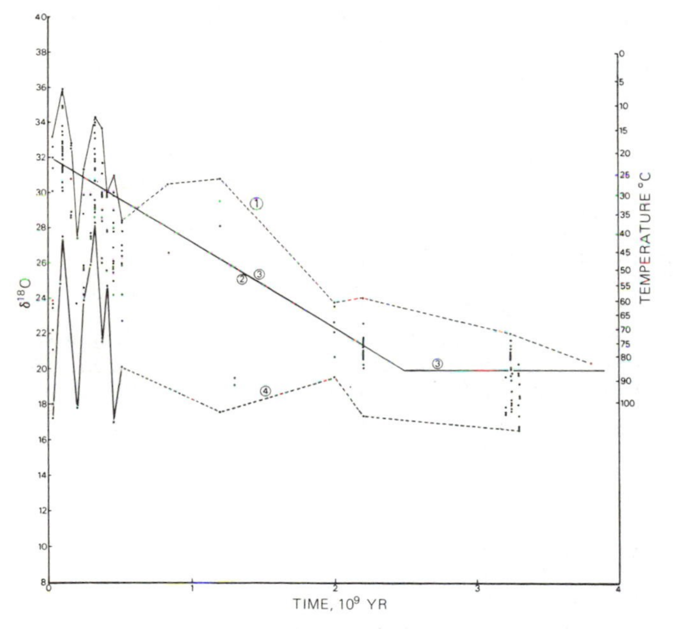

FIGURE 18.4 Compilation of oxygen isotopic compositions of cherts formed during the past 3800 m.y. Lines 1 and 4 form an envelope about the data. Line 3 is a trend line through the data indicating Perry *et al.*'s (1978) estimate of the secular change of the $^{18}O/^{16}O$ ratio of seawater. Temperature scale is based on Knauth and Lowe's (1978) estimate of no secular change in the $^{18}O/^{16}O$ ratio of seawater. From Knauth and Lowe (1978).

THE ISOTOPIC COMPOSITION OF PRECIPITATION ON THE CONTINENTS

Ocean temperature is the parameter of climate most frequently and readily determined from the isotopic study of ocean sediments. Isotopic studies of terrestrial materials for paleoclimate purposes are far less common. The most readily determined climatic parameter obtained from isotopic studies of terrestrial materials is not temperature but the isotopic composition of precipitation. This is in turn dependent largely but not exclusively on temperature of precipitation. Moreover, because precipitation in few areas is uniformly distributed throughout the year, mean temperature of precipitation can be significantly different than mean annual temperature. Still, the $^{18}O/^{16}O$ (or D/H) ratio of precipitation can be a useful climatic variable.

Most studies that have aimed at estimating isotopic composition of precipitation have had their greatest utility in investigations of Quaternary climate and will be mentioned here only briefly. Hanshaw and Hallet (1978) analyzed the $^{18}O/^{16}O$ ratio of subglacial calcite precipitated as a crust on rock surfaces over which glaciers flowed. Because the meltwater from which the calcite precipitated must have a temperature of almost exactly 0°C, the isotopic composition of the meltwater can be obtained from the $^{18}O/^{16}O$ ratio of the calcite. This must be the isotopic composition of precipitation in the glacier's zone of accumulation. There have been no published attempts to apply this approach to pre-Pleistocene subglacial calcites.

Hendy (1971) and Schwarcz, Harmon, and coworkers (e.g., Schwarcz *et al.*, 1976) have estimated both temperature and isotopic compositions from cave deposits. In many instances, the D/H ratio of fluid inclusions trapped in calcite of speleothems can be shown to be representative of that of the seepage waters from which the calcite was precipitated. Because the $^{18}O/^{16}O$ and D/H ratios of precipitation, worldwide, are linearly related (Craig, 1961), the $^{18}O/^{16}O$ ratio of seepage water can be estimated from its D/H ratio. Analysis of the $^{18}O/^{16}O$ ratio of the calcite permits calculation of an isotopic temperature for speleothem formation. This temperature, and the $^{18}O/^{16}O$ ratio of precipitation, should approximate mean annual temperature and isotopic composition of precipitation in the groundwater recharge zone.

Attempts to derive paleoclimatic information from isotopic studies of tree rings (especially of cellulose) appear promising (Epstein and Yapp, 1976; Libby *et al.*, 1976; Epstein *et al.*, 1977; Yapp and Epstein, 1977; Long *et al.*, 1978). There are, however, still problems in the interpretation of these data that need to be resolved. The scarcity of suitable old material for analysis and uncertainties about preservation of original isotope ratios over long periods of time indicate that for the foreseeable future most applications of this method will be to Quaternary samples.

Clay minerals formed during weathering acquire D/H and $^{18}O/^{16}O$ ratios that reflect the temperature and isotopic composition of the weathering environment. Once formed, clay minerals are highly resistant to subsequent alteration of isotopic compositions except when mineralogic reactions also occur (Lawrence, 1970; Savin and Epstein, 1970; Lawrence and Taylor, 1972). Hence, isotopic compositions of ancient soils should yield climatic information. The climatic signal can be obscured in part because each clay mineral has its own isotopic systematics. However, careful mineralogic as well as isotopic study may be sufficient in many cases to resolve the isotopic systematics of different clay minerals within mineralogically complex soils. Lawrence and Taylor (1971) analyzed a large number of Quaternary age soils from the western United States. They found isotopic compositions basically consistent with the modern climatic regime. Lawrence (1970) analyzed D/H and $^{18}O/^{16}O$ ratios of kaolinite-rich soils of Tertiary age from the western United States and found a distribution pattern of D/H generally similar to those of today. However, he found a smaller difference between Tertiary D/H ratios of coastal regions and those of inland regions, suggesting less extreme climatic differences between the two areas than at present.

Studies of soil isotopic composition cannot be expected to provide climatic information of the same precision as can a number of other isotopic paleoclimate tools. However, the method should provide information of value where other approaches may not be applicable. Until now the method has found extremely little application to paleoclimatic problems. It warrants further consideration and, perhaps, development.

SUMMARY

There are a number of techniques whereby paleoclimatic information can be obtained from stable isotope data. Most precise and widely used is the carbonate-water paleotemperature scale. However, its applicability is restricted, for the most part, to the Cenozoic and late Mesozoic record because of lack of preservation of most older samples. The silica-water system has the potential for yielding useful climatic information for Precambrian times, but further developmental work is required before the results can be uniquely interpreted in terms of climate. However, for early Archaen time, even the most approximate estimates of surface temperatures can be useful and the silica-water system has provided these.

Isotopic methods are less useful for yielding information about pre-Pleistocene climates on the continents than they are about marine climates. However, some approaches such as the analysis of paleosols, speleothems, or subglacially precipitated carbonates may be useful in some cases.

ACKNOWLEDGMENTS

Financial support was provided by the National Science Foundation, Grant OCE 79-17017 (CENOP). Alan Hecht and Paul Knauth provided helpful reviews of the original manuscript. Contribution No. 133, Department of Geological Sciences, Case Western Reserve University.

REFERENCES

Arthur, M. (1979). Paleoceanographic events-recognition resolution and reconsideration, *Rev. Geophys. Space Phys. 17*, 1474-1494.

Berger, W. H. (1979). Stable isotopes in foraminifera, *Soc. Econ. Paleontol. Mineral. Short Course 6*, pp. 156-198.

Bowen, R. (1966). *Paleotemperature Analysis*, Elsevier, Amsterdam, 265 pp.

Craig, H. (1961). Isotopic variations in meteoric waters, *Science 133*, 1702-1703.

Craig, H. (1965). The measurement of oxygen isotope paleotemperatures, *Proc. of the Spoleto Conf. on Stable Isotopes in Oceanographic Studies and Paleotemperatures 3*, Cons. Naz. Richerche, Lab. Geol. Nucleare, Pisa, pp. 1-24.

Douglas, R. G., and S. M. Savin (1978). Oxygen isotopic evidence for the depth stratification of Tertiary and Cretaceous planktic foraminifera, *Mar. Micropaleontol. 3*, 175-196.

Douglas, R. G., and F. Woodruff (1981). Deep sea benthic foraminifera, in *The Sea*, Vol. 7, C. Emiliani, ed., Wiley-Interscience, New York.

Duplessy, J. C., C. LaLou, and A. C. Vinot (1970). Differential isotopic fractionation in benthic foraminifera and paleotemperatures reassessed, *Science 168*, 250-251.

Epstein, S., and C. J. Yapp (1976). Climatic implications of the D/H ratio of hydrogen in C-H groups in tree cellulose, *Earth Planet. Sci. Lett. 30*, 252-261.

Epstein, S., R. Buchsbaum, H. A. Lowenstam, and H. G. Urey (1951). Carbonate-water isotopic temperature scale, *Geol. Soc. Am. Bull. 62*, 417-426.

Epstein, S., P. Thompson, and C. J. Yapp (1977). Oxygen and hydrogen isotopic ratios in plant cellulose, *Science 198*, 1209-1215.

Grazzini, C. V. (1976). Non-equilibrium isotopic compositions of shells of planktonic foraminifera in the Mediterranean Sea, *Paleogeogr. Paleoclimatol. Paleoecol. 20*, 263-276.

Gregory, R. T., and H. P. Taylor, Jr. (1981). Oxygen isotope profile in a section of Cretaceous oceanic crust, Samail Ophiolite, Oman: Evidence for $\delta^{18}O$ buffering of the oceans by deep (greater than 5 km) seawater hydrothermal circulation at mid-ocean ridges, *J. Geophys. Res. 86*, 2737-2755.

Hanshaw, B. B., and B. Hallet (1978). Oxygen isotope composition of subglacially precipitated calcite: Possible paleoclimatic implications, *Science 200*, 1267-1270.

Hecht, A. D. (1976). The oxygen isotopic record of foraminifera in deep-sea sediment, in *Foraminifera, Vol. 2*, R. H. Hedley and C. G. Adams, eds., Academic Press, London, pp. 1-43.

Hendy, C. (1971). The isotopic geochemistry of speleothems—I. The calculation of the effects of different modes of formation of the isotopic composition of speleothems and their applicability as paleoclimatic indicators, *Geochim. Cosmochim. Acta 35*, 801-824.

Hudson, J. D. (1977). Oxygen isotope studies on Cenozoic temperatures, oceans, and ice accumulation, *Scottish J. Geol. 13*, 313-325.

Kennett, J. P., and N. J. Shackleton (1976). Oxygen isotopic evidence for the development of the psychrosphere 38 m.y. ago, *Nature 260*, 513-515.

Knauth, L. P., and S. Epstein (1975). Hydrogen and oxygen isotope ratios in silica from the JOIDES Deep Sea Drilling Project, *Earth Planet. Sci. Lett. 25*, 1-10.

Knauth, L. P., and S. Epstein (1976). Hydrogen and oxygen isotope ratios in nodular and bedded cherts, *Geochim. Cosmochim. Acta 40*, 1095-1108.

Knauth, L. P., and D. R. Lowe (1978). Oxygen isotope geochemistry of cherts from the Onverwacht Group (3.4 billion years), Transvaal, S. Africa, with implications for secular variations in the isotopic composition of cherts, *Earth Planet. Sci. Lett. 41*, 209-222.

Kolodny, Y., and S. Epstein (1976). Stable isotope geochemistry of deep sea cherts, *Geochim. Cosmochim. Acta 40*, 1195-1209.

Labeyrie, L. D. (1974). New approach to surface seawater paleotemperatures using $^{18}O/^{16}O$ ratios in silica of diatom frustules, *Nature 248*, 40-42.

Labeyrie, L. D., and A. M. Juillet (1980). Oxygen isotopic exchangeability of biogenic silica, *EOS: Trans. Am. Geophys. Union 61*, 259.

Lawrence, J. R. (1970). $^{18}O/^{16}O$ and D/H ratios of soils, weathering zones, and clay deposits, Unpublished Ph.D. thesis, California Institute of Technology.

Lawrence, J. R., and H. P. Taylor, Jr. (1971). Deuterium and oxygen-18 correlation: Clay minerals and hydroxides in Quaternary soils compared to meteoric waters, *Geochim. Cosmochim. Acta 35*, 993-1003.

Lawrence, J. R., and H. P. Taylor, Jr. (1972). Hydrogen and oxygen isotope systematics in weathering profiles, *Geochim. Cosmochim. Acta 36*, 1377-1393.

Libby, L. M., L. J. Pandolfi, P. H. Payton, J. Marshall, III, B. Becker, and V. Giertz-Sienbenlist (1976). Isotopic tree thermometers, *Nature 261*, 284-290.

Long, A., J. C. Lerman, and A. Ferhi (1978). Oxygen-18 in tree rings: Paleothermometers or paleohygrometers, in *Short Papers of the Fourth International Conference, Geochronology, Cosmochronology, Isotope Geology, 1978*, R. E. Zartman, ed., U.S. Geol. Surv. Open-File Rep. 78-701, pp. 253-254.

Lowenstam, H. A., and S. Epstein (1954). Paleotemperatures of the post-Aptian Cretaceous as determined by the oxygen isotope method, *J. Geol. 62*, 207-248.

Matthews, R. K., and R. Z. Poore (1980). The Tertiary $\delta^{18}O$ record: An alternative view concerning glacio-eustatic sea level fluctuations, *Geology 8*, 501-504.

Mikkelsen, N., L. Labeyrie, and W. H. Berger (1978). Silica oxygen isotopes in diatoms: A 20,000 year record in deep-sea sediments, *Nature 271*, 536-538.

Muehlenbachs, K., and R. N. Clayton (1976). Oxygen isotope composition of the oceanic crust and its bearing on seawater, *J. Geophys. Res. 81*, 4365-4369.

Murata, K. J., I. Friedman, and J. D. Gleason (1977). Oxygen isotope relations between diagenetic silica minerals in Monteray Shale, Temblor Range, California, *Am. J. Sci. 277*, 259-272.

Oskvarek, J. D., and E. C. Perry, Jr. (1976). Temperature limits on the early Archaean ocean from oxygen isotope variations in the Isua supracrustal sequence, West Greenland, *Nature 259*, 192-194.

Perry, E. C., Jr., S. N. Ahmad, and T. M. Swulius (1978). The oxygen isotope composition of 3800 m.y. old metamorphosed chert and iron formation from Isukasia, West Greenland, *J. Geol. 86*, 223-239.

Savin, S. M. (1977). The history of the Earth's surface temperature during the past 100 million years, *Ann. Rev. Earth Planet. Sci. 5*, 319-355.

Savin, S. M., and S. Epstein (1970). The oxygen and hydrogen isotope geochemistry of clay minerals, *Geochim. Cosmochim. Acta 34*, 25-42.

Savin, S. M., and F. G. Stehli (1974). Interpretation of oxygen isotope paleotemperature measurements: Effect of $^{18}O/^{16}O$ ratio of sea water, depth stratification of foraminifera, and selective solution, in *Colloq. Int. CNRS No. 219, Les Methodes Quantitative d'Etudes des Variations au cours du Pleistocene*, pp. 183-191.

Savin, S. M., and H. W. Yeh (1981). Stable isotopes in ocean sediments, in *The Sea*, Vol. 7, C. Emiliani, ed., Wiley-Interscience, New York.

Savin, S. M., R. G. Douglas, and F. G. Stehli (1975). Tertiary marine paleotemperatures, *Geol. Soc. Am. Bull. 86*, 1499-1510.

Schwarcz, H. P., R. S. Harmon, P. Thompson, and D. C. Ford (1976). Stable isotope studies of fluid inclusions in speleothems and their paleoclimatic significance, *Geochim. Cosmochim. Acta 40*, 657-665.

Shackleton, N. J., and J. P. Kennett (1975). Paleotemperature history of the Cenozoic and the initiation of Antarctic glaciation: Oxygen and carbon isotope analyses in DSDP sites 277, 279 and 281, in *Initial Reports of the Deep Sea Drilling Project 29*, U.S. Government Printing Office, Washington, D.C., pp. 743-755.

Shackleton, N. J., J. D. H. Wiseman, and H. A. Buckley (1973). Non-equilibrium isotopic fractionation between seawater and planktonic foraminiferal tests, *Nature 242*, 177-179.

Stevens, G. R., and R. N. Clayton (1971). Oxygen isotope studies on Jurassic and Cretaceous belemnites from New Zealand and their biogeographic significance, *N.Z. J. Geol. Geophys. 14*, 829-897.

Teis, R. V., and D. P. Naidin (1973). *Paleothermometry and Isotopic Composition of Oxygen in Organic Carbonates*, Moscow, 256 pp. (in Russian).

Urey, H. C. (1947). The thermodynamic properties of isotopic substances, *J. Chem. Soc.*, 562.

Urey, H. C., H. A. Lowenstam, S. Epstein, and C. R. McKinney (1951). Measurement of paleotemperatures and temperatures of the upper Cretaceous of England, Denmark, and the southeastern United States, *Geol. Soc. Am. Bull. 62*, 399-416.

van Donk, J. (1970). The oxygen isotope record in deep sea sediments, Ph.D. thesis, Columbia U., New York, 228 pp.

Williams, D. F., M. A. Sommer, and M. L. Bender (1977). Carbon isotopic compositions of recent planktonic foraminifera of the Indian Ocean, *Earth Planet. Sci. Lett. 36*, 391-403.

Woodruff, F., S. M. Savin, and R. G. Douglas (1980). Biological fractionation of oxygen and carbon isotopes by Recent benthic foraminifera, *Mar. Micropaleontol. 5*, 3-11.

Woodruff, F., S. M. Savin, and R. G. Douglas (1981). A high resolution oxygen isotope study of Pacific Miocene bottom temperatures, *Science 212*, 665-668.

Yapp, C. J., and S. Epstein (1977). Climatic implications of D/H ratios of meteoric waters over North America (9500-22,000 yr b.p.) as inferred from ancient wood cellulose C-H hydrogen, *Earth Planet. Sci. Lett. 34*, 333-350.

Cenozoic Variability of Oxygen Isotopes in Benthic Foraminifera

19

THEODORE C. MOORE, JR.
Exxon Production Research Company

NICKLAS G. PISIAS
Oregon State University

L. D. KEIGWIN, JR.
Woods Hole Oceanographic Institution

INTRODUCTION

The long-term (greater than 10^6 yr) character of the Earth's climate appears to exhibit distinct shifts from one state to the next. The succession of such states can be thought of as an evolutionary process with the average characteristics of each successive climatic state fundamentally different from any previous state. Each state may differ in terms of mean condition, in the amount of oscillation around the mean condition, and in the distribution of the amplitude of oscillation as a function of frequency. This long-term evolution of climate appears to be associated with telluric changes (i.e., changes in the geography and topography that form the boundaries to the fluid spheres). The rate of climatic change depends on the nature of the telluric effects. For example, the opening of an ocean gateway to deep and surface flow (such as passage between Antarctica and Australia) may have had a sudden and dramatic effect on average oceanic conditions, whereas the gradual opening of the Atlantic or closing of the Tethyan seaway may have caused longer term shifts in climatic conditions.

The geologic record of the deep sea affords us the opportunity to study the character of global oceanographic conditions over the past 100 million years (Ma) to define the steps in the evolution that have lead from the rather equitable climates of the Cretaceous to the ice ages of the last few million years and to relate these evolutionary steps to the changes in the telluric boundary conditions that are likely to have caused them. Furthermore, we should be able to characterize different parts of the climate system during each stage of this evolution. By studying the way the surface ocean, the deep ocean, the atmosphere, and the cryosphere have changed during each evolutionary stage, we will gain insights into the mechanisms that give rise to long-term climatic change. In addition, the investigation of the climate system under a variety of boundary conditions should give us a better fundamental understanding of how the different elements of this system can, and do, interact.

PREVIOUS WORK

Before such research can proceed, quantitative data on each element of the climate system must be acquired and studied. One of the most extensive quantitative data bases that now exists for the Cenozoic is the record of change in oxygen isotopes of benthic foraminifera (see Table 19.1). These data have been compiled by many investigators (Douglas and Savin, 1973,

1975; Savin *et al.*, 1975; Shackleton and Kennett, 1975a, 1975b). More recent work by Cenozoic Paleo-Oceanography Research Project (CENOP) workers has greatly added to this data base, particularly in the Miocene.

Oxygen isotopes measured on the shells of benthic foraminifers have given us a good representation (Figure 19.1) of the long-term changes that have occurred in the mean conditions of the deep ocean. The deep ocean contains more than 90 percent of the water on the Earth's surface and represents a part of the climate system that is important to the storage and transport of heat. It is a rather slow-moving part of the system, with a response time on the order of 10^3 yr, intermediate between the rapidly responding atmosphere and surface ocean and the much more slowly changing cryosphere and lithosphere. Compared with that of the surface waters, the isotopic composition of the modern deep ocean is relatively homogenous (Craig and Gordon, 1965); thus, an isotopic record of change in the deep ocean from almost any location is likely to give a picture of change in a large and important part of the climate system.

If it is assumed that the shells of benthic foraminifera are deposited in isotopic equilibrium (or that any vital effects can be taken into account), then changes in the oxygen isotopic ratio in the carbonate tests of the deep benthic fauna indicate changes in either the temperature of the bottom waters (with each 1°C equivalent to roughly 0.26 ‰ change in the $^{18}O/^{16}O$ ratio) or in the isotopic composition of deep waters. Changes in the isotopic composition of the deep ocean would most likely be caused by the transfer of a large amount of isotopically light water from the oceans to continental glaciers ($\delta^{18}O$ change of 0.1 ‰ is roughly equivalent to a 10-m glacial sea-level change); however, changes in the mode of formation of deep water that involved a significant change in their salinity would also affect their isotopic composition [with about a 0.1 ‰ change in $\delta^{18}O$ for every 0.2 ‰ change in salinity (Craig and Gordon, 1965)].

Studies of the long-term record of the isotopic record of foraminifera, together with other geologic and geophysical studies, suggest that in the last 100 Ma there were two times when major continental ice caps were formed and extended into the sea: in the Middle Miocene, marking the buildup of the Antarctic ice cap about 14 Ma ago (Shackleton and Kennett, 1975a, 1975b), and in the late Pliocene, marking the formation of continental glaciers in the northern hemisphere about 3 Ma ago (Shackleton and Opdyke, 1977). If the estimates of the effect of ice volume on the isotopic composition of seawater are taken into account, and if it is assumed that changes in salinity have been small, then the record of the long-term changes in the oxygen isotopes can be interpreted as an oceanic temperature record (Figure 19.1). This record suggests that the deep waters have cooled by about 10-13°C in the last 60 Ma. Surface waters followed this trend to the mid-Miocene and then leveled off or warmed slightly (Douglas and Woodruff, 1981). This overall cooling of deep waters is neither monotonic nor gradual. Short reversals in the trend occur, and the major portion of the cooling appears to occur as distinct steps in the record (e.g., in the mid-Eocene, at the Eocene-Oligocene boundary, in the mid-Miocene, and in the Pliocene (Figure 19.1). These sharp drops in the isotopic record are thought to be associated with evolutionary changes in the climate system that give rise to shifts in the mean conditions. It remains to be seen whether other proxy records and oceanic and climatic changes exhibit the same shifts in their records and whether additional evolutionary steps can be identified in these records.

The most recent evolutionary stage as defined in Figure 19.1 is the Quaternary. An important characteristic of the Quaternary climate has been the large degree of variability around the mean climatic state. Most of the Quaternary variability in the benthic oxygen isotope signal is thought to be associated with changes in continental ice volume (Shackleton, 1967; Shackleton and Opdyke, 1973). The variability of both the oxygen isotopes (Shackleton and Opdyke, 1976; Pisias and Moore, 1981) and the planktonic fauna (Ruddiman, 1971; Briskin and Berggren, 1975) changed through the Quaternary and these changes appear to involve both the amplitude and frequency of oscillation. Spectral analyses of one 2-Ma-long record of oxygen isotopes (measured on a planktonic species of foraminifera) indicates that this record can be divided into at least three intervals, each having progressively more variance associated with progressively longer periods of oscillation (Pisias and Moore, 1981). These changes in the spectral character of oxygen isotope records may also indicate evolutionary changes in the climate system. In this example, these changes are thought to result from changing mechanisms of ice-cap growth and decay and may indicate the effects of extensive glacial erosion of continental areas (Pisias and Moore, 1981).

Similar spectral studies of Tertiary records have yet to be undertaken, primarily because (a) few long, relatively undisturbed marine sections have been recovered, and (b) establishing a sufficiently accurate time scale for a detailed spectral analysis is a difficult task. Although it may not be possible yet to investigate the spectral character (i.e., the distribution of variance as a function of frequency of oscillation) of Tertiary oxygen isotopic records, the total variance of such records and how this variance has changed with time and place can be studied.

It is sometimes assumed that the warmer climes of Tertiary and Cretaceous times were more equable than at present, that is, they were less variable. However, little work has been done on the short-term variability of Cenozoic climate. Certainly before the buildup of continental glaciers one might expect to see less variability in the oxygen isotope signal. But was there less oceanographic variability during the Eocene than during the Oligocene or Miocene? Did the variability of the benthic oxygen isotope record increase as the deep-ocean temperature cooled? How did variability in this record change with evolution of the climate system, and what degree of variability is associated with each evolutionary step? Such questions are addressed here in hopes of better defining the true nature of climatic and oceanographic variability through the Cenozoic.

METHODS

An accurate estimate of the total variance in a data set does not require as long a record as a spectral analysis. It does not require that the time scale be known accurately, nor does it re-

TABLE 19.1 Estimated Variance ($^*\sigma^2$) with Linear Trend Removed[a]

Age	Site	Ocean	Water Depth (m)	*N*	Species Group	$^*\sigma^2$	Data Source
L. Quaternary	V19-28	P	2720	142	*Uviger.*	0.168^2	Shackleton, 1977
	V19-29	P	3157	168	*Uviger.*	0.181^2	Shackleton, 1977
	Y6910-2	P	2615	203	*Uviger.*	0.142^2	Shackleton, 1977
	157	P	2591	9	*Uviger.*	0.173^2	Keigwin, 1979a
	E67-135	A	725	31	*Uviger.*	0.184^2	Keigwin, 1979a
	397	A	2900	133	*Uviger.*	0.282^2	Shackleton and Cita, 1979
E. Quaternary	284	P	1068	7	*Uviger.*	0.119^2	Kennett *et al.*, 1979
	310	P	3516	21	*Uviger.*	0.087^2	Keigwin, 1979a
	E67-135	A	725	32	*Uviger.*	0.100^2	Keigwin, 1979b
	397	A	2900	15	*Uviger.*	0.113^2	Shackleton and Cita, 1979
L. Pliocene	V28-179	P	4490	36	*G. subgl.*	0.064^2	Shackleton and Opdyke, 1977
	157	P	2300	16	*Uviger.*	0.052^2	Keigwin, 1979a
	206	P	3110	6	*Uviger.*	0.012^2	Bender (CENOP unpubl.)
	207	P	1360	9	*Uviger.*	0.018^2	Bender (CENOP unpubl.)
	310	P	3510	23	*Uviger.*	0.060^2	Keigwin, 1979a
	503	P	3500	7	*P. wuell.*	0.055^2	Keigwin (CENOP unpubl.)
	281	S(P)	1570	6	*Uviger.*	0.033^2	Bender (CENOP unpubl.)
	E67-135	A	720	25	*Uviger.*	0.068^2	Keigwin, 1979b
	397	A	2890	25	*Uviger.*	0.122^2	Shackleton and Cita, 1979
	502	A	3040	9	*P. wuell.*	0.146^2	Keigwin (CENOP unpubl.)
E. Pliocene	V28-179	P	4480	36	*G. subgl.*	0.018	Shackleton and Opdyke, 1977
	62.1	P	2490	10	*G. subgl.*	0.007	Keigwin *et al.*, 1979
	83A	P	3410	14	*Uviger.*	0.008	Keigwin *et al.*, 1979
	84	P	2700	6	*Uviger.*	0.056	Keigwin *et al.*, 1979
	158	P	1710	11	*Uviger.*	0.050	Keigwin, 1979a
	206	P	3100	8	*Uviger.*	0.006	Keigwin *et al.*, 1979
	207A	P	1340	7	*Uviger.*	0.010	Keigwin *et al.*, 1979
	208	P	1520	19	*Uviger.*; G. subgl.	0.028[1]	Bender (CENOP unpubl.); Keigwin *et al.*, 1979
	284	P	1020	19	*Uviger.*	0.015	Kennett *et al.*, 1979
	310	P	3500	11	*Uviger.*	0.017	Keigwin, 1979a
	503	P	3450	21	*C. kull.*	0.011	Keigwin (CENOP unpubl.)
	E67-135	A	710	24	*Uviger.*	0.025	Keigwin, 1979b
	297	A	2890	20	*Uviger.*	0.090	Shackleton and Cita, 1979
	502	A	3030	17	*P. wuell.*	0.035	Keigwin (CENOP unpubl.)
L. Miocene (postcarbon shift)	77B	P	4250	12	M.B.	0.090	Savin and Weh, 1981
	158	P	1570	27	*Uviger.*; *G. subgl.*	0.030[1]	Keigwin, 1979a
	207A	P	1320	28	*Uviger.*	0.034	Bender (CENOP unpubl.)
	208	P	1480	12	*Uviger.*	0.058	Bender (CENOP unpubl).
	284	P	990	13	*Uviger.*	0.015	Kennett *et al.*, 1979
	289	P	2190	12	*P. wuell.*	0.025	Woodruff *et al.*, 1981
	292	P	2800	27	*Oridos.*	0.046	The Benedum Lab., Brown U. (CENOP unpubl.)
	296*	P	2750	12	*Oridos.*	0.422	The Benedum Lab., Brown U. (CENOP unpubl.)
	310	P	3430	12	*Uviger.*	0.015	Keigwin, 1979a
	503	P	3270	7	*C. kull.*	0.014	Keigwin (CENOP unpubl.)
	278	S(P)	3530	6	*Cibicid.*	0.034	Bender (CENOP unpubl.)
	281	S(P)	1500	6	*Uviger.*	0.014	Bender (CENOP unpubl.)
	329	S(A)	1430	33	M.B.	0.114	Savin *et al.* (CENOP unpubl.)
	238	I	2630	19	*Oridos.*; *P. wuell.*	0.056[1]	Vincent *et al.*, 1980
	357	A	2060	8	*Oridos.*	0.062	The Benedum Lab., Brown U. (CENOP unpubl.)
	397	A	2860	18	M.B.	0.055	Shackleton and Cita, 1979

TABLE 19.1 *(continued)*

Age	Site	Ocean	Water Depth (m)	*N*	Species Group	$^*\sigma^2$	Data Source
	408	A	1190	6	*Oridos.*	0.106	The Benedum Lab., Brown U. (CENOP unpubl.)
	502	A	3020	21	*P. wuell.*	0.035	Keigwin (CENOP unpubl.)
L. Miocene (precarbon shift)	77B	P	4210	16	M.B.	0.149	Savin *et al.*, 1981
	158	P	1410	9	*Uviger.*	0.057	Keigwin, 1979a
	206	P	3100	12	*Uviger.*	0.042	Bender (CENOP unpubl.)
	207	P	1300	13	*Uviger.*	0.018	Bender (CENOP unpubl.)
	208	P	1450	7	*Uviger.*	0.009	Bender (CENOP unpubl.)
	289	P	2180	11	*P. wuell.*	0.046	Woodruff *et al.*, 1981
	296	P	2740	9	*Oridos.*	0.054	The Benedum Lab., Brown U. (CENOP unpubl.)
	310	P	3420	9	*Oridos.*	0.009	Keigwin, 1979a
	503	P	3090	7	*P. wuell.*	0.006	Keigwin (CENOP unpubl.)
	278	S(P)	3510	6	*Cibicid.*	0.045	Bender (CENOP unpubl.)
	281	S(P)	1490	26	*G. subglob.*	0.097	Loutit (CENOP unpubl.)
	329	S(A)	1420	6	M.B.	0.276	Savin *et al.*, (CENOP unpubl.)
	238	I	2640	38	*P. wuell., Oridos.*	0.029[1]	Vincent *et al.*, 1980
	357	A	2050	8	*Oridos.*	0.095	The Benedum Lab., Brown U. (CENOP unpubl.)
	397	A	2850	28	M.B.	0.077	Shackleton and Cita, 1979
	397	A	2850	11	*P. wuell.*	0.011	Bender (CENOP, unpubl.)
	408	A	1170	10	*Oridos.*	0.209	The Benedum Lab., Brown U. (CENOP unpubl.)
	502	A	3000	18	*P. wuell.*	0.020	Keigwin (CENOP unpubl.)
L. Mid. Miocene	77B	P	4180	27	*Cibicid.; G. subgl.*	0.039[1]	Savin *et al.*, 1981; Kennett and Keigwin (CENOP unpubl.)
	206	P	3060	12	*Oridos.*	0.019	Bender (CENOP unpubl.)
	206	P	3060	11	*P. wuell.*	0.065	Bender (CENOP unpubl.)
	207A	P	1240	7	*Uviger.*	0.051	Bender (CENOP unpubl.)
	208	P	1420	11	*P. wuell.*	0.013	Bender (CENOP unpubl.)
	289	P	2160	36	*Cibicid.*	0.053	Woodruff *et al.*, 1981
	310	P	3380	5	*Oridos.*	0.036	Keigwin, 1979a
	281	S(P)	1400	17	M.B.; *Uviger., G. subgl.*	0.004[1]	Shackleton and Kennett, 1975a; Loutit (CENOP unpubl.)
Mid. Miocene (trans.)	77B	P	4090	5	*G. subgl.*	0.086	Kennett and Keigwin (CENOP unpubl.)
	77B	P	4090	23	*Cibicid.; C. kull.*	0.024[1]	Kennett and Keigwin (CENOP unpubl.)
	289	P	2140	37	*Cibicid.*	0.099	Woodruff *et al.*, 1981
E. Mid. Miocene	55	P	2750	6	M.B.	0.048	Savin *et al.*, 1975
	71	P	4270	6	*Cibicid.*	0.080	Savin *et al.*, 1981
	77B	P	4080	17	*Cibicid.*	0.037	Kennett and Keigwin (CENOP unpubl.)
	206	P	3040	18	*P. wuell.*	0.024[1]	Bender (CENOP unpubl.)
	289	P	2100	26	*Cibicid.*	0.033	Woodruff *et al.*, 1981
	281	S(P)	1380	24	*Uviger.*	0.041	Loutit (CENOP unpubl.)
E. Miocene	71	P	4260	37	*Oridos.; Cibicid.*	0.110[1]	Savin *et al.*, 1981
	77B	P	3980	8	*Cibicid.*	0.063	Kennett and Keigwin (CENOP unpubl.)
	206	P	3000	11	*Oridos.*	0.056	Bender (CENOP unpubl.)
	208	P	1350	20	*G. subgl.*	0.036	Bender (CENOP unpubl.)
	289	P	1900	25	*Cibicid.*	0.042	Woodruff *et al.*, 1981
	296	P	2360	18	*Oridos.*	0.070	The Benedum Lab., Brown U. (CENOP unpubl.)

TABLE 19.1 *(continued)*

Age	Site	Ocean	Water Depth (m)	*N*	Species Group	$*\sigma^2$	Data Source
	279	S(P)	2010	35	M.B.; *Gyrod.*	0.055[1]	Bender (CENOP unpubl.); Shackleton and Kennett, 1975
	281	S(P)	1250	17	*Uviger.*	0.015	Loutit (CENOP unpubl.)
	237	I	1440	25	*Oridos.*	0.059	Vincent *et al.* (CENOP unpubl.)
	15	A	2550	17	M.B.	0.060	Savin *et al.*, 1975
	116	A	200	22	*Oridos.*	0.026	The Benedum Lab., Brown U. (CENOP unpubl.)
	366A	A	2700	27	*Oridos.*	0.050	Vincent *et al.*, (CENOP unpubl.)
M.-L. Oligocene	277	S(P)	1222	14	M.B.	0.020	Shackleton and Kennett, 1975a
	366A	A	2860	11	*G. subgl.*	0.035	Boersma and Shackleton, 1977
E. Oligocene	277	S(P)	1222	7	*Oridos.*	0.004	Keigwin, 1980
	292	P	2943	7	*Oridos.*	0.018	Keigwin, 1980
	366A	A	2860	8	M.B.	0.044	Boersma and Shackleton, 1977
L. Eocene	292	P	2943	14	*Oridos.*	0.026	Keigwin, 1980
	277	S(P)	1222	20	M.B., *Oridos.*	0.017[1]	Shackleton and Kennett, 1975a; Keigwin, 1980
M. Eocene	44	P	1478	8	M.B.	0.038	Savin *et al.*, 1975
	277	S(P)	1222	9	M.B.	0.083	Shackleton and Kennett, 1975a
	398*	A	3900	8	M.B.	0.329	Vergnaud-Grazzini, 1979
Paleocene	384	A	3910	13	M.B.	0.065	Boersma *et al.*, 1979

[a]Data sets are arranged according to stratigraphic age and grouped according to ocean basin. Depths given are estimated paleodepths for each stratigraphic age based on standard backtracking techniques. Number of data points (*N*) in individual data sets, species groups used, and data sources are indicated. M.B. indicates data that are based on the mixed benthic assemblage. An asterisk denotes data sets with very large variances, which appear spurious when compared with other data of similar age and location. These data are not shown in Figure 19.2. A "1" denotes those sites in which data are available from two different species groups and have estimated variances that are not significantly different. Their variances are pooled in this table. A "2" indicates those data sets in which all variance estimates from a given ocean basin and age are not significantly different. These variances are pooled for use in Figure 19.2. Site locations and ocean abbreviations are given in Table 19.2.

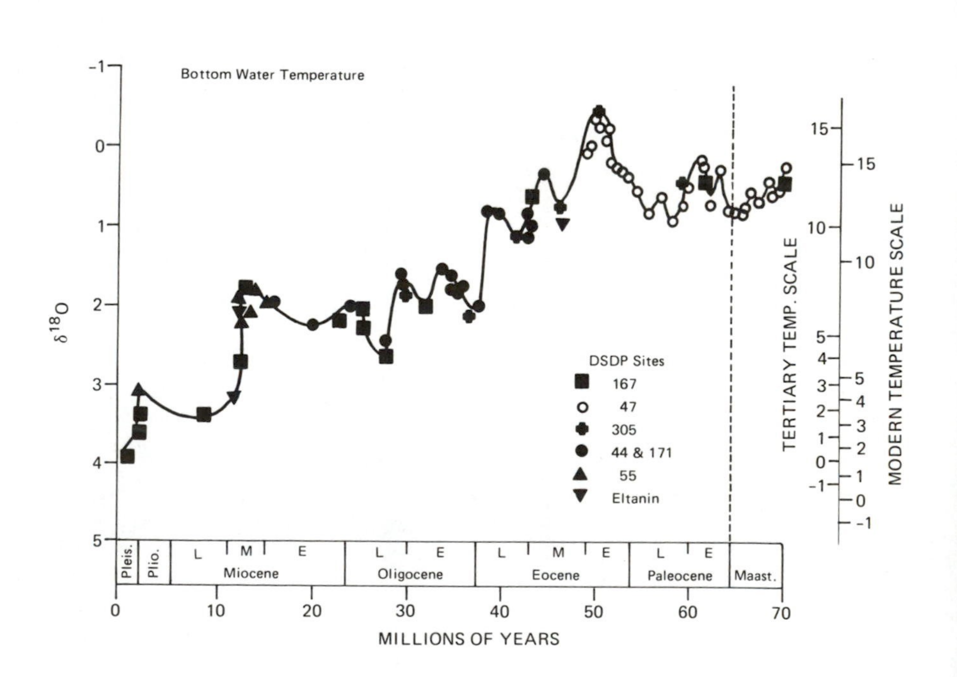

FIGURE 19.1 Benthic oxygen isotope record for the Cenozoic (after Douglas and Woodruff, 1981).

quire that the mean values of two data sets be the same before a comparison of the variances can be made. Thus, virtually all of the short-time series of oxygen isotope data (Table 19.1), regardless of the particular species used to obtain the data, can be used in a comparison of the total variance in different data sets. These variances have units of (‰)^2; however, this notation is omitted from the text discussion.

Most of the samples in the Tertiary data series are rather widely spaced within the recovered sections (usually 100 cm). Accumulation rates of 10-50 meters/million years (m/m.y.) are common in these pelagic sediments and thus would indicate a time spacing of the samples that is often greater than 10^5 yr. Sample spacing in the Miocene isotopic data collected by the CENOP group (Table 19.1) was designed to be between 50,000-100,000 yr. Although such sample spacings are much broader than commonly used in Quaternary studies, they do provide an estimate of the total variance in the long-term record. If the Pleistocene record is sampled at a 50,000-yr spacing, the variance estimate is the same as for a 5000-yr sample spacing. Table 19.1 lists data sources to be used in this analysis and Table 19.2 lists site locations. These data have the following characteristics: (1) Oxygen isotope measurements in each data set were carried out on samples of a single species, a species group, or a mixed benthic assemblage. Data derived from different species groups or size fractions were not combined; however, if their variances were not significantly different (F test), they

TABLE 19.2 Location, Water Depth, and Stratigraphic Age of Benthic Oxygen Isotope Data Used in this Study[a]

Site	Ocean	Latitude	Longitude	Water Depth (m)	Stratigraphic Age Studied
E67-135	A	29°00′ N	87°00′ W	725	Quaternary, Pliocene
V19-28	P	02°22′ S	84°39′ W	2720	L. Quaternary
V19-29	P	03°35′ S	83°56′ W	3157	L. Quaternary
V28-179	P	04°37′ N	139°36′ W	4502	Pliocene
Y6910-2	P	41°16′ N	127°01′ W	2615	L. Quaternary
15	A	30°53′ S	17°59′ W	3927	E. Miocene
44	P	19°18′ N	169°00′ W	1478	M. Eocene
55	P	09°18′ N	142°33′ W	2850	M. Miocene
62	P	01°52′ N	141°56′ W	2591	Pliocene
71	P	01°26′ S	125°49′ W	4419	M. Miocene, E. Miocene
77	P	01°39′ S	127°52′ W	4290	L. Miocene, M. Miocene, E. Miocene
83	P	04°03′ N	95°44.2′ W	3632	Pliocene
84	P	05°45′ N	82°53′ W	3096	Pliocene
116	A	57°56′ N	15°56′ W	1161	E. Miocene
157	P	01°46′ S	85°54′ W	2591	L. Quaternary, Pliocene
158	P	06°37′ N	85°14′ W	1953	Pliocene, L. Miocene
206	P	32°01′ S	165°27′ E	3196	Pliocene, L. Miocene, M. Miocene, E. Miocene
207	P	36°58′ S	165°26′ E	1389	Pliocene, L. Miocene, M. Miocene
208	P	26°07′ S	161°13′ E	1545	Pliocene, L. Miocene, E. Miocene
237	I	07°05′ S	58°07′ E	1640	E. Miocene
238	I	11°09′ S	70°32′ E	2844	L. Miocene
277	S(P)	52°13′ S	166°11′ E	1222	L. Oligocene, L. Eocene, M. Eocene
278	S(P)	56°33′ S	160°04′ E	3698	L. Miocene
279	S(P)	51°20′ S	162°38′ E	3371	E. Miocene
281	S(P)	48°00′ S	147°46′ E	1591	Pliocene, L. Miocene, M. Miocene, E. Miocene
284	P	40°30′ S	167°41′ E	1068	E. Quaternary, Pliocene, L. Miocene
289	P	00°30′ S	158°31′ E	2206	L. Miocene, M. Miocene, E. Miocene
292	P	15°49′ N	124°39′ E	2943	L. Miocene, E. Oligocene, L. Eocene
296	P	29°20′ N	133°32′ E	2920	L. Miocene, E. Miocene
310	P	36°52′ N	176°54′ E	3516	E. Quaternary, Pliocene, L. Miocene, M. Miocene
329	S(A)	50°39′ S	46°06′ W	1519	L. Miocene
357	A	30°00′ S	35°34′ W	2109	L. Miocene
366	A	05°41′ N	19°51′ W	2860	E. Miocene, L. Oligocene, E. Oligocene
384	A	40°22′ N	51°40′ W	3910	Paleocene
397	A	26°51′ N	15°11′ W	2900	Quaternary, Pliocene, L. Miocene
398	A	40°58′ N	10°48′ W	3900	M. Eocene
408	A	63°23′ N	28°55′ W	1634	L. Miocene
502	A	11°29′ N	79°23′ W	3052	Pliocene, L. Miocene
503	A	04°03′ N	95°38′ W	3672	Pliocene, L. Miocene

[a]Ocean locations: A, Atlantic; P, Pacific; I, Indian, S(P), Southern Ocean, Pacific sector; S(A), Southern Ocean, Atlantic sector.

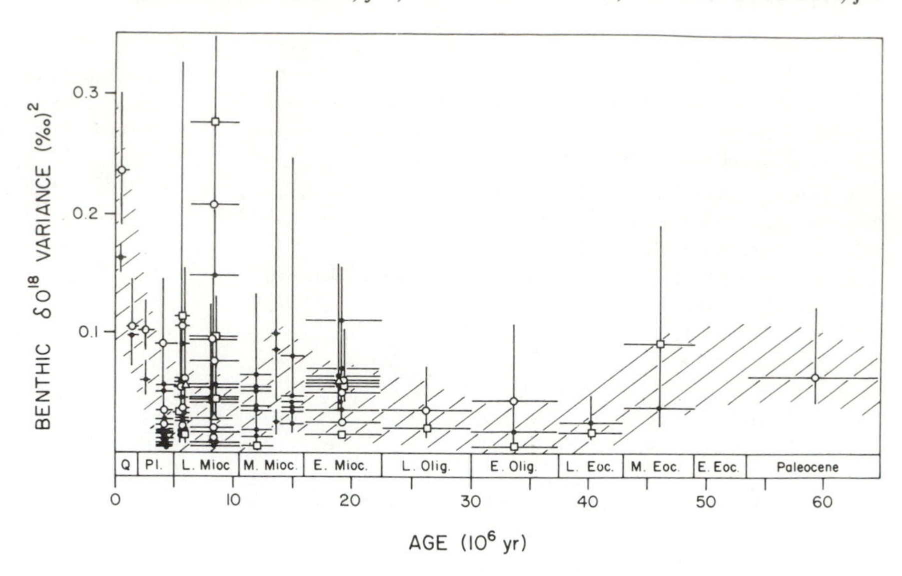

FIGURE 19.2 Variance in benthic oxygen isotopes as a function of age in the Cenozoic. Data sources are listed in Table 19.1. Horizontal lines indicate stratigraphic range represented by the data sets, vertical lines indicate 80 percent confidence limits of the variance estimates. Closed circles represent Pacific Ocean data sets; open circles, Atlantic Ocean data sets; squares, Southern Ocean data sets; triangles, Indian Ocean data sets. Shaded area encompasses most of the data from the Pacific and Southern Oceans; excluded are the very deep Pacific sites in the Early and Late Miocene and the highly variable sites of the Atlantic and Southern Oceans sites in the Late Miocene.

could be pooled. (2) Each data set comes from a single site. (3) Each group of measurements is associated with a known stratigraphic age.

Stratigraphic intervals over which the variance was calculated were kept as short as possible to lend more detail to the long-term record of variability. Data were selected to avoid any clear jumps or shifts in the record that might be associated with evolutionary changes in the system. Such shifts in the data would lead to an abnormally large estimate of the variance. To guard further against more gradual shifts in the data, trends were removed from each data set using a simple linear regression. The estimate of variance used in this study is the variance around this regression line.

For each stratigraphic age the variances of all data sets were compared using the *M* test (Thompson and Merrington, 1944) for homogeneity of variances. If the difference in the variance estimates were nonsignificant ($p \geq 0.95$), then the variances of all the data sets of the age could be pooled. This procedure allows the combination of several Quaternary time series. If variances were nonhomogeneous, they were subdivided according to ocean basin locations, and again tested for homogeneity. Data sets from the Atlantic, Pacific, and Indian Oceans were compared; when sufficient data were available, their variability as a function of water depth was contrasted.

In a few cases the variability in individual data sets is high, greatly exceeding that found in other sites at the same time and depth interval. Such data might result from unrecognized diagenic or stratigraphic problems. They are considered spurious here, and, although included in Table 19.1, they are not plotted in the figures. Sites having extremely low variability might result from severe disturbance and mixing of rotary-drilled sections; however, such sites appear to show some degree of spatial and temporal coherence and are not excluded from the figures.

THE LONG-TERM RECORD

The record of the variance in benthic oxygen isotopes is shown in Figure 19.2, with vertical lines giving the 80 percent confidence limits of the estimate and horizontal lines indicating the range of the stratigraphic age of the individual data sets. Each symbol is located at the midpoint of the stratigraphic range. Different symbols are used for Pacific, Atlantic, Antarctic, and Indian Ocean data sets.

Although the amount of data varies greatly through the Cenozoic (Figure 19.2), there appear to be several important changes in the character of isotopic variance as a function of time: (a) a decrease in the estimated variance from mid-Eocene to Oligocene time, (b) a slight Middle Miocene maximum in variance, (c) an increase in the variance of Atlantic sites relative to Pacific sites in post-Early Miocene times, and (d) a sharp increase in the variance at both Atlantic and Pacific sites beginning in the late Pliocene. The background variability above which the intervals of high variance rise is surprisingly consistent. It averages about 0.04 and is four times greater than the variance associated with laboratory error.

In the latest part of the record the high variance in benthic oxygen isotopes increases from values of 0.06-0.1 in the late Pliocene to approximately 0.2 in the late Quaternary. Most of this high degree of variability is related to changes in the isotopic composition of the oceans as continental glaciers waxed and waned (Shackleton, 1967; Shackleton and Opdyke, 1973).

There is some indication that the pooled variance in Atlantic sites is slightly higher than in Pacific sites. The difference is not significant in the Quaternary (*M* test, $p \geq 0.95$); however in the Pliocene and Late Miocene the Atlantic and Pacific do appear to show different degrees of variability. Shackleton and Cita (1979) noted the generally higher degree of varability in Atlantic sites during the latest Miocene. Such differences are also seen in the Pliocene and Late Miocene data presented here (Figure 19.2), where several (but not all) Atlantic and Southern Ocean sites show a higher variance than most Pacific sites, with the highest variances measured in the earlier part of the Late Miocene. During the Early Miocene and Oligocene, the variance of benthic oxygen isotopes in Atlantic sites were also slightly higher than those of the Pacific; however, the differences are not statistically significant (F test, $p \geq 0.9$).

In the Pliocene and Miocene, measurements from deep sites (>4000 m) are included in the data set. These Pacific sites (DSDP 71 and DSDP 77 of the Deep Sea Drilling Project) show generally higher variance than others from the Pacific Ocean basin during both the Early and Late Miocene. During the Middle Miocene, however, detailed studies of a mid-depth site (DSDP 289—Woodruff *et al.*, 1981) have an equally high degree of variability. This interval spans the time of glacial buildup in Antarctica (Shackleton and Kennett, 1975a) and appears as a major shift in the benthic isotopic values (Figure 19.1). The maximum in variance associated with this shift in isotopic values is clearly seen in the original data (Woodruff *et al.*, 1981). To remove the effect of this shift in isotopic mean values on estimates of variance, the Middle Miocene data are subdivided into three groups (Table 19.1, Figure 19.1): pretransition, transition, and posttransition. Trends within these data subsets were removed using a linear regression.

There are few accurate estimates of oxygen isotopic variance in the Paleogene; however, the data that are available indicate a minimum in variability during Late Eocene through Oligocene times. Discounting the high mid-Eocene variance of DSDP 398 (which may be spurious), the variance in benthic oxygen isotopes of the early Paleogene was near 0.06, whereas those in the Late Eocene through Oligocene were significantly lower (pooled variance = 0.027). This apparent decrease in variance parallels a general cooling trend through the mid- to late-Paleogene (Figure 19.1) when bottom-water temperatures are estimated to have dropped from almost 15°C in the mid-Eocene to 6°C in the Oligocene (Figure 19.1). This late Paleogene minimum in variability is also significantly less than that measured in the Miocene, when the variance was usually near 0.04.

Many of the changes in the variance of the oxygen isotopic data can be associated with major evolutionary transitions in the Cenozoic climate. The maxima in variance during the Quaternary and mid-Miocene are both correlated with the growth of major ice caps. It is presumed that like those of the Quaternary, the mid-Miocene intervals of high variance are associated with instability in the climate system and that most of the variation is due to changes in the isotopic composition of the oceans.

The next older major step in climatic evolution indicated by the oxygen isotopes (Figure 19.1) occurs at the Eocene-Oligocene boundary. Although the data are more sparse around and across this boundary, the existing data (Kennett and Shackleton, 1976; Keigwin, 1980) indicate a rapid, monotonic shift in isotopic values. This shift is observed in both planktonic and benthic species in high latitudes but only in the benthics at low-latitude Pacific sites (Keigwin, 1980). Thus this evolutionary step, which is thought to be associated with the opening of the Australian-Antarctic seaway (Kennett and Shackleton, 1976), appears to be related to a marked cooling of deep waters and high-latitude surface waters. However, there does not appear to have been a change in the variability of deep-ocean waters on either side of this boundary. Nor was there a marked instability associated with the transition from one climatic state to the next, as observed with the growth of continental ice in the Middle Miocene and Quaternary.

The increase in benthic isotopic variability that occurred in the mid-Eocene is based on sparse data (Table 19.1); but if further work supports its existence, it is the third maximum in variability associated with a marked shift in the mean ^{18}O content of benthic tests (cf., Figure 19.1). The cause of the mid-Eocene shift in mean isotopic values is not certain. It could have been caused by a telluric change, such as the opening of a passage between South America and Antarctica (Norton and Sclater, 1979), which might have led to a marked cooling of the ocean waters. It might also have been caused by an early buildup of fairly large mountain glaciers in Antarctica (Matthews and Poore, 1980) or by some combination of temperature and ice volume effects.

Sufficient data have been gathered from the Miocene and Pliocene (Table 19.1) to allow a view of how benthic isotopic variability is distributed in space as well as time. In the Late Miocene there is an increased oxygen isotopic (temperature) contrast between deep and bottom waters (Douglas and Woodruff, 1981). This is also the time when marked differences between the isotopic variability of the Atlantic and Pacific Oceans are first noted. This divergence of Atlantic and Pacific estimates of isotopic variance in the Late Miocene suggests the development of different source regions for deep waters in the two basins.

The general pattern of change with depth seen in the Miocene and Pliocene of the Pacific Ocean is from low variance in shallower sites to higher variance in deeper sites. In both the Early and Late Miocene, the variance in the deepest sites (DSDP 71 and DSDP 77) is higher than in any other depth zone in the Pacific Ocean and is exceeded in magnitude only by the Quaternary data and the Late Miocene data from the Atlantic and Southern Ocean. In the Atlantic, Late Miocene variability of benthic isotopes at shallow depths is high (up to about 0.25). The variance decreases with depth, so that at 3000 m all oceans show approximately the same rather low degree of variability. The marked difference in the variance of Atlantic and Pacific benthic isotopic data has been noted previously (Shackleton and Cita, 1979); however, this difference does not appear to occur prior to the mid-Miocene.

DISCUSSION AND CONCLUSIONS

The data presented here indicate that the variability of benthic oxygen isotopes have changed with time and that these changes have often been associated with major steps in the evolution of the oceans. Although the data from the Paleogene are sparse, they indicate that the variability in benthic oxygen isotopes of the Oligocene and Late Eocene was significantly less than at any other time in the Cenozoic. The Oligocene was a time of relatively cool and equable climate (Kennett, 1978; Fischer and Arthur, 1977) and low eustatic sea levels (Vail *et al.*, 1977). The low isotopic variability of this interval may be attributable to homogeneous deep waters derived from a single source region. The higher variability of the early Paleogene was associated with much warmer high-latitude and bottom-water temperatures. These conditions might have given rise to deep waters with a wider range of isotopic compositions. Different source regions and the initial buildup of sizable glaciers on Antarctica could also have served to introduce variability in the isotopic composition of the deep waters; however, the data are not sufficient to explore these possibilities.

The well-documented isotopic shift across the Eocene-Oligocene boundary is interpreted as cooling of the high-latitude ocean and deep waters by about 3°C (Keigwin, 1980). There is no maximum in isotopic variability associated with the buildup of continental ice during the mid-Miocene and Pliocene-Pleistocene. Rather the record indicates a sudden, monotonic shift from one relatively stable oceanic state to another.

In the Early Miocene the average variability increased again, but not up to the levels of the early Paleogene. This change is not readily associated with major oceanographic changes (see Figure 19.1), and the data are not sufficient to tell whether this shift to increased variance was relatively sudden (as in the mid-Miocene and Quaternary) or more gradual. The gradual shoaling of the calcite compensation depth and increase in the carbonate dissolution gradients (Heath *et al.*, 1977) through the Oligocene and Early Miocene suggest a slow, long-term change in the character of the deep waters during this time interval.

An apparent maximum in variability is associated with the growth of the Antarctic ice cap during the mid-Miocene and suggests that some degree of instability in this ice cap may have existed during its early growth phase (Woodruff *et al.*, 1981). In the Pacific Ocean, variation in the benthic oxygen isotopes was approximately the same before and after growth of the Antarctic ice cap; however, variability in data from the Atlantic Ocean greatly increased by Late Miocene times.

The development of northern hemisphere ice sheets in the Late Pliocene is associated with an increase in oxygen isotopic variability that continues into the Quaternary and reaches a maximum in the late Pleistocene. This increase in variance is readily associated with the fluctuations in the isotopic composition of seawater caused by the growth and decay of continental ice sheets.

There are several keys to assessing the changes in the Cenozoic oceans that may have led to changes in the variability of benthic isotopes. Changes in the global ice volume (which caused a 1.6 ‰ change in the oxygen isotopic composition of the oceans during the late Quaternary) is clearly associated with two of the maxima noted in the historical record (Figure 19.2). Such compositional changes may also be associated with the high variance of the early Paleogene (Matthews and Poore, 1980).

By itself the range of oxygen isotopic compositions of modern deep waters (0.57 ‰) is close to the range of variation in the Cenozoic record of benthic oxygen isotopes; however, the temperatures of these modern watermasses are such that the isotopically heaviest waters [North Atlantic deep water (NADW) at +0.12 ‰] are also comparatively warm (about 4°C), and the ^{18}O-depleted waters [Antarctic bottom water (AABW) at −0.45 ‰] are cold (about 2°C). Thus, the temperature-fractionation effects of calcite precipitation on oxygen isotopes tend to offset the compositional differences and result in benthic foraminiferal tests with similar isotopic compositions.

The temperature and isotopic differences between modern AABW (at about 5°C and −0.15 ‰) and NADW (at 4°C and +0.12 ‰) tend to enhance the differences measured in the benthic foraminifera. If over long periods of time a site were alternately bathed by NADW and AABW, the isotopic record measured on benthic foraminifera would have a variance close to that measured in most of the Cenozoic sites studied. This suggests that the range of isotopic compositions of modern deep waters is at least sufficient to account for most of the long-term Cenozoic variation in benthic oxygen isotopes (excluding the large compositional changes associated with development of the cryosphere).

How such a degree of long-term variation actually takes place remains unresolved. There are four mechanisms that seem plausible: (1) changes in the isotopic composition of the deep waters at their area of formation; (2) changes in the temperature of salinity characteristics of the deep water masses [which might also be associated with (1) above]; (3) changes in the number of source areas producing deep waters of dissimilar isotopic composition; and (4) changes in vertical structure of the oceans that would lead to fluctuations in hydrographic boundaries between waters of different physical and/or isotopic character.

To evaluate the likelihood of any of these mechanisms operating at a particular time, data are needed from sites that sample a wide depth range in several ocean basins. Such data are not available for the Paleogene but are now being produced for the Neogene. The most common pattern seen in the Miocene data is that of relatively low variability at shallow depths (<1500 m), moderate variability between about 1500 and 4000 m, and increased variability below 4000 m.

High variability in the deepest sites is not seen in Pliocene times. The Early Pliocene has relatively low variability (0.01-0.03) at all depths except between 2000-3000 m. In the Late Pliocene, the data are rather homogeneous within the Atlantic and Pacific Oceans. There is some indication that variability increases slightly with depth and that the Atlantic is more variable than the Pacific. These tendencies are statistically significant, but the data base is not large. During this time interval, northern hemisphere glaciations are thought to have begun (Shackleton and Opdyke, 1977). Although variance estimates for the Pacific Ocean are only slightly less than those of the mid-Miocene ice buildup in Antarctica, the amount of variability estimated for this interval of northern hemisphere glacial buildup is no greater than that found in the shallower waters of the Atlantic Ocean in the Late Miocene and is nearly the same as that estimated for the Pacific Ocean during the early Paleogene. Thus, relatively high isotopic variability may be closely associated with times when major changes occur in the average oxygen isotopic composition of the oceans (such as during glacial buildup), but they may also occur in association with the creation of new, isotopically different types of deep and intermediate water masses. If the later mechanism applies, large differences in the variability of benthic oxygen isotopes may be found between different oceans and between different depths in the same ocean.

The oxygen isotopic data presented here are discussed in terms of only one simple statistical characteristic—its variability. Even with this simple tool, major changes in the deep waters of the oceans can be discerned. Clearly, the record of paleo-oceanographic change is dependent not only on its position in time but also on its geographic and depth location. A more thorough understanding of the variability of the oceans awaits an evaluation of the spectral character of oceanic oxygen isotopic variability that could take into account the relative importance of long-term and short-term oscillations. Questions concerning how such variance spectra have changed with time and the likely mechanisms of such change await the gathering of longer time series and the further refinement of the geologic time scale.

ACKNOWLEDGMENTS

We wish to express our appreciation to the CENOP project scientists for their thoughtful comments and criticisms of this work. Discussions with Sam Savin, Michael Bender, Nick Shackleton, and David Graham were particularly helpful. We also thank Fay Woodruff, Edith Vincent, Robley Matthews, Mike Sommers, Michael Bender, and Sam Savin for making their unpublished data available to us. This research was supported by National Science Foundation grants to the CENOP project, including OCE 79-14594 at the University of Rhode Island.

REFERENCES

Boersma, A., and N. J. Shackleton (1977). Tertiary oxygen and carbon isotopic stratigraphy, Site 357 (mid latitude South Atlantic), in *Initial Reports of the Deep Sea Drilling Project 39*, U.S. Government Printing Office, Washington D.C., pp. 911-924.

Boersma, A., N. Shackleton, M. Hall, and Q. Given (1979). Carbon and oxygen isotope records at DSDP site 384 (North Atlantic) and some Paleocene paleotemperatures and carbon isotope variations in the Atlantic Ocean, in *Initial Reports of the Deep Sea Drilling Project 43*, U.S. Government Printing Office, Washington, D.C., pp. 695-717.

Briskin, M., and W. A. Berggren (1975). Pleistocene stratigraphy and quantitative paleo-oceanography of tropical North Atlantic core V16-205, Late Neogene Epoch Boundaries, *Micropaleontology*, 167-198.

Craig, H., and L. I. Gordon (1965). Isotopic oceanography: Deuterium and oxygen-18 variations in the ocean and marine atmosphere, in *Symposium on Marine Chemistry*, D. R. Schinek and J. T. Corless, eds., Occas. Publ. No. 3 of Narragansett Mar. Lab., Graduate School of Oceanogr., U. of Rhode Island, pp. 277-374.

Douglas, R. G., and S. M. Savin (1973). Oxygen and carbon isotope analysis of Cretaceous and Tertiary foraminifera from the central North Pacific, in *Initial Reports of the Deep Sea Drilling Project 17*, U.S. Government Printing Office, Washington, D.C., pp. 591-605.

Douglas, R. G., and S. M. Savin (1975). Oxygen and carbon isotope analyses of Tertiary and Cretaceous microfossils from Shatsky Rise and other rise sites in the North Pacific, in *Initial Reports of the Deep Sea Drilling Project 32*, U.S. Government Printing Office, Washington, D.C., pp. 509-520.

Douglas, R. G., and F. Woodruff (1981). Deep sea benthic foraminifera, in *The Sea*, Vol. 7, C. Emiliani, ed., Wiley-Interscience, New York.

Fischer, A. G., and M. A. Arthur (1977). Secular variations in the pelagic realm, in *Deep Water Carbonate Environment*, H. E. Cook and P. Enos, eds., Soc. Econ. Paleontol. Mineral. Spec. Publ. No. 25, pp. 19-50.

Heath, G. R., T. C. Moore, and T. H. van Andel (1977). Carbonate accumulation and dissolution in the equatorial Pacific during the past 45 million years, in *The Fate of Fossil Fuel CO_2 in the Oceans*, N. R. Anderson and A. Malahoff, eds., Plenum, New York, pp. 627-640.

Keigwin, L. D., Jr. (1979a). Late Cenozoic stable isotope stratigraphy and paleontology of DSDP sites from the eastern equatorial and central Pacific, *Earth Planet. Sci. Lett. 45*, 361-382.

Keigwin, L. P., Jr. (1979b). Cenozoic stable isotope stratigraphy, biostratigraphy, and paleoceanography of deep-sea sedimentary sequences, Ph.D. dissertation, U. of Rhode Island, 188 pp.

Keigwin, L. D., Jr. (1980). Paleoceanographic change in the Pacific at the Eocene-Oligocene boundary, *Nature 287*, 722-725.

Keigwin, L. D., Jr., M. L. Bender, and J. P. Kennett (1979). Thermal structure of the deep Pacific Ocean in Early Pliocene, *Science 205*, 1386-1388.

Kennett, J. P. (1978). The development of planktonic biogeography in the southern ocean during the Cenozoic, *Mar. Micropaleontol. 3*, 301-346.

Kennett, J. P., and N. J. Shackleton (1976). Oxygen isotopic evidence for the development of the psychrosphere 38 Myr ago, *Nature 260*, 513-515.

Kennett, J. P., N. J. Shackleton, S. V. Margolis, D. E. Goodney, W. C. Dudley, and P. M. Kroopnick (1979). Late Cenozoic oxygen and carbon isotopic history and volcanic ash stratigraphy: DSDP site 284, South Pacific, *Am. J. Sci. 279*, 52-69.

Matthews, R. K., and R. Z. Poore (1980). Tertiary $\delta^{18}O$ record and glacial-eustatic sea-level fluctuations, *Geology 8*, 501-504.

Norton, I. O., and J. G. Sclater (1979). A model for the evolution of the Indian Ocean and the breakup of Gondwanaland, *J. Geophys. Res. 84*, 6803-6830.

Pisias, N. G., and T. C. Moore (1981). The evolution of Pleistocene climate: A time series approach, *Earth Planet. Sci. Lett. 52*, 450-458.

Ruddiman, W. F. (1971). Pleistocene sedimentation in the equatorial Atlantic: Stratigraphy and faunal paleoclimatology, *Geol. Soc. Am. Bull. 81*, 283-302.

Savin, S. M., and H. W. Yeh (1981). Stable isotopes in ocean sediments, in *The Sea*, Vol. 7, C. Emiliani, ed., Wiley-Interscience, New York.

Savin, S. M., R. G. Douglas, and F. G. Stehli (1975). Tertiary paleotemperatures, *Geol. Soc. Am. Bull. 86*, 1499-1510.

Shackleton, N. J. (1967). Oxygen isotope analyses and Pleistocene temperatures re-assessed, *Nature 215*, 15-17.

Shackleton, N. J. (1977). The oxygen isotope stratigraphic record of the Late Pleistocene, *Phil. Trans. R. Soc. London B 280*, 169-182.

Shackleton, N. J., and M. B. Cita (1979). Oxygen and carbon isotope stratigraphy of benthic foraminifers at site 397: Detailed history of climatic change during the Late Neogene, in *Initial Reports of the Deep Sea Drilling Project 47*, U.S. Government Printing Office, Washington, D.C., pp. 433-445.

Shackleton, N. J., and J. P. Kennett (1975a). Paleotemperature history of the Cenozoic and the initiation of Antarctic glaciation: Oxygen and carbon isotope analyses of DSDP sites 277, 279, 281, in *Initial Reports of the Deep Sea Drilling Project* 29, U.S. Government Printing Office, Washington, D.C., pp. 743-755.

Shackleton, N. J., and J. P. Kennett (1975b). Late Cenozoic oxygen and carbon isotopic change at DSDP site 284: Implications for glacial history of the Northern Hemisphere and Antarctica, in *Initial Reports of the Deep Sea Drilling Project* 29, U.S. Government Printing Office, Washington, D.C., pp. 801-807.

Shackleton, N. J., and N. D. Opdyke (1973). Oxygen isotope and paleomagnetic stratigraphy of equatorial Pacific core V28-238: Oxygen isotope temperatures and ice volumes on a 10^5 and 10^6 year scale, *Quat. Res. 3*, 39-55.

Shackleton, N. J., and N. D. Opdyke (1976). Oxygen-isotope and paleomagnetic stratigraphy of Pacific core V28-239 late Pliocene to latest Pleistocene, in *Investigations of Late Quaternary Paleoceanography and Paleoclimatology*, R. M. Cline and J. D. Hays, eds., Geol. Soc. Am. Mem. 145, pp. 449-464.

Shackleton, N. J., and N. D. Opdyke (1977). Oxygen isotope and paleomagnetic evidence for early northern hemisphere glaciation, *Nature 270*, 216-219.

Thompson, C. M., and M. Merrington (1944). Tables for testing the homogeneity of a set of estimated variances, *Biometrika 33*, 296-304.

Vail, P. R., R. N. Mitchum, Jr., and S. Thompson (1977). Global relative changes of sea-level, in *Seismic Stratigraphy—Application to Hydrocarbon Exploration*, C. E. Payton, ed., Am. Assoc. Petrol. Geol. Mem. 26, Part IV, pp. 83-98.

Vergnaud-Grazzini, C. (1979). Cenozoic paleotemperatures at site 398, eastern North Atlantic: Diagenetic effects on carbon and oxygen isotopic signal, in *Initial Reports of the Deep Sea Drilling Project 47*, U.S. Government Printing Office, Washington, D.C., pp. 507-512.

Vincent, E., J. S. Killingley, and W. H. Berger (1980). The magnetic Epoch 6 carbon shift: A change in the oceans $^{13}C/^{12}C$ ratio 6.2 million years ago, *Mar. Micropaleontol. 5*, 185-203.

Woodruff, F., S. Savin, and R. Douglas (1981). A detailed study of the Mid-Miocene isotopic record and its paleoclimatic implications, *Science 212*, 665-668.

Seasonality and the Structure of the Biosphere

20

JAMES W. VALENTINE
University of California, Santa Barbara

INTRODUCTION

This paper traces the major effects of seasonality on organisms, particularly in the sea, from both theoretical and observational data. The tilting of the Earth's axis of rotation away from the axis of the ecliptic produces a regular annual pattern of variation in the solar radiation received on Earth, giving rise to seasons. The seasonality in solar radiation, greatest near the poles and least under the Sun's track, entrains variation in a vast array of physical environmental parameters including temperature, rainfall, atmospheric and oceanic circulation, salinity regimes and nutrient upwelling in the sea, and humidity on land.

These parameters are highly important in the adaptation of organisms, and they affect in turn many other factors of great biological significance. The present pattern of seasonality is reflected in patterns in the biosphere. Patterns of seasonality must have been different in the past; the fossil record of patterns in ancient biospheres can be useful in understanding climatic change and its biological consequences.

POPULATION RESPONSES TO SEASONALITY

Intrinsic Population Features

The logistic equation of population growth is a convenient starting point from which to evaluate seasonal effects on populations. It is usually written

$$dN/dt = rN(1 - N/K),$$

where N is population size, r is the intrinsic population growth rate, and K is the carrying capacity of the environment for that population. A population that is very small with respect to the carrying capacity will grow at a rate set by r until its density approaches K, when growth may be damped and the population size stabilized at an equilibrium value (Figure 20.1A). However, a wide range of other population behaviors is possible (May, 1975, 1979; May and Oster, 1976). If there is a time delay in response by a population to the operative factors in K (Hutchinson, 1948, 1965; May, 1973), the population becomes less likely to attain a stable equilibrium as the time lag in-

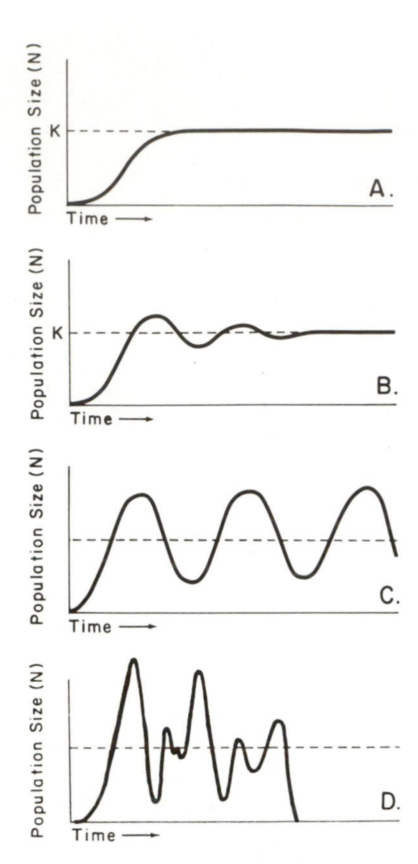

FIGURE 20.1 Diagrammatic representation of the dynamics displayed by populations of overlapping generations with a time delay in the regulatory mechanism. A, If time delay is sufficiently short a population may grow to a stable equilibrium level, K. B, With a larger time delay and sufficiently high r a population may at first penetrate the level of K, to which it may be returned via damped oscillations. C, With still larger time delay and sufficiently high r, a population may oscillate indefinitely in a limit cycle rather than seeking stability at an equilibrium level. D, As time delays increase, and if r is still sufficiently high, the limit cycle breaks down into random oscillations that can lead to extinction of the population. These concepts are reviewed by May (1979).

creases. Most time delays are on the order of a year and are related to seasonality. When r is sufficiently high and the factors in K are sufficiently slow to act, the population size can be driven above K temporarily. K may then be approached by damped oscillations that finally settle on the equilibrium value (Figure 20.1B). If the time lag is greater still, the population may penetrate K to such an extent that the subsequent crash carries it considerably below K; r may then return the population to its former peak level from which it will again crash, and the oscillations may continue indefinitely (rather than being damped) in a stable limit cycle (Figure 20.1C). Finally, if the time lag continues to increase, the stable cycle will break down into chaotic population density variations, the bounds of which increase so that at the lower bound the population size is below some critical low level from which it cannot recover; extinction ensues. Thus, intrinsic population properties even in a constant environmental regime can lead a population to range in behavior from stable to chaotic. Real populations must limit their downward fluctuations above some critical extinction point, and indeed to a point from which they may recover sufficiently to cope with subsequent inclement conditions.

Seasonal variation in environmental factors further complicates population dynamics. In general, the greater such intrinsic variation, the less the population, which can be driven through the same wide range of behavior as discussed above (see May, 1979). All real environments vary to some degree, evoking appropriate adaptive responses on the part of populations. Seasonality is the most regular pervasive source of environmental variation. It is logical to ask, what sorts of adaptations do populations evolve in order to cope with seasonality, and what are the consequences? Adaptations to seasonality are referred to here as *seasonal strategies*.

Extrinsic Density-Independent Factors

To endure in seasonal environments, populations must respond in ways that maintain their oscillations within some limited range. Two sorts of extrinsic factors are recognized (Smith, 1935). *Density-independent* factors affect organisms without regard to whether there are few or many individuals present. Most effects of physical climatic factors are of this sort. The lethality of a temperature change, for example, does not ordinarily depend on population size. *Density-dependent* factors, on the other hand, have increasing effects as population size increases. They form components of K in the logistic equation. Many biological and some physical factors are of this sort; commonly they involve a factor, such as food supply or habitat space, that can be used up. However, as emphasized by Andrewartha and Birch (1954), any factor that is density-independent can become density-dependent in some circumstances. It is better to speak of density-dependent and -independent *processes* or *effects*.

Density-independent processes play an important role in shaping the individual tolerances of organisms and therefore the modal niches of populations. This must be particularly true of seasonal factors because they impose annually recurrent effects, although with somewhat variable intensities. Selection arising from density-independent processes tends to act most severely near the margins of species' distributions, where even small departures from normal conditions may exceed the tolerances of most individuals. It causes "stabilizing" or "centripetal" evolution. If the environmental regime changes, selection must adjust to tolerances and rates so as to be adaptive to the new conditions; this causes "directional" evolution. If the changes are too large or abrupt for evolution to track, extinction ensues.

Although they act independently of population size, it is possible for density-independent factors to control population densities at levels below those at which density-dependent effects occur (Andrewartha and Birch, 1954). For example, density-independent mortality may occur so frequently that populations never reach their carrying capacities. In this case it is postulated that selection favors a high-reproductive potential (r selection). When environments are relatively stable and populations at their carrying capacities, on the other hand, it is postulated that selection favors adult efficiency in utilizing resources (*K* selection) (MacArthur and Wilson, 1967).

Extrinsic Density-Dependent Factors

The strategies that can be employed in response to seasonality have been investigated by Boyce (1979). He modeled the relative fitness of individuals with different combinations of three properties that seem particularly relevant to density-dependent selection: (1) intrinsic reproductive potential (r); (2) level of resource demand (D); and (3) population decay when resources are inadequate (z)—almost the opposite of (r). Boyce calculated the relative fitness of individuals with contrasts in these properties in different regimes of seasonality and at different population densities. The results are displayed in Figure 20.2.

In case A (Figure 20.2), individuals with higher r have the higher resource demand but resistance to inclement conditions is equal, compared with low-r individuals. The high-r individuals have the higher fitness over a range from low to high seasonality, except at very high population densities, when their high resource demand becomes a significant liability. Case D is similar in that the high-r individuals have the higher resource demand, but they have more resistance to inclement conditions than do low-r individuals. High-r individuals increase their margin of fitness in more highly seasonal conditions at low population densities. At very high densities, however, the high resource demand still poses a significant liability and individuals with the lower r become the more fit.

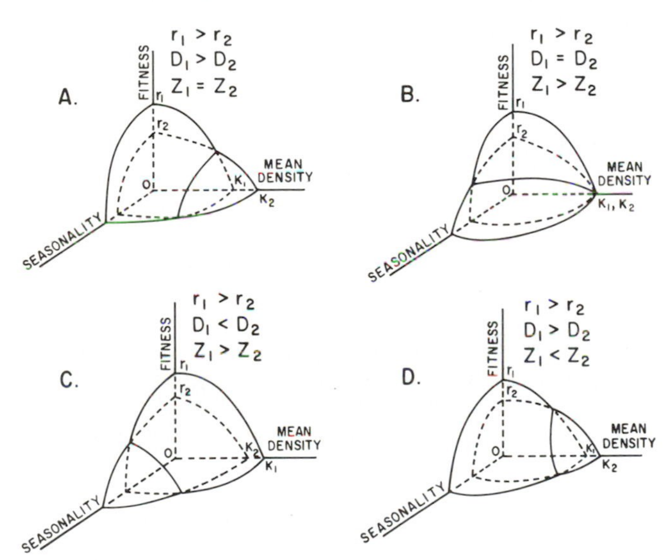

FIGURE 20.2 Models of the reproductive potential (r) and carrying capacity (K) of populations over a range of seasonalities and resource supplies (resource supplies are inversely proportional to mean densities). D is resource demand, Z is rate of population decrease. The population with the higher reproductive potential is always the fitter at low seasonalities, and it is also the fitter at high seasonalities if Z is equal to or lower than in the other population (cases A and D). However, the population with the greater resistance to decrease (smaller Z) is the fitter under highly seasonal conditions (after Boyce, 1979, copyright U. of Chicago Press).

Cases B and C have rather different outcomes. In both of these cases the high-r individuals have less resistance to inclement conditions than do low-r individuals. These high-r individuals remain most fit in the less seasonal environments but are least fit in highly seasonal conditions except at high densities. In case B both kinds of individuals have equal resource demands and equal fitness at the highest density. In case C the low-r individuals have the higher resource demands and therefore have less fitness at high densities.

It is evident from Figure 20.2 that adaptation is more difficult in environments with greater seasonality. Fitness always falls with increasing seasonality. Similarly, the carrying capacity is always lower in highly seasonal environments (as indicated by the curves of the fitness dimension of the seasonality-fitness faces in Figure 20.2). Similarly, the carrying capacity is always lower in highly seasonal environments (as indicated by the curves of the "mean density" dimension on the basal seasonality-mean density faces in Figure 20.2). The two main seasonal strategies are (1) to increase reproduction and (2) to fortify the populations against decrease during inclement conditions. The reproductive strategy has no special damping effect on population oscillations but prevents extinction by returning the population size rapidly toward K or beyond. It might result in rather large size fluctuations, but if successful, constrains them within limits (the lower significantly above zero) to give rise to a limit cycle. In the model these strategies (reproduction and fortification) seem about equally effective, although when greater resource demands are involved the fortification strategy is least affected.

Observation of species in highly seasonal environments reveals that these strategies are in operation in nature (Boyce, 1979). Mammals provide many terrestrial examples. Some have larger litter sizes in more seasonal (high-latitude) environments (Lord, 1960), following the strategies indicated by cases A or D. Other mammals in seasonal environments follow the fortification strategies. Some, such as hibernating forms, put energy into resource storage rather than into reproduction, following the strategy of case B; others grow to larger body sizes in seasonal regimes, buffering the effects of seasonality and thus following the strategy of case C.

Marine Populations and Seasonality

The fossil record of the marine shelf and shallow seas contains the most diverse fauna and embraces the greatest time span of all environmental realms. It is thus of interest to determine the seasonal strategies of shelf invertebrates. In the shelf realm there is a trend from high-reproductive potentials among species in regions with little seasonality to low-reproductive potentials in seasonal regions (Thorson, 1950; Mileikovsky, 1971)—just the opposite of what is expected of the reproductive strategies in cases A or D. Plausible explanations have been proposed for this trend in r. In less seasonal regions (such as the tropics) there is a relatively stable resource base in the water column—photosynthesis goes on at about the same intensity all year—and most shelf species have planktonic larvae that feed in the water column (planktotrophic larvae). Such species can produce many eggs, because nourishment need not be furnished to the larvae. Planktotrophy is probably favored by selection because it aids dispersal and enhances recruitment in an environment densely populated and rich in rather specialized species and therefore patchy in opportunities. Invertebrates in stable shallow-water environments tend to be both *r*- and *K*-selected (Valentine and Ayala, 1978). In highly seasonal regimes (such as in high latitudes) most benthic invertebrates have a relatively small number of large yolky eggs; the young feed in the egg but hatch as nonfeeding larvae or as miniature adults. Each egg requires a considerable investment in energy and therefore fecundity is low, but clearly the chances of reproductive success per egg are higher than for planktotrophic forms, provided as they are with food and more or less protected during development, often by brooding. In terms of Boyce's models, highly seasonal marine invertebrates follow case B or C strategies; the difference depends on whether their reproductive effort is equal (B) or greater (C) than invertebrates in less seasonal environments. It thus appears that marine invertebrates in seasonal environments are neither *r*- selected nor *K*-selected (Valentine and Ayala, 1978) but follow a strategy of developmental fortification.

Why this particular strategy has been selected, rather than an *r* strategy, is probably explained by the energetics of fluctuating populations. Imagine a population in a seasonal environment that is subject to such an inclement season as to suffer fairly heavy mortality—*x* percent. This must be made up via reproduction and (assuming deaths are random with respect to age) growth. In a more extreme (more highly seasonal) environment, mortality might be $2x$ percent. To make up for this heavier mortality, reproduction and growth must be correspondingly greater, and more energy must be consumed to support this greater effort. In each case the population has only until the next inclement season to return to a condition sufficient to cope with the accompanying wave of mortality. In general, then, under conditions of seasonally imposed mortality, the more a population fluctuates, the higher the cost in terms of energy use.

Because carrying capacities and fitnesses are low in highly seasonal environments (Figure 20.2), additional properties may be required of successful species. One strategy is to become ecologically generalized, occurring over a broad region in a wide variety of habitats and indulging in a catholic diet. In this way a species in a seasonal regime may maintain a large population but will cover the functional range of number of specialized species in stable environments. In a sense such adaptations represent a fortification strategy, since during inclement periods generalists are likely to be able to subsist on such food items as happen to be available and may occur in areas and in habitats that are least severely affected by the adverse conditions. It seems likely that a generalist strategy would maximize fitness under the circumstances.

DIVERSITY AND SEASONALITY

Diversity (species richness) patterns have attracted wide attention and controversy. They are often interpreted as climate related and if so would be useful in paleoclimatology for they represent one aspect of biotic structure that can be inferred from the fossil record. They are also of potential interest in predicting biotic response to future climatic change.

The strategies outlined above permit the survival of permanent communities in the "boom-and-bust" economies of highly seasonal environments, even with their low carrying capacities. At base, species persist by minimizing the effects of the seasonal fluctuations. The seasonal strategies, however, are resource-intensive. The available trophic resources must be apportioned in large shares. This places a limit on the number of species that can be accommodated in communities in seasonal regimes. In regions of low seasonality, resources are continuously available or nearly so, and carrying capacity is much higher per unit of resource. The resources may therefore be partitioned finely among numerous different species, which can have small populations and specialized habits since they can rely on a steady resource supply and need not regularly undergo energetically expensive fluctuations in population size.

Insofar as can be told from available data, the correlation of seasonality and diversity seems high in modern oceans (Valentine, 1971, 1973). The well-known latitudinal gradient in diversity parallels the latitudinal gradient in seasonality. Additionally, patterns of trophic resource seasonality that are created by hydrographic factors—alternations of nutrient-rich and nutrient-poor waters—are particularly important in ocean climates, because perturbations of the oceanic water column lead directly to nutrient supply (net upwelling) or removal (net downwelling). Horizontal currents also change nutrient-poor for nutrient-rich water on a seasonal basis in some regions (as along the northeastern American coast).

It would obviously be much to the point to compare the modern pattern of marine diversity, normalized to some standard measure, with the pattern of seasonality of productivity in the oceans today. Diversity data, though spotty, are improving and are probably sufficient to determine broad global patterns. The pattern of seasonality of productivity, however, is not known from observations, which are restricted to only a few scattered localities (there is a great lack of winter data). It must be inferred from general principles and from the hydrographic patterns. Not to be possessed of actual observations sufficient to produce a map of global marine seasonality of productivity is a serious gap in our knowledge of the oceans and of the global ecosystem. Generally, diversity displays an inverse relation to nutrient seasonality, just as it does to solar seasonality, whenever trends can be inferred (Valentine, 1973).

Factors other than seasonality affect local patterns of species richness. Some factors act to restrict the sizes of populations so as to release resources than can be utilized by additional populations. Predators may have this effect (Paine, 1966), and physical disturbances can also free resources, often by increasing spatial patchiness (Dayton, 1971). There is no evidence that global trends in predation or in disturbance regulate global diversities, however. Other possible diversity controls have been suggested, but when all are considered it seems reasonable to conclude that the effects of seasonality have the largest claim to both theoretical and empirical bases (so far as the data are available) sufficient to explain global diversity patterns.

SEASONALITY PAST AND FUTURE

If seasonality is indeed the major determinant of global diversity patterns, then the major diversity trends as registered in the fossil record should indicate past patterns of seasonality. Diversity trends can be observed when the record is studied on a hemispheric or global scale (for example, Stehli, 1970). Furthermore, techniques are available to test hypotheses of seasonality based on diversity patterns. Seasonal growth rhythms are commonly recorded by tree rings and by shell-growth increments in shallow marine invertebrates. Indeed, rhythmic patterns in the growth of stromatolites (Panella, 1976) and in the deposition of iron formations (Trendall and Blockley, 1970) are interpreted as suggestive of seasonality in rocks over 2 billion years old.

Although seasonal patterns have not yet been systematically followed through time, it is clear from the data available that seasons have varied considerably. This has probably been in large part due to effects of plate tectonic processes, to the changing geographies of land and sea, to mountain ranges, to ocean currents, and to sea-level fluctuations. Biogeographic patterns are strongly affected by such changes, as indicated by other chapters in this volume. Seasonal patterns and their biotic reflections should also be affected.

In the geologic record there are indications that some periods were characterized by relatively broad, equable climates, when tropical (or at least low-latitude) faunas penetrated much further poleward than they do today, and latitudinal provinciality was relatively low. Indeed, during much if not all of the pre-Pleistocene, the record suggests a weaker latitudinal marine provinciality than today's (Valentine *et al.*, 1978). At some times, as in the Jurassic (see Chapter 17), there seem to have been only one or two marine provinces between equator and pole. Three or four may be inferred at other times (perhaps during the late Cretaceous). Today, by contrast, there are six or seven separate marine-shelf provinces along the northeastern Pacific shelf between the equator and pole (Valentine, 1966).

An important question is, how are patterns of marine seasonality affected when climates broaden? Because we can probably assume that the gradient in seasonality of solar radiation has been about the same during the Phanerozoic, the question becomes, how are patterns of seasonality of nutrient supplies affected when climates broaden? It is possible to develop hypotheses that suggest a general reduction of nutrient seasonality with warmer poles and broader climates, although the subject is so complicated (see Chapter 14) that the opposite notion, relating higher nutrient seasonality to a broader climate, cannot yet be ruled out. The question might be solved by contrasting diversity patterns of more broadly zoned with those of more narrowly zones periods. Additional important questions concern other factors that can significantly alter the nutrient regime and therefore the seasonal pattern of productivity in the world ocean. Such factors may include the pattern of continentality, of narrowness or breadth of oceans, of the width of continental shelves (the height of sea level), the prevalence of east-west versus north-south coastlines, and the geography of ocean gateways. Many such questions can probably be tested, with care, from fossil-diversity data.

In conclusion, seasonality appears to be a significant parameter in structuring the biosphere, and it would seem reasonable to encourage research efforts in two main areas. One is in recording and interpreting the pattern of seasonality of productivity in the present oceans; this is one of the more important and basic pieces of information about how the present world operates and its collection is quite technically feasible. The second area is in determining the global patterns of marine diversity of the past. When local effects are filtered out, these patterns should reflect the patterns of marine seasonality. Their relations can then be tested against the breadth of past climates and other relevant factors and employed in constructing models of pre-Pleistocene climates.

REFERENCES

Andrewartha, H. G., and L. C. Birch (1954). *The Distribution and Abundance of Animals*, U. of Chicago Press, Chicago, 789 pp.

Boyce, M. S. (1979). Seasonality and patterns of natural selection for life histories, *Am. Nat. 114*, 569-583.

Dayton, P. K. (1971). Competition, disturbance, and community organization: The provision and subsequent utilization of space in a rocky intertidal community, *Ecol. Monogr. 41*, 351-389.

Hutchinson, G. E. (1948). Circular causal systems in ecology, *Ann. N.Y. Acad. Sci. 50*, 221-246.

Hutchinson, G. E. (1965). *The Ecological Theatre and the Evolutionary Play*, Yale U. Press, New Haven, Conn., 139 pp.

Lord, R. D. (1960). Litter size and latitude in North American mammals, *Am. Midl. Nat. 64*, 488-499.

MacArthur, R. H., and E. O. Wilson (1967). The theory of island biogeography, *Princeton Monogr. Pop. Biol. 1*, 1-203.

May, R. M. (1973). Time-delay versus stability in population models with two and three trophic levels, *Ecology 54*, 315-325.

May, R. M. (1975). Stability and complexity in model ecosystems, *Princeton Monogr. Pop. Biol. 6*, 1-265.

May, R. M. (1979). The structure and dynamics of ecological communities, in *Population Dynamics*, R. M. Anderson, B. D. Turner, and L. R. Taylor, eds., Blackwell Science Publishers, Oxford, pp. 385-407.

May, R. M., and G. F. Oster (1976). Bifurcations and dynamic complexity in single ecological models, *Am. Nat. 110*, 573-599.

Mileikovsky, S. A. (1971). Types of larval development in marine bottom invertebrates, their distribution and ecological significance: A re-evaluation, *Mar. Biol. 10*, 193-213.

Paine, R. T. (1966). Food web complexity and species diversity, *Am. Nat. 100*, 65-75.

Panella, G. (1976). Geophysical inferences from stromatolite lamination, in *Stromatolites*, M. R. Walter, ed., Elsevier, Amsterdam, pp. 673-685.

Smith, H. S. (1935). The role of biotic factors in the determination of population densities, *J. Edon. Entomol. 28*, 873-898.

Stehli, F. G. (1970). A test of the Earth's magnetic field during Permian time, *J. Geophys. Res. 75*, 3325-3342.

Thorsen, G. (1950). Reproductive and larval ecology of marine bottom invertebrates, *Biol. Rev. 25*, 1-45.

Trendall, A. G., and J. G. Blockley (1970). The iron formations of the Precambrian Hamersley Group, Western Australia, *Geol. Surv. Western Australia Bull. 199*, 1-365.

Valentine, J. W. (1966). Numerical analysis of marine molluscan ranges on the extratropical northeast Pacific shelf, *Limnol. Oceanogr. 11*, 198-211.

Valentine, J. W. (1971). Plate tectonics and shallow marine diversity and endemism, an actualistic model, *Syst. Zool. 20*, 253-264.

Valentine, J. W. (1973). *Evolutionary Paleoecology of the Marine Biosphere*, Prentice-Hall, Englewood Cliffs, N.J., 511 pp.

Valentine, J. W., and F. J. Ayala (1978). Adaptive strategies in the sea, in *Marine Organisms: Genetics, Ecology and Evolution*, B. Battaglia and J. A. Beardmore, eds., Plenum, New York, pp. 323-346.

Valentine, J. W., T. C. Foin, and D. Peart (1978). A provincial model of Phanerozoic marine diversity, *Paleobiology 4*, 55-66.

21 Paleozoic Data of Climatological Significance and Their Use for Interpreting Silurian–Devonian Climate

ARTHUR J. BOUCOT
Oregon State University

JANE GRAY
University of Oregon

INTRODUCTION

In principle, enough geologic data of climatological import have been available for many years to enable climatologists to attempt global climatic reconstructions for the Paleozoic. However, only in the past few decades, particularly with the advent of plate tectonic concepts, has this information begun to be synthesized in a manner that makes it usable to the climatologist. Lithofacies data with climatic implications, for example, are now available for the Paleozoic on a global scale, and biogeographic data are being synthesized for the major time intervals. But most classes of information are not adequately compiled, and others for which syntheses exist are for such relatively broad time intervals that they will confuse attempts to gain climatological understanding. Our purpose in this paper is to review categories of information useful in reconstructing Paleozoic climates. We begin with those most readily available and most important and conclude with some that, while little exploited, may provide important supplementary data. We present some examples of how these data have been used climatologically in the Paleozoic, and finally, using several categories of information, we provide an example of how these data might be used to interpret Silurian-Devonian climates.

PALEOZOIC CLIMATIC INFORMATION

Biogeography

The data of Paleozoic biogeography are readily available in several volumes, edited by Middlemiss *et al.* (1971), Hallam (1973), Hughes (1973), Ross (1974), and Gray and Boucot (1979). These volumes include papers covering many Paleozoic organisms but chiefly shallow-water, marine invertebrates. The coverage is by no means encyclopedic, but it is extensive enough to provide an approximate biogeography for the Paleozoic Periods.

What data does biogeography provide for the climatologist? In view of what we understand of modern biogeography we can conclude that the animals and plants of any biogeographic unit were in reproductive contact and that the environment within the unit had a certain level of uniformity. Thus, a shallow-water biogeographic unit of continental-shelf depth based on sea shells (as are most of the Paleozoic units) indicates a water mass having relatively uniform properties and a shallow-water circulation system permitting reproductive communication. Although the larval stages of some marine benthos are capable of long-range transport within the plankton (teleplanic species), it appears that most species do not belong to this category. The taxa characterized by long-range transport and consequently more cosmopolitan distribution do, however, yield information about some aspects of long-range, shallow-current circulation.

In the nonmarine environment, Paleozoic data are almost exclusively for post-Devonian woody plants. Animals, vertebrate and invertebrate, of the nonmarine environment are too rare as fossils to be of practical value, although future study may alter this condition. With plants we deal chiefly with megascopic structures, but global studies of pollen and spores are beginning to provide supporting data.

Additionally, Paleozoic biogeography provides direct information about climates and climatic gradients for the Periods. Taxonomic diversity at all levels from the species through the superfamily tends to be significantly lower in cold or cool (temperate to glacial) climates than in warm (warm temperate to tropical) climates (Stehli *et al.*, 1967; Stehli, 1968). Boucot (1975) summarized diversity data of this type for the Silurian and Devonian. Warm and cold or cool biogeographic units may also be indicated by the diversity present in benthic marine animal communities. Cold- to cool-water communities tend to have a much smaller number of species per community than do warm-water communities. There is also a tendency for the shells of organisms of cold- or cool-water units to be smaller and thinner than those of warm water units (Nicol, 1967).

The climatologist should recognize that levels of provincialism increase and decrease during the Paleozoic (see Boucot and Gray, 1979, for a brief summary of provincialism from the Cambrian through the Permian). These changes are probably the results of many interacting forces such as changes in global climatic gradients, paleogeography, and shallow-water circulation systems.

Carbonate-Noncarbonate Lithofacies

Boucot and Gray (1979, 1980) emphasized in general terms the climatic importance of carbonate and noncarbonate sedimentary rock sequences. Heckel and Witzke (1979, pp. 99-103) provided a detailed discussion of how carbonate rocks may be used for purposes of climatic analysis in the Paleozoic. Evidence indicates that carbonate rocks (limestone and dolomite of sedimentary origin) usually denote tropical to warm temperate conditions and that regions lacking such rock types in the marine facies probably were temperate to cooler. We have summarized above some biologic data that parallel the lithofacies evidence. Cool-climate, noncarbonate sedimentary rocks commonly are richer in unweathered mica (which imparts a glitter to the bedding planes) than warm climate noncarbonate rock sequences. When noncarbonate sequences are found in warm climates they are commonly associated with at least some redbeds (marine and nonmarine) and other evidence favoring an interpretation of warm climate.

However, climatic interpretation of a noncarbonate rock sequence requires a large sample; it cannot be based on single rock samples, rocks of a single roadcut, or rocks of a single limited geographic area. When an entire region has been studied, however, the climatic conclusions have a high degree of reliability.

Coal

The climatic significance of coal has been discussed briefly by Boucot and Gray (1980), who cited previous reviews by Krausel (1964) and Schopf (1972) dealing with the distribution and paleoclimatic significance of coal. Coal deposits indicate high humidity in association with *either* high or low temperature, as well as conditions adequate for the preservation of organic material. Similar observations were made by Davis (1913).

In the Paleozoic, coals occur in both types of climatic regimes—in the cool, high-humidity climate of the Gondwana Realm of the Late Carboniferous-Early Permian and in the major warm, high-humidity climate of the extra-Gondwanic Realms (Gray and Boucot, 1979). Heckel (1977) showed that the North American distribution of major warm-climate Pennsylvanian coals is well removed from contemporary evaporites. The areal distribution of coal belts relative to evaporite belts as well as scattered paleosols can thus provide climatic information. Heckel's approach assumed that local orographic, rain-shadow effects will probably not affect the geologic record enough to prevent interpretation of broad climatic belts. However, some Paleozoic coal deposits will probably have to be explained in terms of local orographic perturbations. Heckel (The University of Iowa, personal communication, 1981) suggested, for example, that thin, nonminable coals and evaporites of Pennsylvanian age in Colorado were deposited in the rain shadow of the ancestral Rockies, although other interpretations are possible. Knowledge of local phenomena should make an orographic appeal rational in some cases.

Meyerhoff (1970) provided an excellent global summary of the distribution of Paleozoic coal deposits. Coals are virtually absent prior to the Late Devonian, when, presumably for the first time, enough land plant debris became available in suitable preservational sites to provide material for extensive coal deposits.

Evaporites

Evaporites, derived chiefly from marine water, form today in arid regions with ready access to the sea. In addition to an arid climate in which evaporation exceeds freshwater input, it is necessary that there be limited circulation between the evaporating basin and the oceanic reservoir lest the waters recombine before those of the evaporating basin approach salinities compatible with precipitation of evaporite minerals. Thus, evaporite deposits may be absent in regions believed from other evidence to have been arid during the Paleozoic. Moreover, the presence of evaporites during every Paleozoic Period does not imply that all parts of the Period were arid. For example, during the Silurian, an interval of about 30 million years (m.y.), evaporites are widespread in middle latitudes during the Late Silurian but are rare during the Early Silurian. In addition, the width of past evaporite belts is partly a function of the global climatic gradients present during each time interval. For a further discussion of the climatic significance of evaporites see Chapter 10.

Evaporite deposition is incompatible with the climatic conditions characteristic of coal formation. Therefore, the distribution of coeval coal and evaporite deposits can provide an index to humid and arid climatic belts (Heckel, 1977; Heckel and Witzke, 1979), although coals are not sufficiently abundant prior to the Late Devonian for this purpose. Meyerhoff (1970) provided a useful Period-by-Period summary of Paleozoic evaporite distribution.

Redbeds

In the geologic record, a high correlation exists between warm climate indicators and the presence of both marine and nonmarine redbeds. Redbeds are defined here as detrital sedimentary sequences containing many red interbeds, as well as yellow and orange rocks. Geologic-mineralogic studies indicate that much of the red, yellow, and orange pigment consists of iron minerals oxidized under near-surface terrestrial conditions before being incorporated in the sedimentary record (Walker, 1967, 1975). However, older redbeds may be eroded and redeposited under climatic conditions entirely different from those under which they formed. Modern intertidal-shallow subtidal red sediments found near the head of the Bay of Fundy, Nova Scotia, and derived from the weathering and redeposition of Triassic beds are a good example of this phenomenon. In addition, there are diagenetic processes capable of producing red, yellow, and orange minerals that have little to do with the ordinary surface climatic condition responsible for the formation of these pigments.

Gray and Boucot (1979) reviewed some of the previously synthesized data relating to the distribution of Paleozoic redbeds. Global summaries for many of the Periods do not exist, although many raw data are available in the geologic literature. Nor are syntheses available for the distribution of redbeds within the Paleozoic Periods, even from secondary sources, such as regional geologies. What data have been synthesized, however, are consistent with the conclusion that redbeds are indicators of warm climates both past and present. For additional discussion of climate and redbeds, see Chapter 11.

The presence of redbeds in one region for a part of any Paleozoic Period does not necessarily indicate that conditions suitable for redbed formation characterized that region during the entire Period. The Paleozoic Periods are lengthy, and conditions changed from place to place within as well as between Periods.

Plant Morphological Features

A variety of gross morphological and anatomical characters recognizable in Paleozoic plant megafossil remains represent adaptations to environmental parameters that can be climatically interpreted (see Potonie, 1911; White, 1913, 1925, 1931; Noe, 1931; Krassilov, 1975, for Paleozoic examples). The basis for such interpretations are the autecological adaptations found among living plants, many of which are summarized by Daubenmire (1974), Richards (1979), and others. To use characters found in fossil plants as a basis for paleoenvironmental interpretations, it is necessary to assume that plants growing under similar circumstances and stresses have always adapted in the same morphological and anatomical manner. It is also necessary to assume that analogous features in fossil plants carry the same environmental connotations that they do in living plants. It must be borne in mind, however, that the adaptive significance of morphological characters of largely extinct plants, unrelated or only distantly related to living plants, may be different, particularly if there is conflicting biological and/or physical information. Moreover, the precise climatic variable—for example, temperature or moisture—to which the adaptation is a physiological response is not always clear, even when interpreting varied morphological features in modern plants (see Dolph and Dilcher, 1979). Finally, the physiological response to different climatic variables, perhaps even the opposite climatic variable, may lead to the same morphological adaptation (White, 1931, p. 272). However, when the varied morphological and anatomical data reinforce one another, and when they reinforce and bolster other data from the physical environment, they can prove an important source of paleoclimatic information.

Leaves are often regarded as among the most sensitive of plant structures to climatic conditions as they are the most exposed; roots the least as they are seldom exposed. There are specific types of root adaptations, however, that provide definite environmental information. Among the varied morphological and anatomical sources of environmental data in the Paleozoic are diverse features related to leaves (size, texture, gross morphology, and anatomy), tree rings, and other anatomical features connected with woody stems, varied adaptations connected with reproductive structures and roots, and growth habits.

Specific attributes of leaves that are regarded as environmental adaptations are texture (soft, delicate, thick, leathery, or coriaceous), size, surface hairs, scales and glands, the arrangement, number and position of stomates and varied anatomical adaptations related to cell size and cell wall thickness, intercellular spaces, and the development of certain tissue types. Two examples illustrate how this information may be interpreted environmentally. Many of the leaves of the Hermit Shale (Permian, Arizona) are very thick and leathery and have a scaly covering. White (1929, p. 21) interpreted such information, which reinforces other data bearing on the climate of that area, as indicating "a semi-arid climate with a long dry season. . . ." The "reduced, coriaceous and generally densely villous leaves" of Mississippian plants from Illinois suggested to White (1931, p. 272) similar unfavorable conditions of growth with a climate "characterized by severe droughts," or with soils possibly "overdrained during dry seasons," providing in either case evidence of seasonality.

Tree rings or growth rings in woody plants have been used extensively as a basis for making climatic deductions throughout geologic time. Compilations of tree-ring records for various intervals of geologic time are provided by Goldring (1921), Antevs (1925), and Chaloner and Creber (1973). Such rings are known to be common in living woody plants in environments with marked seasonality, whether wet-dry seasonality or a seasonality that reflects temperature changes. In these circumstances, tree rings represent annual increments of wood laid down following a period of growth dormancy. Conversely, the absence of growth rings is believed to imply a more equable climate, in which growth occurred throughout the year.

In deducing information about ancient climates from tree rings, it is important to recognize that rings of a nonannual kind may be correlated with a variety of factors including fire, drought, disease, frost, floods, and defoliation among others. Antevs (1925), Tomlinson and Craighead (1972), and Chaloner and Creber (1973) discussed possible limitations to the use of tree rings in environmental interpretation. Tomlinson and Craighead (1972) and Chaloner and Creber (1973) noted a variety of nonclimatic factors that influence growth ring formation. In addition, it is essential to recognize that the presence or absence of tree rings in woody stems may be an unreliable guide in some circumstances to present climatic conditions, and by implication to climatic conditions of the past. For example, Tomlinson and Craighead (1972) noted that in the distinctive woody flora of subtropical south Florida, which mixes tropical and temperate plants, there is so much variety and variability in growth ring formation that it is difficult to find any climatic correlation. As a result of their studies in this region they write ". . . trees within a single climatic zone and vegetation type may or may not exhibit growth rings. . . . One is forced to the conclusion that the ability to develop growth rings is primarily determined by the genetic make-up of the individual species and only in a limited number of species is there a correlation with climate such that one distinct ring per year is produced" (Tomlinson and Craighead, 1972, p. 49).

The Paleozoic examples that follow presume that the presence of growth rings implies seasonality and the absence of growth rings lack of seasonality. The possible complications in this interpretation, noted above, should be borne in mind, especially where there may be conflicting biological and/or physical information. Arnold (1947, p. 391) provided a striking illustration of a stem of *Callixylon* from the Upper Devonian of New York State that shows what appear to be well-developed growth rings. However, Chaloner and Creber (1973) discussed other examples of *Callixylon* from Upper Devonian strata, including materials from the eastern and central United States and Europe, that show no growth rings or only very obscure growth rings. Chaloner and Creber (1973) concluded from their limited survey of Devonian woody plants that some show growth rings, but they suggest that the rings are less pronounced than would be expected from woody plants at comparable latitudes at this time. White (1931) used the slight development or even "obscurity of annual rings" in Carbondale age floras of Indiana and Illinois together with other features of the vegetation to suggest general equability of temperature. Plumstead (1963) reported an example of silicified wood from Lashly Mt. (Antarctica) of Middle to Late Devonian age that shows "closely set annual rings indicating slow growth and marked seasons." Examples of Permian gymnospermous woods from Antarctica that show broad and well-marked growth rings are provided by Schopf (1972; see also Plumstead, 1965).

Other anatomical features of fossil wood with possible climatic significance are discussed by Noe (1931).

The habit of cauliflory (bearing of reproductive organs directly on the stem) is typical today of many tropical rainforest trees. Although some examples of cauliflory are known for the Paleozoic (see White, 1913, 1931; Noe, 1931), Potonie (1953, as cited in Krassilov, 1975, pp. 118-119) suggested that the habit in Paleozoic plants may have been an adaptation specifically for protection of the reproductive organs against heavy rain, rather than having any temperature significance.

Roots are regarded as among the least sensitive of the major plant organs to climatic variables. There are special cases, however, where they have environmental significance as, for example, in the occurrence of pneumatophores or breathing roots found in swamp plants. The presence of these structures or of subaerial roots of the type found growing flat near the ground surface suggest poorly aerated soil and plants growing in bogs or swamps with a shallow but permanent water cover. The anatomical structures of such roots (i.e., the presence of air chambers) may confirm their function of regulating air supply in a swamp environment. Aerial or stilt roots of the type found among modern mangroves have also been reported for some Paleozoic plants. The complex of anatomical and morphological features found in the roots of *Amyelon*, a Pennsylvanian cordaitalean, for example, can only be matched among modern mangroves. This conclusion led Cridland (1964, p. 201) to suggest that this gymnosperm must have occupied a similar habitat in "tropical or subtropical saline swamps of sheltered marine shores and estuaries. . . ." Additional examples of the use of roots in interpreting the environment of Carboniferous plants are provided by Potonie (1911), White (1913, 1925, 1931), and Noe (1931).

Luxuriance of growth as predicted from size and abundance of plants has also been used as a basis for climatic conclusions. The small size of plants from the Hermit Shale (Permian, Arizona) reinforces sedimentary data relative to the unfavorable climate under which these plants lived (White, 1929). The stunted appearance of Chester age plants from Illinois and Indiana, together with the general poverty of the flora and the presence of xerophytic characters, led White (1931) to suggest unfavorable conditions of growth. By contrast, the "lush" Carbondale floras of Indiana and Illinois together with the large size of the leaves and trunks were "proof of ample rainfall." Dilation of tree bases, a condition found among modern swamp plants, also occurs in certain Carboniferous plants (*Calamites* and Sigillarians) and has been used to reinforce interpretation of their growth in permanent swamps.

An increase in plant size during the Paleozoic (as a general index to luxurious growth) might be interpreted to indicate more favorable conditions for growth relative to the availability of soils, oxygen content of the atmosphere, and insulation from ultraviolet radiation. Chaloner and Sheerin (1979) summarized data related to the maximum observed axis diameter of Late Silurian and Devonian plants. They noted that the potential for tree size was achieved at least by the end of the Devonian. Chaloner and Sheerin also pointed out that more complex plant communities would have been possible as a result of an increase in plant size, because stratification possible with plants of different sizes would have led to "different micro-environments at different levels of light intensity and humidity."

Land Plants

During the Early Paleozoic we have the first evidence for the presence of higher, green land plants and, following them, land animals. Although the time interval for the initial advent of land plants remains in question, the earliest known vascular plant megafossils have been found in the Wenlock (early Late Silurian) of Great Britain (Edwards and Feehan, 1980) and possibly in the late Llandovery or Wenlock of North Africa (Boureau *et al.*, 1978); the earliest specimens attributed to nonvascular land plants in the Llandovery (Early Silurian; Pratt *et al.*, 1978) and records of varied land plant microfossils (trilete spores, spore tetrads, and cuticle remains) have been found from Pridoli (latest Silurian) to the Caradocian portion of the Ordovician (Middle Ordovician; Gray and Boucot, 1971, 1978; Gray *et al.*, 1974, 1980). From this information it can be concluded that the land estate was possibly gained as early as the Middle Ordovician. This benchmark has varied meaning to the climatologist with regard to the oxygen budget, CO_2 content of the atmosphere, incidence of ultraviolet radiation, soil formation, and other factors.

Paleosols

A great deal is known about the climatic significance of various soil types, although Paleozoic paleosols have largely escaped attention except for studies related to the significance of the mineralogy of Permo-Carboniferous underclays in coal deposits.

Boucot and Gray (1980) synthesized data on paleosols and soil products currently recognized in the Paleozoic, including kaolins and bauxites, as well as gibbsite, boehmite, and emery deposits, the last being the regionally metamorphosed equivalent of alumina-rich soils. Additional data on Paleozoic paleosols is summarized by Retallack (in press).

At present, Paleozoic paleosol data are too limited for independent climatic conclusions, although the data may provide useful constraints. The distribution of Paleozoic calcretes, for example, provides insight into regions of very moderate seasonal rainfall. Such information may be combined with similar data for evaporite and coal distribution to make climatic conclusions more realistic. Pre-Carboniferous Paleozoic calcretes are largely restricted to the Devonian. Dineley (1963) and McKerrow *et al.* (1974) described Late Ludlovian or Early Pridolian age calcretes from the upper Red Member of the Moydart Formation of Nova Scotia. Allen (1974) described Downtonian (Pridolian) calcretes from England and has documented a number of occurrences of Lower and Upper Devonian calcretes in nonmarine sequences in England, Wales, and Scotland. Dineley and Hickox (1974) described calcareous nodules and conglomerates from the Lower Devonian Knoydart Formation of Nova Scotia that are similar in all respects to the same age calcretes from Britain described by Allen (1974). Woodrow *et al.* (1973) found similar calcretes in the Middle and Upper Devonian of New York. McPherson (1979) notes Upper Devonian calcrete from Antarctica. D. L. Woodrow (Hobart and William Smith Colleges, personal communication, 1979) mentioned that Brian G. Jones, University of Wollongong, finds Upper Devonian calcretes in Australia. Loope and Schmitt (1980) reported calcrete from the Pennsylvanian of Wyoming. We have recently observed Lower Devonian calcrete in eastern Yunnan Province, southwestern China.

Yaalon (The Hebrew University, personal communication, 1979) wrote that "calcrete develops best under semiarid conditions, with an optimum [rainfall] of about 300-350 mm, i.e., when the soil absorbs all the moisture and there is practically no recharge to groundwater. Above 600 mm leaching is too strong and too frequent to form secondary $CaCO_3$. In fully arid regions, calcrete forms slowly and at shallow depths and gypcrete is more common, provided sulphate is available . . . laterite plus bauxite . . . require much more humid conditions, but . . . seasonal aridity (about 2 months) is frequently a prerequisite, to enable groundwater level changes and/or the drying of seepages that lead to the irreversible hardening of the accumulated sesquioxides."

Widespread Lower and Upper Devonian calcretes in the Caledonian Belt of Britain, and from Nova Scotia to New York in the Appalachian Belt, as well as in the Upper Devonian of Antarctica and the Lower Devonian of southwestern China help to extend the arid zone into those regions, although the absence of evaporites in all of them previously provided somewhat ambiguous climatic information. The occurrence of Upper Devonian coal in North America (Heckel and Witzke, 1979) and in Spitsbergen and Bear Island far to the north of the Devonian calcretes indicates a consistent climatic difference.

Phosphorites

Possible evidence regarding ancient oceanic circulation patterns of climatic importance can be obtained from the distribution of marine phosphorite deposits. Boucot and Gray (1980) summarized the Paleozoic occurrence of these deposits and discussed some of the principal possibilities for their genesis. There is reasonable evidence that at least a few of the Paleozoic phosphorites, such as those of the Permian in western North America, are best considered to indicate upwelling from deep water on the western margins of mid- to low- but not lowest-latitude major land areas. However, not all phosphorite deposits of the Paleozoic can be regarded as having had such an origin.

Glacial Deposits

Boucot and Gray (1979, 1980) and Crowell (Chapter 6) synthesized the major sources of data pertaining to Paleozoic glacial deposits. The latest Ordovician provides evidence for extensive southern hemisphere Gondwana-region glaciation; so does the Late Carboniferous-Early Permian. The regional limits of the areas affected differ. Evidence for the Late Paleozoic suggests that a number of glacial centers were present but that they were not simultaneously active. No evidence exists for extensive glacial activity at sea level during the Paleozoic at times other than those specified above. It is reasonable to assume, however, that many Paleozoic mountain belts were glaciated at elevations greater than 1000 m above sea level, although evidence is lacking because of the destruction of high-elevation regions by erosion.

Mountain Belts

Mountain belts are a common feature for many intervals of the Paleozoic; these belts are now preserved only as their eroded roots. The distribution in time and space of the widespread Paleozoic orogenic belts would provide an additional set of climatic clues. The time interval of orogeny must be accurately dated to know when such orogenic belts presumably had climatologically significant elevations.

Both mountain building and erosion may be relatively rapid geologic processes, each occupying no more than a few million years. Although the average or maximum elevation of Paleozoic mountain belts may be estimated from the volume of debris eroded from them, from their area, and from the interval of time during which the mountain belt was being uplifted and eroded, such estimates are too imprecise to be of value to the climatologist. Nevertheless, orographic effects may be useful in helping to explain some climatological anomalies, and this possibility should be kept in mind.

Summaries of mountain-belt locations are readily available in the literature for the Paleozoic, period by period. The reliability of the data decreases with time. For the Cambrian particularly, there are some large uncertainties in the currently available data.

Regression-Transgression

For many Paleozoic time intervals, relatively detailed information on changing shoreline positions is available. This is the principal data of paleogeographic maps. When synthesized on a global scale, the maps provide a measure of the percentage of the continental areas covered with shallow seawater. These data should provide insight into changing albedo values.

The climatologist should understand that the relative amount of seawater cover on the continents varies from time interval to time interval and from continent to continent. North America, for example, had a large part of its area covered by shallow seas during the Paleozoic. Africa in contrast had a large part of its area above sea level during the Paleozoic. The Early Cambrian had far more overall continental area above sea level than did the mid-Silurian. A time sequence of paleogeographic maps (i.e., Boucot and Gray, 1979, 1980) is necessary, therefore, to evaluate this regression and transgression of the Paleozoic seas from the continents.

Paleowinds

There have been several attempts to use sedimentary structures formed by wind as a measure of wind direction of the past. For the later Paleozoic, Poole (1964), Runcorn (1964), Glennie (1972), and Van Veen (1975) provided good summaries of the kinds of data employed. Although aeolian deposits are neither common nor widespread during the Paleozoic, advantage should be taken of the data to provide information on average wind direction, bearing in mind that a large sample is necessary before conclusions have any real significance.

Krinsley and Wellendorf (1980) suggested that microsculpturing on quartz grains provides evidence about the last wind velocity to which the grain was subject. If these data prove reliable, they should provide wind velocities for some Paleozoic aeolian deposits, as well as velocities for "floating" quartz grains found here and there in marine sediments.

SUMMARY OF SILURIAN AND DEVONIAN CLIMATES

We have discussed categories of data that have proved useful, or are potentially useful, in reconstructing Paleozoic climates. We now provide an example of how the data can be used to determine something about the climates of the Silurian and Devonian.

The reconstruction of past climates is intimately involved with the reconstruction of past geographies. We begin with a brief consideration of one possible interpretation of Early Paleozoic paleogeography—the Pangaeic. The climatologist may wish to test this and the many alternative reconstructions for the Paleozoic generated by interest in plate tectonic concepts.

Figures 21.1 and 21.2 present a possible interpretation of Late Silurian and Early Devonian paleogeography within the Pangaeic framework. The canons on which this interpretation is based and the data used to support it have been discussed in detail elsewhere and need not be repeated here (Boucot and Gray, 1979, 1980). We would like to comment, however, about one facet of physical geology that may influence acceptance or consideration of the Pangaea—we refer to the assumption of some geologists that all mountain belts, tectonic zones, and ophiolite occurrences represent suture zones for areas previously separated by thousands of kilometers. Such an assumption often sets up situations that are biogeographically and oceanographically unworkable in terms of oceanic circulation patterns necessary to explain the available data. For example, Scotese *et al.* (1979) provided a set of period-by-period paleogeographic maps for the Paleozoic that assume that most orogenic and ophiolite belts indicate the location of ancient plate boundaries to either side of which there was a large amount of movement. They also assumed that their selected paleomagnetic data are reliable. Their maps suggested a fragmentation of Asia during the Paleozoic for which good geologic and paleontological evidence to the contrary exists. Chang (1981, p. 184) comments ". . . Ziegler *et al.* (1977) placed Chinese Tibet on the equator contiguous with the rest of western China and directly opposite Australia across a seaway. Two years later the same group of authors [Scotese *et al.* (1979)] placed Chinese Tibet adjacent to India but very distant from China." No justification for this major change was provided. Their interpretation was also inconsistent with what we now conclude about surface ocean-current circulation based on the biogeography of the Cambrian through the Devonian.

Hall (1980) indicated why one such tectonic belt, located in the Mediterranean region, had nothing to do with significant

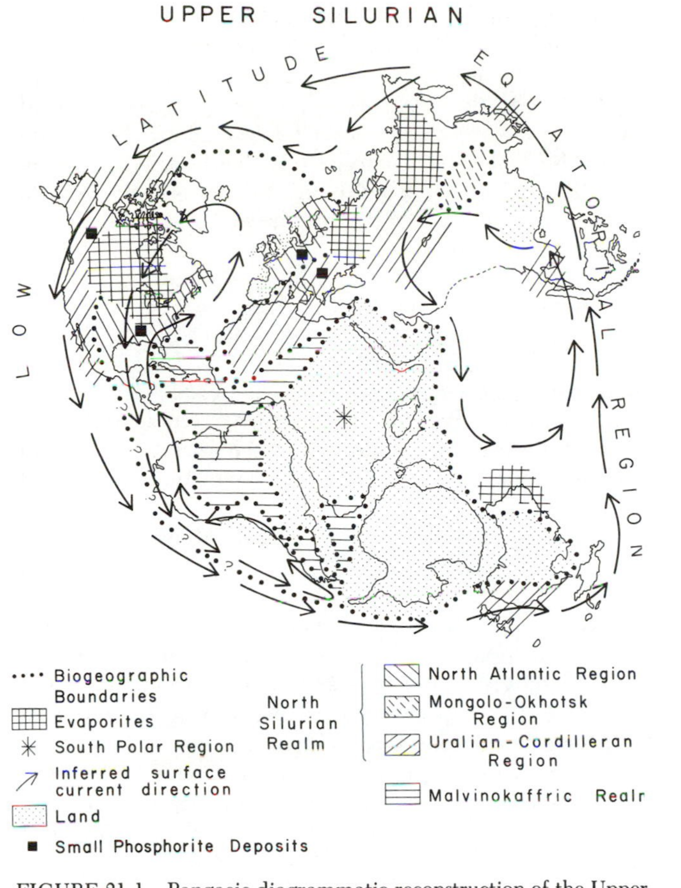

FIGURE 21.1 Pangaeic diagrammatic reconstruction of the Upper Silurian (from Boucot and Gray, 1980).

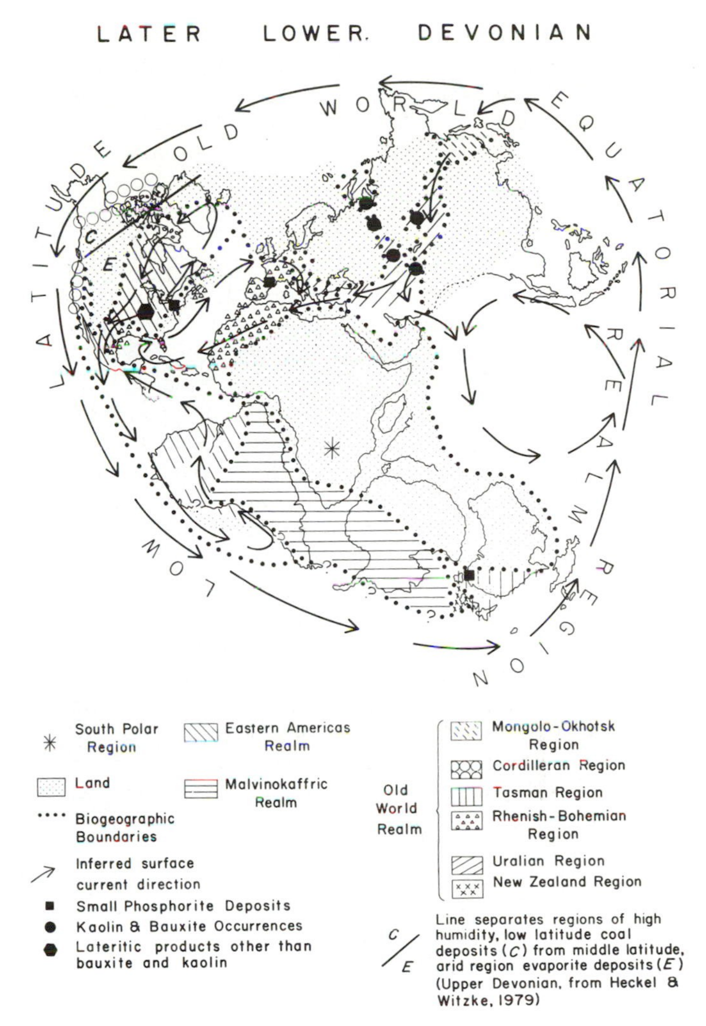

seafloor spreading. Similarly, the orogenic belts of later and mid-Paleozoic age situated along the eastern margins of North America (the Appalachian system), and corresponding units of the western margins of Europe (the Caledonide systems in particular, plus some parts of the Hercynian), provide us with little solid data with regard to their Paleozoic geographic relations to each other. Although a water body separated much of eastern North America from western Europe and northern Africa during the Late Silurian and Early Devonian (the "Iapetus" of some writers), the physical data are unconvincing whether it was of the magnitude of the Mediterranean, the Pacific, or something in between. Thus, each belt or zone must be considered on its own merits.

Although we do not regard Figures 21.1 and 21.2 as maps in the strict sense, we believe that they provide a more rational synthesis of the varied pre-Carboniferous lithofacies and biofacies data on which they are principally based than do other types of pre-Carboniferous paleogeographic reconstructions. Engel and Kelm (1972) employed a similar Pangaeic reconstruction for the Precambrian, although basing it on data different than ours because of the earlier time interval. In addition to biogeography and lithofacies information, our re-

FIGURE 21.2 Pangaeic diagrammatic reconstruction of the later Early Devonian (from Boucot and Gray, 1980). No coals are known for this time interval. Coals are present in the Middle Devonian of Gaspe and Pirate Cove between the La Garde and Pirate Cove Formations (P. J. Lesperance, Université de Montreal, personal communication, 1981). Coals are widespread in the modern Arctic and Subarctic regions in beds of Late Devonian Age. Recent work in China has shown that a major land area is present only in the so-called Sino-Korean platform area of east-central China and Korea to the east. Wang Yu *et al.* (in press) have provided the evidence for the presence of a major biogeographic unit, the South China Region of the Old World Realm, which takes in most of South China plus adjacent North Korea as far south as the Red River Valley. In the Eifeleian, the South China Region is distinct from coeval faunas present in the Shan States of Burma. The closest affinities of the latter are with the Rhenish-Bohemian Region. The Late Devonian biofacies and lithofacies data necessitate a major shift of the Devonian continental units from the position shown here to one with a closer approach to the geography of the Carboniferous. There is good evidence for the presence in the Silurian and Devonian of a single Uralian-Mongolian volcanic-rich geosyncline situated between a Siberian-Kolyma Platform to the north and a Chinese block to the south of the Mongolian region part of the geosyncline. The present fragmented nature of this east Asian region is a post-Mesozoic phenomenon.

constructions outline surface oceanic circulation patterns that will reproductively connect the same biogeographic unit present in two or more areas, i.e., surface current patterns that are consistent with the present known biogeographic data.

For the Silurian, the chief source of climatic information, in addition to biogeography, is the distribution of evaporites and carbonate (including reefs) and noncarbonate rock sequences. The carbonate and reef facies, denoting tropical to warm temperate conditions, are confined to the North Silurian Realm. Silurian coals have not been recognized. Paleosol data of climatic significance are limited: calcretes, indicative of semiarid or seasonally arid regions, are confined to the latest Silurian where their distribution complements known areas of major Late Silurian evaporites. The significance of phosphorites is too poorly understood within the Silurian to employ them as clues or even as definite indicators of oceanic circulation.

Paleoclimatic data for the Devonian are more varied. Figure 21.2 shows biogeographic subdivisions for the Early Devonian as well as some lithofacies data. As in the Silurian, carbonate rocks continue to be absent from the Malvinokaffric Realm. For the Late Devonian the distribution of coals (humid indicators) is shown latitudinally separated from that of evaporites (arid indicators) in North America. Similar data are available for the Devonian of Eurasia and Australia for both the Middle and Late Devonian (Oswald, 1968). As in the Silurian, the distribution of Devonian calcretes complements the distribution of evaporites. Lacustrine beds in northern Scotland yield authigenic aegirine compatible with a semiarid, possibly seasonal climate (Fortey and Michie, 1978). These data are consistent with low-latitude humid climates and middle- to high-latitude dry or seasonally dry climates. Paleosols of seasonally humid-arid implications (bauxites, kaolin, and other lateritic products) are prominent in the Uralian region and are also found in Siberia and Iran. Their distribution implies a regional climatic anomaly inconsistent with a latitudinal, completely parallel distribution of humid climate in low and middle latitudes and relatively arid and seasonal climate in higher latitudes.

In sum, lithofacies and biofacies evidence for most of the Silurian and Devonian indicate the presence of a major southern hemisphere high-latitude region (the Malvinokaffric Realm) lacking carbonate rocks, reefs, redbeds, evaporites, coal deposits, and other evidences of warm climate. During the Silurian and Devonian there are widespread evaporite belts deduced to have been present in low (but not lowest) to middle latitudes. These are best preserved during the Late Silurian, the Middle Devonian, and the Frasnian (lower half of the Late Devonian) but are poorly represented in both the Early Silurian and Early Devonian. Varied paleosols geographically complement the distribution of evaporites for the most part. Silurian-Devonian redbeds are widespread in regions ringing the Malvinokaffric Realm in consort with other evidences of warm climate. Coal deposits are present in the Late Devonian in regions deduced to represent low latitudes.

During most of the Silurian and Devonian the climate may be characterized as having had a high latitudinal gradient, less than during intervals of extensive glaciation such as the Permian and Quaternary but higher than that of the Lower Carboniferous and much of the Mesozoic. Continental glaciation in the southern hemisphere also marks the end of the Ordovician (the Ashgillian) as a time of highest climatic gradient. The absence of positive evidence for continental glaciation in the Silurian and Devonian leads us to conclude that the overall climatic gradient was lower than in the Late Ordovician. Carozzi (1979) mentioned Upper Devonian tillite on the north side of the Amazon Basin, but no paleontologic evidence is provided to document that this tillite could not be of Permo-Carboniferous age. Sometime during the Late Devonian, however, the climate of the high-latitude Malvinokaffric Realm appears to have ameliorated, and the Realm no longer existed as a biogeographic unit. These changes are consistent with the presence of a globally low climatic gradient similar to that which characterized the Early Carboniferous on a global scale.

SUMMARY AND CAVEATS

To understand Paleozoic climates, detailed time-sequence maps are necessary on which shoreline positions for the continents are indicated. On these should be plotted the boundaries of biogeographic units, lithologic data (distribution of redbeds, carbonate-noncarbonate rock sequences, evaporite and coal deposits, paleosol and glacial deposits, and phosphorites of the upwelling type), mountain belts, and biological information of environmental significance. Only then can climatic deductions and constraints be suggested for the Paleozoic. Maps such as Figures 21.1 and 21.2 and those by Boucot and Gray (1979, 1980) are a beginning but include only a fraction of the usable data. Maps for the Devonian by Heckel and Witzke (1979) are well-documented attempts to synthesize more varied climatic indicators, with emphasis on sedimentary data.

Probably the chief warning to the climatologist attempting to reconstruct Paleozoic climate and climatic events is to employ as short a time interval as possible but one lengthy enough to provide sufficient global data to permit informed speculation. For the Paleozoic, time intervals of 10-15 m.y. are usable, and it is improbable that units much shorter than that can be synthesized in the near future. In geologic terms this means that a geologic Period, measured in tens of millions of years, is commonly too large a unit for which to collect meaningful data because there is good evidence that globally significant shifts in the position of climatic belts, as well as changes in the global climatic gradient, have taken place during these lengthy intervals. For example, White (1931) contrasted the plant remains of the Late Mississippian (Chester) with those of the earlier Pennsylvanian of the Eastern Interior, noting that the former have a "generally stunted" appearance and "more or less distinctly xerophytic characters," whereas the general luxuriance of the Pennsylvanian vegetation together with its large size and other features clearly indicates not only equable climates but ample rainfall. In the Carboniferous of the Northern Appalachian region (New England through Newfoundland) the Early Carboniferous (Mississippian) is characterized by the deposition of evaporites such as those of the Windsor Group, whereas the Late Carboniferous (Pennsylvanian) was sufficiently humid to permit the accumulation and preservation of enough plant material to form economically important coal beds. In a recent publication Habicht (1979) confused this paleoclimatic evidence by plotting evaporites and coals together on the Carboniferous map of the Northern Appalachian region without regard for the fact that they accumulated at different times. Habicht (1979) presented a similar anomaly for the Cambrian of North Africa, where the Early Cambrian includes many marine carbonate beds, indicators of warm climatic regime, whereas the Middle and Late Cambrian lacks carbonates and is characterized by a cool climate marine fauna. We have pointed out earlier (Boucot and Gray, 1979, 1980) that in the Antarctic Devonian, the Early Devonian is of cold climate, Malvinokaffric Realm type, whereas the Late Devonian is of warm type and even includes a recently discovered calcrete (McPherson, 1979).

Recognition that the time correlation of the features being discussed here have different values is crucial. Marine carbonate rocks rich in fossils can probably be correlated with a precision within 2-3 m.y. The dating of nonmarine bauxites and varied paleosols may be much more approximate and in some instances no more precise than 10-15 m.y. Therefore, the climatologist should continually consult with the biostratigrapher-paleontologist to make certain about the reliability of the correlation and the precision of the data being employed.

REFERENCES

Allen, J. R. L. (1974). Sedimentology of the Old Red Sandstone (Siluro-Devonian) in the Clee Hills Area, Shropshire, England, *Sed. Geol. 12*, 73-167.

Antevs, E. (1925). The climatologic significance of annual rings in fossil woods, *Am. J. Sci. 9 (5th series)*, 296-302.

Arnold, C. A. (1947). *An Introduction to Paleobotany*, McGraw-Hill, New York, 433 pp.

Boucot, A. J. (1975). *Evolution and Extinction Rate Controls*, Elsevier, New York, 427 pp.

Boucot, A. J., and J. Gray (1979). Epilogue: A Paleozoic Pangaea? in *Historical Biogeography, Plate Tectonics, and The Changing Environment*, J. Gray and A. J. Boucot, eds., Oregon State U. Press, Corvallis, pp. 465-482.

Boucot, A. J., and J. Gray (1980). A Cambro-Permian pangaeic model consistent with lithofacies and biogeographic data, in *The Continental Crust and its Mineral Deposits*, D. W. Strangway, ed., Geol. Assoc. Canada Spec. Pap. No. 20, pp. 389-419.

Boureau, E., A. Lejal-Nicol, and D. Massa (1978). A props du Silurien et du Devonien en Libye. Il faut reporter au Silurien la date d'apparition des plantes vasculaires, *C. R. Acad. Sci. Paris 286*, ser. D, pp. 1567-1571.

Carozzi, A. V. (1979). Petroleum geology in the Paleozoic clastics of the Middle Amazon Basin, Brazil, *J. Petrol. Geol. 2*, 55-74.

Chaloner, W. G., and G. T. Creber (1973). Growth rings in fossil woods as evidence of past climates, in *Implications of Continental Drift to the Earth Sciences*, D. H. Tarling and S. K. Runcorn, eds., Academic, New York, pp. 425-437.

Chaloner, W. G., and A. Sheerin (1979). Devonian macrofloras, in *The Devonian System*, M. R. House, C. T. Scrutton, and M. G. Bassett, eds., Palaeontol. Assoc. Spec. Pap. in Palaeontol. No. 23, pp. 145-161.

Chang, W.T. (1981). On the northward drift of the Afro-Arabian and Indian Plates, *R. Soc. Victoria Proc. 92*, 181-185.

Cridland, A. A. (1964). *Amyelon* in American coal-balls, *Palaeontology* 7, 186-209.

Daubenmire, R. F. (1974). *Plants and Environment: A Textbook of Plant Autecology* (3rd ed.), Wiley, New York, 422 pp.

Davis, C. A. (1913). Origin and formation of peat, in *The Origin of Coal*, D. White and R. Thiessen, Bureau of Mines Bull. 38, pp. 165-186.

Dineley, D. L. (1963). The "Red Stratum" of the Silurian Arisaig Series, Nova Scotia, Canada, *J. Geol. 71*, 523-524.

Dineley, D. L., and C. F. Hickox (1974). Arisaig Group: Knoydart Formation, in *Geology of the Arisaig Area, Antigonish County, Nova Scotia*, Geol. Soc. Am. Spec. Pap. 139, pp. 68-76.

Dolph, G. E., and D. L. Dilcher (1979). Foliar physiognomy as an aid in determining paleoclimate, *Palaeontographica 170*, 151-172.

Edwards, D., and J. Feehan (1980). Records of *Cooksonia*-type sporangia from late Wenlock strata in Ireland, *Nature 287*, 41-42.

Engel, A. E. J., and D. L. Kelm (1972). Pre-Permian global tectonics: A tectonic test, *Geol. Soc. Am. Bull. 83*, 2325-2340.

Fortey, N. J., and U. McL. Michie (1978). Aegirine of possible authigenic origin in Middle Devonian sediments in Caithness, Scotland, *Mineral. Mag. 42*, 439-442.

Friend, P. F., and M. Moody-Stuart (1970). Carbonate sedimentation on the river floodplains of the Wood Bay Formation (Devonian) of Spitsbergen, *Geol. Mag. 107*, 181-195.

Glennie, K. W. (1972). Permian Rotliegendes of northwest Europe interpreted in light of modern desert sedimentation studies, *Am. Assoc. Petrol. Geol. Bull. 56*, 1048-1071.

Goldring, W. (1921). Annual rings of growth in Carboniferous woods, *Bot. Gaz. 72*, 326-330.

Gray, J., and A. J. Boucot (1971). Early Silurian spore tetrads from New York: Earliest New World evidence for vascular plants? *Science 173*, 918-921.

Gray, J., and A. J. Boucot (1978). The advent of land plant life, *Geology 6*, 489-492.

Gray, J., and A. J. Boucot, eds. (1979). *Historical Biogeography, Plate Tectonics, and The Changing Environment*, Oregon State U. Press, Corvallis, 500 pp.

Gray, J., S. Laufeld, and A. J. Boucot (1974). Silurian trilete spores and spore tetrads from Gotland: Their implications for land plant evolution, *Science 185*, 260-263.

Gray, J., D. Massa, and A. J. Boucot (1980). Caradocian land plant microfossils from Libya: The oldest known land plant spores, *Geol. Soc. Am. Abstr. Programs 12*, p. 436.

Habicht, J. K. A. (1979). *Paleoclimate, Paleomagnetism, and Continental Drift*, Am. Assoc. Petrol. Geol. Studies in Geology No. 9, Tulsa, Okla., 31 pp.

Hall, R. (1980). Contact metamorphism by an ophiolite peridotite from Neyriz, Iran, *Science 208*, 1259-1262.

Hallam, A., ed. (1973). *Atlas of Palaeobiogeography*, Elsevier, New York, 531 pp.

Heckel, P. H. (1977). Origin of phosphatic black shale facies in Pennsylvanian cyclothems of mid-continent North America, *Am. Assoc. Petrol. Geol. Bull. 61*, 1045-1068.

Heckel, P. H., and P. J. Witzke (1979). Devonian world palaeogeography determined from distribution of carbonates and related lithic palaeoclimatic indicators, in *The Devonian System*, M. R. House, C. T. Scrutton, and M. G. Bassett, eds., Palaeontol. Assoc. Spec. Pap. in Paleontol. No. 23, pp. 99-123.

Hughes, N. F., ed. (1973). *Organisms and Continents Through Time*, Palaeontol. Assoc. Spec. Pap. in Palaeontol. No. 12, 334 pp.

Krassilov, V. A. (1975). *Paleoecology of Terrestrial Plants: Basic Principles and Techniques*, Wiley, New York, 283 pp. (translated from Russian by H. Hardin).

Krausel, R. (1964). Introduction to the palaeoclimatic significance of coal, in *Problems of Palaeoclimatology*, A. E. M. Nairn, ed., Wiley-Interscience, New York, pp. 53-56, 73-74.

Krinsley, D., and W. Wellendorf (1980). Wind velocities determined from the surface textures of sand grains, *Nature 283*, 372-373.

Loope, D. B., and J. G. Schmitt (1980). Caliche in the Late Paleozoic Fountain Formation: Rediscovery and implications, *Geol. Soc. Am. Abstr. Programs 12*, p. 473.

McKerrow, W. S., A. J. Boucot, and C. F. Hickox (1974). Arisaig Group: Beechhill Cove Formation, Ross Brook Formation and others, in *Geology of the Arisaig Area, Antigonish County, Nova Scotia*, Geol. Soc. Am. Spec. Pap. 139, pp. 34-68.

McPherson, J. G. (1979). Calcrete (caliche) palaeosols in fluvial redbeds of the Aztec Siltstone (Upper Devonian), southern Victoria Land, Antarctica, *Sed. Geol. 22*, 267-285.

Meyerhoff, A. A. (1970). Continental drift: Implications of paleomagnetic studies, meteorology, physical oceanography, and climatology, *J. Geol. 78*, 1-51.

Middlemiss, F. A., P. F. Rawson, and G. Newall, eds. (1971). *Faunal Provinces in Space and Time*, Seel House Press, Liverpool, 236 pp.

Nicol, D. (1967). Some characteristics of cold-water marine pelecypods, *J. Paleontol. 41*, 1330-1340.

Noe, A. C. (1931). Evidences of climate in the morphology of Pennsylvanian plants, *State of Ill., State Geol. Surv. Bull. 60*, 282-289.

Oswald, D. H., ed. (1968). *International Symposium on the Devonian System, Calgary 1967*, Calgary, Alberta Soc. Petrol. Geol., Vol. 1, 1055 pp.; Vol. 2, 1377 pp.

Plumstead, E. P. (1963). Palaeobotany of Antarctica, in *Antarctic Geology*, SCAR Proc., XI Palaeontology, pp. 637-654.

Plumstead, E. P. (1965). Glimpses into the history and prehistory of Antarctica, *Antarktiese Bull. 9*, 1-5.

Poole, F. G. (1964). Palaeowinds in the western United States, in *Problems in Paleoclimatology*, A. E. M. Nairn, ed., Wiley-Interscience, New York, pp. 394-405, 423-424.

Potonie, H. (1911). Die Tropen-Sumpfflachmoor-Natur der Moore des Productiven Karbons, *Kgl. Preuss. geol. Landesanst. Jahrb. (1909) 30*, 389-443.

Potonie, R. (1953). Zür Palaobiologie der karbonischen Pflanzenwelt, *Naturwissenschaften 40*, 119-128.

Pratt, L. M., T. L. Phillips, and J. M. Dennison (1978). Evidence of non-vascular land plants from the Early Silurian (Llandoverian) of Virginia, U.S.A., *Rev. Palaeobot. Palynol. 25*, 121-149.

Retallack, G. (in press). Fossil soils—indicators of ancient terrestrial environments, in *Evolution, Paleoecology and the Fossil Record*, K. Niklas, ed., Praeger, New York.

Richards, P. W. (1979). *The Tropical Rain Forest: An Ecological Study*, Cambridge U. Press, Cambridge, 450 pp.

Ross, C. A., ed. (1974). *Paleogeographic Provinces and Provinciality*, Soc. Econ. Paleontol. Mineral. Spec. Publ. No. 21, 233 pp.

Runcorn, S. K. (1964). Paleowind directions and palaeomagnetic latitudes, in *Problems in Palaeoclimatology*, A. E. M. Nairn, ed., Wiley-Interscience, New York, pp. 409-421, 424.

Schopf, J. M. (1972). Coal, climate and global tectonics, in *Implications of Continental Drift to the Earth Sciences*, Vol. 1, D. H. Tarling and S. K. Runcorn, eds., Academic, New York, pp. 609-622.

Scotese, C. R., R. K. Bambach, C. Barton, R. Van der Voss, and A. M. Ziegler (1979). Paleozoic base maps, *J. Geol. 87*, 217-277.

Stehli, F. G. (1968). Taxonomic diversity gradients in pole location: The Recent model, in *Evolution and Environment*, E. T. Drake, ed., Yale U. Press, New Haven, Conn., pp. 163-227.

Stehli, F. G., A. L. McAlester, and C. E. Helsley (1967). Taxonomic diversity of Recent bivalves and some implications for geology, *Geol. Soc. Am. Bull. 78*, 455-466.

Tomlinson, P. B., and F. C. Craighead (1972). Growth-ring studies on the native trees of sub-tropical Florida, in *Research Trends in Plant Anatomy* (Chowdhury Commemorative Volume), Ghouse and Yunus, eds., McGraw-Hill, New York, pp. 39-51.

Van Veen, F. R. (1975). Geology of the Lemen gas-field, in *Petroleum and the Continental Shelf of Northwest Europe*, Vol. 1, A. W. Woodland, ed., Halsted, New York, pp. 223-231.

Walker, T. R. (1967). Formation of red beds in modern and ancient deserts, *Geol. Soc. Am. Bull. 78*, 353-368.

Walker, T. R. (1975). Red beds in the western interior of the United States, in *Paleotectonic Investigations of the Pennsylvanian System in the United States, Pt. II*, L. D. McKee, E. J. Crosby, and others, U.S. Geol. Surv. Prof. Pap. 853, pp. 593-606.

Wang Yu, A. J. Boucot, Rong Jia-Yu, and Yang Xue-Chang (in press). Silurian and Devonian biogeography of China, *Geol. Soc. Am. Bull.*

White, D. (1913). Physiographic conditions attending the formation of coal, in *The Origin of Coal*, D. White and R. Thiessen, Bur. of Mines Bull. 38, pp. 52-84.

White, D. (1925). Environmental conditions of deposition of coal, *Trans. Am. Inst. Min. Metal. Eng. 71*, 3-34.

White, D. (1929). *Flora of the Hermit Shale, Grand Canyon, Arizona*, Carnegie Institution of Washington Publ. No. 405, 119 pp.

White, D. (1931). Climatic implications of Pennsylvanian flora, *State of Ill., State Geol. Surv. Bull. 60*, 271-281.

Woodrow, D. L., F. W. Fletcher, and W. F. Ahrnsbrak (1973). Paleogeography and paleoclimate at the depositional sites of the Devonian Catskill and Old Red facies, *Geol. Soc. Am. Bull. 84*, 3051-3064.

Ziegler, A. M., K. S. Johnson, M. E. Kelly, C. R. Scotese, and R. Van der Voo (1977). Silurian continental distributions, paleogeography, climatology, and biogeography, *Tectonophysics 40*, 13-51.